THE MEDITERRANEAN GENETIC CODE - GRAPEVINE AND OLIVE

Edited by **Danijela Poljuha** and **Barbara Sladonja**

The Mediterranean Genetic Code - Grapevine and Olive

http://dx.doi.org/10.5772/3442
Edited by Danijela Poljuha and Barbara Sladonja

Contributors

Stefano Meneghetti, Zohreh Rabiei, Sattar Tahmasebi Enferadi, José Eiras-Dias, Jorge Cunha, Pedro Fevereiro, Margarida Teixeira-Santos, João Brazão, Massimo Muganu, Marco Paolocci, Mirza Musayev, Zeynal Akparov, Lidija Tomić, Branka Javornik, Nataša Štajner, Rosa Adela Arroyo-Garcia, Eugenio Revilla, Denis Rusjan, Jernej Jakše, Rotondi Annalisa, Catherine Marie Breton, André Berville, Anthony Ananga, Vasil Georgiev, Joel W. Ochieng, Bobby Phills, Violetka Tsolova, Devaiah Kambiranda

Published by InTech

Janeza Trdine 9, 51000 Rijeka, Croatia

Publishing Process Manager Iva Simcic
Typesetting InTech DTP Team
Cover InTech Design Team

Additional hard copies can be obtained from orders@intechopen.com

The Mediterranean Genetic Code - Grapevine and Olive,
Edited by Danijela Poljuha and Barbara Sladonja
p. cm.
ISBN 978-953-51-1067-5

For EU product safety concerns:
IN TECH d.o.o., Prolaz Marije Krucifikse Kozulić 3, 51000 Rijeka, Croatia,
info@intechopen.com or visit our website at intechopen.com.

Meet the editors

Dr. Danijela Poljuha graduated in molecular biology at the Faculty of Science, University of Zagreb, Croatia, where she also obtained a Master's degree (MSc) and a Doctoral Degree (PhD) in Natural Sciences. During the first years of her professional carrier in grape and olive genetics she installed a brand new Genetic laboratory and introduced molecular techniques at the Institute of Agriculture and Tourism. She spent two months as a visiting researcher at Centro di ricerca per la viticoltura (CRA-VIT) in Conegliano, Italy, and collaborated on many national and international scientific projects. She now works as Associate Research Scientist at the Istrian Development Agency in Pula, Croatia, where she established a new Biotechnical Department.

Dr. Barbara Sladonja graduated in biology – ecology at the Faculty of Science, University of Zagreb, Croatia, and obtained a Doctoral degree (PhD) in Natural Sciences at the University of Udine and University of Florence, Italy. She organised the international scientific conference "Mediterranean Malvasias" in Poreč, Croatia, in 2005, and the International Scientific Congress GESCO (Groupe d'Etude des Systemes de Conduite de la Vigne) in Poreč (Croatia) in 2007. She acted as a collaborator or leader on many national and international scientific projects covering the grape and olive field. She works at the Institute of Agriculture and Tourism Poreč.

Contents

Preface IX

Section 1 Molecular Insight into Variability 1

Chapter 1 **Characterization of Grapevines by the Use of Genetic Markers 3**
Lidija Tomić, Nataša Štajner and Branka Javornik

Chapter 2 **Application of Microsatellite Markers in Grapevine and Olives 25**
Jernej Jakše, Nataša Štajner, Lidija Tomić and Branka Javornik

Chapter 3 **The Current Status of Wild Grapevine Populations (Vitis vinifera ssp sylvestris) in the Mediterranean Basin 51**
Rosa A. Arroyo García and Eugenio Revilla

Chapter 4 **Inter- and Intra-Varietal Genetic Variability in Vitis vinifera L. 73**
Stefano Meneghetti, Luigi Bavaresco, Antonio Calò and Angelo Costacurta

Section 2 Genetics in Service of National Germplasms Preservation 97

Chapter 5 **Centuries-Old Results of Cultivation and Diversity of Genetic Resources of Grapes in Azerbaijan 99**
Mirza Musayev and Zeynal Akparov

Chapter 6 **Portuguese Vitis vinifera L. Germplasm: Accessing Its Diversity and Strategies for Conservation 125**
Jorge Cunha, Margarida Teixeira-Santos, João Brazão, Pedro Fevereiro and José Eduardo Eiras-Dias

Chapter 7 **Genetic and Phenotypic Diversity and Relations Between Grapevine Varieties: Slovenian Germplasm 147**
Denis Rusjan

Section 3 From Genotype to Product 177

Chapter 8 **Italian National Database of Monovarietal Extra Virgin Olive Oils 179**
Annalisa Rotondi, Massimiliano Magli, Lucia Morrone, Barbara Alfei and Giorgio Pannelli

Chapter 9 **Challenges for Genetic Identification of Olive Oil 201**
Sattar Tahmasebi Enferadi and Zohreh Rabiei

Section 4 And All Begins with Genetics 219

Chapter 10 **Adaptation of Local Grapevine Germplasm: Exploitation of Natural Defence Mechanisms to Biotic Stresses 221**
Massimo Muganu and Marco Paolocci

Chapter 11 **Production of Anthocyanins in Grape Cell Cultures: A Potential Source of Raw Material for Pharmaceutical, Food, and Cosmetic Industries 247**
Anthony Ananga, Vasil Georgiev, Joel Ochieng, Bobby Phills and Violeta Tsolova

Chapter 12 **From the Olive Flower to the Drupe: Flower Types, Pollination, Self and Inter-Compatibility and Fruit Set 289**
Catherine Breton and André Bervillé

Preface

Grapes and olives were once a symbol and an exclusive trademark of the Mediterranean. Nowadays these cultures are present on all continents and their cultivation is increasing constantly, becoming an important economical branch. Therefore, the science based on these two cultures involves scientists from all over the globe.

The book "The Mediterranean Genetic Code – Grapevine and Olive" collects relevant papers documenting the results of research in grapevine and olive genetics, as a contribution to overall compendium of the existing biodiversity for both species with insight into molecular mechanisms responsible for their desirable and important traits. Book encompasses a broad and diverse palette of different topics related to grapevine and olive genetics, with no areal or any other strict limitation, keeping the title as a loose frame for borderless science. Divided in four sections it takes us for a "molecular walk" through different levels of genetic variability, uncovering the remains of still existing wild populations and treasures of neglected local peculiarities, weaving the network from plant to product and back to the beginning, to the hearth of all questions asked and answers hidden in genetics.

The first section gives an overview of genetic markers used in grapevine research, with special emphasis on microsatellite markers and their application in grapevine and olive, accompanied by practical examples. Since wild grapevines are endangered in their natural habitats, conservation priority is given to these populations. This section provides also a detailed insight in the current status of the remaining wild grape populations around the Mediterranean basin and their relationships with cultivated varieties obtained by molecular genetics approach. Many researches worldwide have tried to clarify origin and phylogenetic relationships of a great number of today known grapevine varieties. Here we present a molecular strategy applied in inter- and intra-varietal genetic variability studies with the aim of ascertaining relationships between molecular profiles, environmental parameters and morphological traits in grapevine.

A special accent is given on the preservation of autochthonous grapevine biotypes and supporting a targeted propagation of local genetic material, selected for centuries and adapted to locally specific environment. This is elaborated in detail on the examples of national collections and germplasms preservation in Azerbaijan, Portugal and Slovenia given in the second section.

Third part articulates peculiar connection and traceability between plant genotype and final product – olive oil. The example of efficient strategy of valorization and promotion of local and national olive genetic heritage presented on the case of Italian National Database of Monovarietal Extra Virgin Olive Oils and supplemented with recent advances in application of

DNA markers in olive oil authentication and traceability, implies olive biodiversity preservation, olive oil quality improvement as well as consumers' education and interest protection.

The last section discusses molecular mechanisms responsible for important traits of both grapevine and olive, comprising natural defense mechanisms and responses to abiotic stress, anthocyanin biosynthesis and finally closing with the description of main phases and steps from blossoming to harvest in olive, from both physiological and genetic point of view.

The book is aimed at researchers interested in molecular methods, growers and producers of olives, olive oil, grapes and wine, agricultural experts, biotechnical students, olive oil and wine educated consumers and marketing operators for agricultural products.

By accepting the challenge of this book adventure we hoped to provide answers to some questions deeply rooted in genetics. We honestly believe we succeeded in this mission.

The book has come to fruition thanks to the efforts and expertise of the contributing authors, as well as of good friends and colleagues. We hope that this shared effort will be the start of more collaboration possibilities in the future, and also an impulse for new questions and answers in some future journey aimed to reveal secrets hidden in molecules.

Danijela Poljuha
Research Centre Metris, Istrian Development Agency
Croatia

Barbara Sladonja
Institute of Agriculture and Tourism Poreč
Croatia

Section 1

Molecular Insight into Variability

Chapter 1

Characterization of Grapevines by the Use of Genetic Markers

Lidija Tomić, Nataša Štajner and Branka Javornik

Additional information is available at the end of the chapter

http://dx.doi.org/10.5772/52833

1. Introduction

Grapevine (*Vitis vinifera* L.), used worldwide for producing wine, table grapes and dried fruits, is an important horticultural species; the total number of grapevine cultivars in ampelographic collections worldwide is estimated to be 10,000 [1]. Grapevine cultivars have traditionally been characterized and identified by standard ampelographic descriptors. In order to establish comparable evaluation of grapevines, a unique system for cultivar description was introduced. In 1873, the International Ampelography Committee was established in Vienna, which prepared the first international standards for the classification of grapevines based on morphological traits. Ampelogrpahy is based on visual observation of certain traits, while ampelometry developed as a method that relies on precise measurement of the phenotypic characteristics of grapevines, mainly based on leaf traits. Today, the ampelographic description of cultivars includes 150 descriptors. The Office International de la Vigne et du Vin (OIV), the Union International pour la Protectione des Obtentions Végétales (UPOV) and the International Board for Plant Genetic Resources (IBPGR) agreed to establish a common methodology for the ampelographic description of cultivars, which is used for the characterization and evaluation of cultivars in order to identify them, characterize their traits, to protect authors' rights and for the needs of gene banks. Ampelographic descriptions enable the identification of cultivars taking into account the development stage of the plants, their health status and environmental conditions [2]. Standard ampelographic methods can sometimes result in misunderstandings because the expression of morphological characters depends on the developmental phase of the plant (sample), health status of the sample and environmental conditions. At the same time, the vast number of different established cultivars makes it hard to differentiate them all by morphological characteristics [3]. In parallel, genetic erosion in grapevine germplasm has been observable, due to the worldwide predominance of few successful cultivars in all major wine producing regions. There is a significant shift in varietal structure in favour of modern cultivars and thus a decrease or even disappearance of regionally typical or local cultivars. Accurate identification is needed for numerous such cultivars, as well as systematic characterization of identified cultivars in terms of their sustainable use and breeding for future needs and conservation. Modern viticulture must be innovative and of high quality but, at the same time, must also take environmental protection into consideration. Grape growers and wine producers need to have access to a variety of grape genetic resources, in order to be able to create new varieties and new wine tastes. Growers also need to be able to certify their products, so the accurate names of local, potentially valuable grapevine varieties, and their genetic and geographic origins, need to be available. Biochemical characterization of grapevines was developed as a supplementary method to ampelographic characterization but issues associated with enzyme extraction, the general lack of a discriminating enzyme system and inconsistency in assaying enzymes have hindered the wider application of this method. Characterization of grapevines has today been complemented by the use of molecular markers, providing a different set of data, which enables more accurate identification and extended characterization. The introduction of molecular markers has allowed more accurate identification, since the results are independent of environmental factors. DNA based markers have enabled a new approach to genetic characterization and to the assessment of diversity within an analyzed set of samples, which is important for evaluation of the range and distribution of genetic variability. In grapevines, diverse marker techniques, such as RFLP or PCR based RAPD, SSR or AFLP and, recently, SNP have been widely used during recent decades. Among them microsatellites, or SSR (simple sequence repeat) markers, have become molecular markers of choice, since they offer some advantages over other molecular markers, including their co-dominant inheritance, hyper-variability and, once they are developed, they are easy to use and the data can be readily compared among laboratories. Microsatellites have also become favoured molecular markers for identifying grapevine cultivars, and their properties enable a wide range of applications, since they are ubiquitous, abundant and highly dispersed in genomes, with high variability at most loci. In *Vitis,* a large number of markers have been developed by individual groups and these markers have been very successfully applied for genetic studies. The suitability of *Vitis* SSR markers for assessing genetic origin and diversity in germplasm collections, cultivar identification, parentage analysis and for genetic mapping is well documented.

2. Biochemical methods

Isoenzyme analysis was an important tool in the characterization of grapevines during the nineties, thus preceding the wide use of molecular marker technologies. Biochemical characterization of grapevines was developed as a supplementary method to ampelographic characterization. The biochemical approach includes analysis of isoenzymes, phenolic and aromatic compounds, as well as serological analysis of pollen proteins.

During the nineties, various studies applied isoenzymes in the characterization of grapevines. Bachmann [4] developed simplified and improved isolation of active cytoplasmatic enzymes in grapevines. The polymorphism of peroxidase isoenzyme activity in phloem and dormant canes in 313 cultivars and species in *Vitis* has been evaluated. Single polymorphic isoenzyme peroxidase was sufficient to group cultivars and to discriminate between two samples. Royo *et al.* [5] characterized eight Spanish grapevine varieties and their clones by analysis of the polymorphism of isozyme activities carried out for esterases, peroxidises, catechol oxidase, glutamate oxalacetate transaminase and acid phosphatase. In the analyses, the zymograms varied in relation to the time of sampling, phenophase and origin of the plant tissues. In this case, it was concluded that two or more repetitions of sampling and isoenzyme analysis are needed for the generation of repetitive zymogram patterns. Isoenzyme analyses were also used to assess differentiation among table grapevine cultivars. A combination of four isoenzyme zymograms (peroxidises, catechol oxidase, glutamate oxalacetate transaminase and superoxide dismutase) allowed differentiation of 31 cultivars out of 43. The catechol oxidase system showed the highest level of polymorphism. This methodology was recommended for the differentiation of grapevine cultivars by Sanchez-Escribano *et al.* [6]. Analysis of isoenzymes of catechol-oxidase and acid phosphatase also allowed differentiation of the additional cultivars Kéknyelű and Picolit, considered to be synonymic [7]. Cultivars have been reported as synonyms in the Vitis International Variety Catalogue, despite differences in leaf morphology and type of wine produced. Cabernet Sauvignon and Chardonnay were used as reference cultivars for isoenzyme analysis, in which the same zymograms were obtained as with previous studies while Kéknyelű and Picolit differed in both studied enzyme systems.

Isoenzymes have mostly been used in biochemical characterisation for differentiation between cultivars but issues related to the success of enzyme extraction, lack of zymogram repeatability between repeated reactions, as well as the lack of a general discriminating enzyme have hindered wider application of this method [2].

3. Molecular methods

Ampelographic and biochemical methods for genotype characterization have been shown to be dependent on environmental conditions and sample status (developmental stage of plant and health status), resulting in a lack of repeatability and reproducibility in the analyzed set of parameters. In recent decades, classical methodologies have been supplemented by molecular techniques using various marker systems for the detection of DNA polymorphism.

4. Restriction Fragment Length Polymorphism (RFLP)

Restriction Fragment Length Polymorphism (RFLP) was the first widely used marker technique for molecular characterization of grapevines. Digestion of genomic DNA by restriction enzymes results in the production of numerous DNA fragments, and RFLP markers are detected by the hybridization of known probes to these fragments. Point mutations, insertions and deletions that occur within or between restriction sites can result in an altered length of RFLP fragments, revealing polymorphism among the analyzed genotypes. The main advantage of RFLP markers is their co-dominance and high reproducibility but they require a high amount of relatively pure DNA and a high labour input.

RFLP markers in grapevines have been used to differentiate between genotypes and for cultivar or rootstock identification, as well as for studying polymorphism within an analyzed set of cultivars and for verifying known relationships.

Bourquin *et al.* [8] used RFLP markers for the identification of grapevine rootstocks. Sixteen *Vitis* rootstocks were differentiated by means of RFLP analysis by the combination of the *HinfI* restriction endonuclease and probes obtained from DNA sequences of cv. Chardonnay. Additionally, 5 clones of SO 4 (*V. berlandieri* × *V. riparia*) and 3 clones of 41 B Mgt (*V. berlandieri* × *V. vinifera*) were analysed however no difference within clones of a same hybrid were found, since no polymorphism appeared using different probes. These analyses were a successful continuation of the study by Bourquin et al. [9], in which rootstocks of cultivars were differentiated by RFLP analysis with the restriction enzymes *Alu*-I and *Hinf*-I, using 9 different *Pst*-I inserts from *E. coli* recombinant clones derived from cv. Chardonnay as probes. Bourquin et al. [10] analyzed 46 grapevine cultivars by RFLP markers and detected significant polymorphism among all of them. As with rootstocks, RFLP markers could not identify cultivars belonging to the Pinot, Gewuerztraminer and Gamay group of cultivars. Forty six cultivars could be defined as belonging to six taxonomic groups, which were partially in accordance with relationships assessed from ampelographic data.

The RFLP technique showed high reproducibility but it is very demanding in terms of labour. Bourquin et al. [11] therefore reported PCR primers developed from four cloned *Pst*I DNA fragments of the cultivar Chardonnay, which had been shown to be the most informative RFLP probes from previous studies. PCR products were then digested by *Dde*I, *HinfI* and *Alu*I. This method was shown to be suitable for rapid differentiation among the majority of commercialized rootstocks (22 rootstocks), either by direct amplification or by RFLP analysis of the amplified products but they were not able to discriminate between clones of the same hybrid (rootstock 3309 C).

Versatile techniques have been developed based on polymerase chain reaction (PCR), which is more sensitive for germplasm characterization in terms of the ability for fast generation of a huge number of markers. PCR based techniques are less laborious than RFLP and require small amounts of DNA. Randomly Amplified Polymorphic DNA (RAPD), microsatellites (SSR, simple sequence repeats) and Amplified Fragment Length Polymorphism (AFLP) have proved to be most useful for grapevine germplasm analysis.

5. Random Amplified Polymorphic DNA (RAPD)

The RAPD technique is based on a PCR reaction and the use of short primers of an arbitrary nucleotide sequence, which results in amplification of an anonymous fragment (RAPD markers) of genomic DNA. The most important advantages of the RAPD technique are its technical simplicity and the fact that there is no need for advance knowledge of the DNA sequence. RAPD reproducibility among different laboratories and the requirement for strict experimental conditions are hard to achieve, which are the main disadvantages of this technique [12]. This technically least demanding method (RAPD) became popular during the nineties and due to its ease of application, it is also used nowadays.

Collins and Symons [13] used a sensitive and reproducible RAPD technique to establish a unique fingerprint of grapevine cultivars and for assessing polymorphism within the cultivars analyzed. They demonstrated that distinguishing between cultivars is already possible using single primer or by a mixture of two primers. Jean-Jaques et al. [14] confirmed this possibility by using RAPD markers in identity analysis of eight cultivars. Among 50 RAPD primers that were used in the analysis, reliable identification of analyzed cultivar was found by comparison between the RAPD patterns obtained by at least two primers (OPA 01 and OPA 18). Grando et al. [15] used 44 RAPD primers in order to assess the genetic diversity existing between wild and cultivated grapevines. The amplification patterns of the primers used did not differentiate between cultivated and wild grapevines but this RAPD approach enabled the analysis of genetic relationships within *V. vinifera* L. species.

Stavrakakis et al. [16] analyzed 8 grapevine cultivars grown on the island of Crete with the use of 15 RAPD decamer primers. Each grape cultivar showed a unique banding pattern for 5 or more primers used. Genetic similarity was calculated and a dendrogram of the 8 cultivars was constructed. The obtained results demonstrated that RAPD is a reliable method for the identification, discrimination and genomic analysis of grapevine cultivars. RAPD analysis of genetic diversity has been performed for cultivars from the Carpathian Basin [17], Turkish grape cultivars [18], Indian cultivars [19], and many others.

RAPD markers have also been shown to be very efficient in distinguishing between grapevine rootstocks. This et al. [20] demonstrated a high level of polymorphism among 30 grapevine rootstock cultivars by the use of 21 decamer primers. Using three primers (OPA09, OPA20 and OPP17), it was possible to identify each of the analyzed rootstock.

RAPD marker analysis has been shown to be advantageous since it is cheaper and easier to perform than RFLP analysis or isoenzyme characterization.

RAPD markers have been successfully applied in genetic mapping. Lodhi et al. [21] constructed one of the first genetic linkage maps using population derived from a cross between Cayuga White and Aurore. The map was based on 422 RAPD markers and also included some RFLP and isozyme markers. The seedlessness of grapevines, defined through various traits (mean fresh weight of one seed, total fresh weight of seeds per berry, perception of seed content, seed size categories evaluated visually, degree of hardness of the seed coat, degree of development of the endosperm and degree of development of the embryo) were assessed in 82 offsprings from of a cross between Early Muscat and Flame Seedless [22]. One hundred and sixty RAPD decamer primers were used, among which 12 molecular markers were identified that correlated with the seven traits of seedlessness. Identified markers can be used in a marker assisted selection to exclude seeded offsprings at an early stage breeding process. Luo et al. [23] used 280 RAPD primers to construct linkage map and found marker tightly linked to a major gene for resistance to downy mildew (*Plasmopara viticola)* (*RPv*-1). Similarly, Merdinoglu et al. [24] used 151 RAPD primers for linkage analysis related to downy mildew resistance.

6. Amplified fragment length polymorphism (AFLP)

The AFLP technique is the selective amplification of DNA fragments generated by restriction enzyme digestion. The AFLP approach enables simultaneous analysis of a large number of loci in a single assay, providing stable and reproducible marker patterns. AFLP, just as RAPD, are dominant markers, so are not suitable for parentage analysis. In grapevine germplasm analysis, the AFLP technique has mainly been used to assess genetic similarities among different varieties and to study genetic relationships among grapevines. Fanizza et al. [25] studied genetic relationships among aromatic grapevines varieties by the use of AFLP markers. The result of cluster analysis showed a separation between Moscato and Malvasia varieties but no grouping of *V. vinifera* varieties into aromatic and non-aromatic grapevines could be made, as had been done by some ampelographers in the past. AFLP markers were used for the characterization of a collection of 35 table grapevine varieties [26]. They detected that genetic similarity among them varied between 0.65 and 0.90, while sibling varieties derived from the same cross showed a genetic similarity over 0.80. AFLP analysis enabled distinction of all 35 analyzed cultivars and can be a powerful technique in identifying variety specific polymorphic fragments for distinguishing table grapevine cultivars.

AFLP markers have also been applied for assessing intra-varietal variability and for differentiation between clones of the same variety. The variety Flame Seedless, characterized by earlier bud burst, was differentiated from its parental genotype by analysis of 64 AFLP primer combinations. Two markers were identified, which were unique either only to the mutant or only to the parental line [27]. Cervera et al. [28] analyzed the intra-varietal diversity of 31 accessions called Tempranillo or described as a synonym of this Spanish cultivar. Two AFLP primer combinations generated 95 markers, indicating that the cultivar Tempranillo consists of various clones, with a genetic similarity over 0.97. Tomić [29] analyzed 56 samples from 5 locations of the Bosnian and Herzegovinian cultivar Žilavka by AFLP markers in order to assess intra-cultivar heterogeneity in the Herzegovina region. No clustering of Žilavka samples in relation to the location or names of the samples was detected. AFLP results showed high intra-varietal variability of cultivar Žilavka, expressing average polymorphism above 50.

AFLP have been used together with microsatellite markers in various studies in order to analyze genetic diversity within a single cultivar [30,31]; to evaluate genetic relatedness [32,33] or to identify and characterize grapevine rootstocks [34].

AFLP markers have also been used a great deal for the construction of genetic linkage [35-40], primarily aimed at mapping markers closely linked to important grapevine traits. For example, resistance to powdery mildew is controlled by single locus *Run* derived from *M. rotundifolia.* Pauquet et al. [41] identified 13 AFLP markers linked to *Run*1 and constructed a local map around the gene. Three markers out of 13 were shown to be always present in all resistant genotypes (absent in susceptible), which makes them a good diagnostic tool for selection for resistance.

7. Short sequence repeats (SSRs) – microsatellites

Microsatellites have become widely used genetic markers for the characterization of grapevine germplasm. Microsatellites are short (1-5 bp), tandemly repeated DNA sequences that are ubiquitous, abundant and highly dispersed in genomes. The variability of length of microsatellites is caused by changes in the number of repeats units, which can be easily detected by PCR, thus providing highly informative markers. The advantage of microsatellite markers is their co-dominant inheritance, as well as high polymorphism in terms of size due to the variable number of tandem repeats. Reproducibility and standardization of the SSR technique is easy to achieve but this marker system requires prior knowledge of primer binding, which increases the cost inputs for markers development. SSR markers are used for the identification of cultivars, revealing synonyms and homonyms, pedigree reconstruction and genetic relatedness, as well as population genetic studies, genome mapping and for marker assisted selection [3,12].

Large microsatellite sets of data in grapevines have been generated by numerous studies worldwide. Many of them are available in published papers and various on-line databases. The public availability of microsatellite genetic profiles of genotyped grapevine cultivars enables comparison of the obtained data, thus allowing even wider characterisation by confirmation of trueness-to-typeness and elimination of duplicates.

Vitis microsatellites markers have been developed within various laboratories [42-49]. Microsatellite primer sequences from these studies are available in the literature. Thomas and Scott [42] identified 26 grapevine cultivars, 6 *Vitis* species and *Muscadinia rotundifolia* L. by means of microsatellites. They established five microsatellite loci (VVS1, VVS2, VVS3, VVS4 and VVS5) from the genomic library of *V. vinifera* L. cultivar Sultana, of which VVS2 and VVS5 were shown to be the most polymorphic ones. Thomas et al. [43] and Cipriani et al. [2] used the same microsatellites for accurate and reliable identification of 80 and 16 grapevine cultivars, respectively. Bowers et al. [44] developed four new microsatellite loci (VVMD5, VVMD6, VVMD7 and VVMD8) from the genomic library of *V. vinifera* L. cultivar Pinot Noir. Seventy-seven cultivars of *V. vinifera* L. were analyzed and all four loci showed high polymorphism, with PIC values over 75%. Bowers et al. [45] developed an additional 22 VVMD loci for CT repeat motifs, initially cloned from the genomic library of Pinot Noir and Cabernet Sauvignon. They analyzed 51 to 347 cultivars, respectively, and twelve markers out of 22 proved to be polymorphic (VVMD6, VVMD8, VVMD17, VVMD21, VVMD24, VVMD25,

VVMD26, VVMD27, VVMD28, VVMD31, VVMD32 and VVMD36). An Austrian research group developed 15 markers from *Vitis riparia* [46, 50]. Two out of 15 loci did not amplify in *V. vinifera,* while the remaining 13 (ssrVrZAG7, ssrVrZAG15, ssrVrZAG21, ssrVrZAG25, ssrVrZAG29, ssrVrZAG30, ssrVrZAG47, ssrVrZAG62, ssrVrZAG64, ssrVrZAG67, ssrVrZAG79, ssrVrZAG83 and ssrVrZAG112) were successively analyzed in 120 cultivars. Four to fifteen alleles per locus were detected and expected heterozygosity ranged between 0.37 and 0.88. The highest information content was provided by locus ssrVrZAG79 (PI 0.05) because of the even distribution of the frequencies of the 13 alleles found. The remaining most informative markers were ssrVrZAG47, ssrVrZAG62, ssrVrZAG64 and ssrVrZAG67. Microsatellite loci from previous research with the highest values of polymorphic content are mainly used in microsatellite studies of grapevines. Loci VVS2 [42], VVMD5 and VVMD7 [44], VVMD27 [45], ssrVrZAG62 and ssrVrZAG79 [46] were chosen as a standard set of alleles for cultivar identification and distinction among cultivars [51], while loci VVMD25, VVMD28 and VVMD32 [45] have recently been used as additional microsatellite DNA markers for grapevines. Once microsatellite markers have been developed, they can be used for the analysis of different genotypes within a species and transferred between two different species within the same genus. Lefort et al. [52] designed primers for seven microsatellite loci (ssrVvUCH2, ssrVvUCH11, ssrVvUCH12, ssrVvUCH19, ssrVvUCH29, ssrVvUCH35 and ssrVvUCH40) from a microsatellite enriched genomic DNA library from the grapevine cultivar Syrah. These loci proved to be highly polymorphic for genotyping analysis of various *Vitis* species and hybrids used as rootstocks. These seven markers display high heterozygosity, all of them having a high number of amplified alleles, which makes them useful for genotype identification. Goto-Yamamoto et al.[49] also used cv. Syrah for development of new microsatellite markers. They developed 9 microsatellite primer pairs which have been successfully used for analysis of oriental and occidental cultivars, as well as for characterization of non-vinifera species (*V. labrusca, V. riparia* and *V. rotundifolia*).

Microsatellite studies of grapevines have many practical implications. The generation of unique cultivar profiles and assessment of true identity enables the genetic fidelity of planting material to be tested and offer solution to errors occurring through a long period of vegetative propagation. Identification and characterization of genetic material helps the selection of parents in breeding programmes and the sustainable management of germplasm collections. Microsatellite data obtained for a single genotype provide the microsatellite profile of that cultivar [3]. Since microsatellites have been shown to be a reliable tool for genotype identification, many research groups have adopted the technology and sets of microsatellite profiles have been increasing rapidly. This has enabled comparison of newly studied cultivars with those already genotyped. Comparison of genotypes of cultivars has revealed unique profiles of cultivars, as well as many cases of synonyms and homonyms. Microsatellites have been used for the identification of Portuguese cultivars [53], Greek cultivars [54], Spanish autochthonous grapevine varieties [55], Albanian [56] and Turkish varieties [57], old Slovenian varieties [58, 59]; Macedonian autochthonous varieties [60]; Algerian grapevine cultivars [61], Bulgarian cultivars [62], Romanian cultivars [63] and Bosnia and Herzegovina cultivars [64]. Microsatellites have proved to be reliable tools for identification and differentiation of grapevine rootstock [34, 50, 65].

In terms of the identification of grapevine cultivars, the question has been raised of the minimum sufficient number of loci required for accurate analysis of identity. In theory, five unlinked markers, each with five equally frequent alleles, could produce over 700,000 different genotypes [44]. In practice, this is not always easy to achieve and so the markers that are most informative should be selected for reliable discrimination [3]. Calculation of different genetic parameters has been used for assessing the informativeness of specific microsatellite loci. Counting alleles can overestimate the value of a given microsatellite locus due to the unequal distribution of alleles. Calculations that are based on allele frequencies are a more reliable measure of the informativeness of a locus. Two measures that are based on allele frequencies and genotype frequencies are probability of identity (probability of identical genotypes) (PI) and discrimination power (D) [3]. They describe the probability that two unrelated cultivars can be differentiated by a particular marker.

Discovering parentage and kinship analysis in grapevines is important for revealing the origin of particular cultivars. Selection of grapevines started almost seven centuries ago but reconstruction of the events that have led to the creation of specific cultivars is difficult. Many ancestors that could have provided evidence of the origin of grapevine cultivars have probably already become extinct [66]. Microsatellites have proved to be a reliable tool for parentage analysis, allowing the reconstruction of crosses. The origins of the widespread and best known grapevine cultivars from northeastern France were discovered by microsatellite analysis of 300 cultivars by 32 markers showing that Chardonnay, Gamay noir, Aligoté and Melon are the progeny of a single pair of parents, Pinot and Gouais blanc, dating from the Middle Ages [45]. Using 25 polymorphic microsatellite markers, Piljac et al. [67] analyzed possible parent progeny relationships within fourteen Croatian cultivars. Crespan [68] confirmed that the cultivar Muscat of Hamburg, which is a fine black table grape variety with a muscat flavour, is the progeny of Schiava Grossa × Muscat of Alexandria, which had been previously assumed in the literature. In this case, parentage was determined by analysis of chloroplast microsatellite loci. Since cytoplasm is inherited from the maternal side, it is possible to deduce the female parent. Microsatellites have been used to determine parent-offspring relationships among many grapevine cultivars. The cultivar Vitouska, which is grown in north-eastern Italy and western Slovenia, was shown to be the progeny of Prosecco and Malvasia Bianca Lunga, with one allele derived from each parent at 37 microsatellite loci [69]. The Italian important cultivar Sangiovese was shown to be the progeny of Ciliegiolo and Calabrese di Montenuovo confirmed by the high likelihood value [70]. Cardinal is one of the most successful table grapes and, after many decades, has remained the most used table grapevine variety grown worldwide, accounting for 20% of total production. This cultivar is a Californian grapevine created by E. Snyder and F. Harmon in 1939 and should have be derived from the cross between Flame Tokay and Alphonse Lavaleé, however microsatellite analyses did not confirm Flame Tokay as a maternal parent [71]. Cipriani et al. [72] analyzed a set of grapevines consisting of 1005 international, Italian national and local varieties. Altogether, 211 putative trios (2 parents and their offspring) were determined, of which 94 were designated with high confidence (95%), 19 with relaxed confidence (80%) and the remainder with an assigned confidence level. The assigned confidence level was due to an inability to select one parent of the pair, amongst a number of candidates with equal probability. Finally, 74 complete pedigrees were found, some of which were already known and some newly revealed. Recently, a total of 138 grapevine cultivars collected in five countries from the Balkan Peninsula were analyzed using 22 microsatellite loci. Kinship analysis resulted in various trios. Some were false trios because the apparent parent-offspring relationship was a result of near synonyms (clones or siblings). In the set of 138 samples, one unknown parentage [Furmint (Knipperlé, Ortlieber) = Pinot Noir × Rebula Stara] was revealed and one pedigree related to Serbian cultivars already reported in the literature (Župljanka = Pinot Noir × Prokupac) was confirmed. The microsatellite analysis also gave the first evidence of the origin of cv. Žilavka, most widespread autochthonous cultivar in Bosnia and Herzegovina. However, the pedigree of Serbian cultivar Petra was found to be false as the origin of cv. Godominka [73].

Microsatellites can be also used for determining the parentage of grapevine rootstock. For example, microsatellite analysis confirmed that the rootstock Fercal, which is important due to its high tolerance to limestone chlorosis, is the progeny of B.C.n°1B and 31 Richter [74]. Pedigree analysis should usually be confirmed by ampelographic observations, since misnaming and mislabeling of samples cannot be entirely excluded. Successful reconstruction of many pedigrees depends on the availability of ancient cultivars and pedigree data of cultivars.

The first genetic map based on microsatellite markers was developed by Riaz et al. [75]. The mapping population was represented by 153 progeny plants from a cross of Riesling and Cabernet Sauvignon and 152 microsatellite markers were mapped to the 20 linkage groups (LG), with an average distance between markers of 11.0 cM. Adam Blondon et al. [76] developed a second microsatellite reference map, consisting of 245 SSR markers, which was derived from the progeny of Syrah and Grenache. This map was more saturated, with 6.5 new markers per linkage group. These reference microsatellite genetic linkage maps have been further used for the fine mapping and QTL analysis.

Resistance locus *Run*1 was located by the microsatellite marker VMC4f3.1 [77], placed within LG12. A single dominant allele, designated *Ren*1, represents another source of resistance to powdery mildew (resistance to *Erysiphe necator* 1). Hofmann et al. [78] deduced that the closest markers to the *Ren*1 locus were microsatellite loci VMC9H4-2, VMCNG4E10-1 and UDV-020, assigned to LG13. Downy mildew resistance is inferred by the unique major gene *Rpv*1 and was found to be closely linked to *Run*1. Microsatellite loci that were mapped on the same linkage group have been shown to have a high correlation with the *Rpv*1 [24]. In relation to the presence of different flower types in grapevines (female, hermaphroditic and male), a cross between male and hermaphroditic plants was performed. The segregating ratio was 1:1 of these two types, assuming a single-locus hypothesis. The microsatellite locus VVS3 was shown to be close to the sex locus, which was mapped on LG2 [35]. Fernandez et al. [79] discovered the microsatellite locus linked to the fleshless berry mutation (*flb* locus) on LG18 (VMC2A3), while a seed development inhibitor, the *Sdl* locus, related to seedlessness, was also mapped on LG18, close to microsatellite VMC7F2 [39, 40]. Microsatellite maps have also been used for QTL mapping as for example, microsatellite markers VVS2 and VMC6G1 showed tight linkage to the magnesium deficiency QTL [80].

8. Single nucleotide polymorphism (SNP)

Advanced sequencing technologies have made available ever more sequence data, which can be used for marker development, particularly single nucleotide polymorphism (SNP). SNPs are sites in genomes where mutations naturally occur as a single nucleotide exchange (base substitutions), as a consequence of either transition or transversion events [12]. One locus of an SNP can comprise two, three or four alleles [12] but SNPs are rather biallelic markers, representing two alleles that may differ in a given nucleotide position in a diploid genome. SNPs are highly abundant, their density depends on the genome region and they differ among organisms. They are usually categorized according to their position in the genome and their effect on coding or regulatory sequences. Exonic SNPs that do not cause a change in the amino acid composition in the coded protein are synonymous SNPs, while SNPs causing a change in amino acid are non-synonymous SNPs. Non-synonymous SNPs that affect the protein function, thus influencing the phenotype, are called diagnostic SNPs. Diagnostic SNPs may be linked to specific important traits and their detection is one of the most important aims of discovering and developing SNPs.

A number of methods for SNP discovery and genotyping are available, although not all of them are equally useful nor it is clear which is the most suitable and most efficient [81]. The discovery of SNPs can usually be done by either a database search or an experimental approach. Most SNPs are extracted from expressed sequence tag (EST) databases [12]. In the experimental approach, candidate genes or genome regions are screened for the presence of SNPs by a series of techniques, such as microchip hybridization, direct sequencing or electrophoresis of PCR fragments containing candidate sequences on DNA single strand conformation polymorphism (SSCP) or denaturing gradient (DGGE) gels [12, 81]. SNP genotyping techniques can be classified into various groups: direct sequencing, cleaved amplified polymorphic sequences (CAPS), allele-specific PCR, allele specific primer extension, allele specific oligonucleotide hybridization etc. [12].

In *Vitis*, the identification and detection of single nucleotide polymorphisms for the development of molecular marker systems have recently dramatically increased with the publication of whole genome sequences [82, 83]. Previously, Salmaso et al. [84] scanned grapevine genes (sugar metabolism, cell signalling, anthocyanin and defence related pathways) to explore the possibility of developing an SNP marker system. Seven *V. vinifera* L. cultivars, the American species *V. riparia* L. and one complex hybrid were analysed for the distribution of SNPs along the gene fragments in order to assess the frequency and type of SNPs, nucleotide diversity, haplotypes and polymorphic information content using SSCP on none-denaturing gel electrophoresis and DNA re-sequencing of PCR amplicons. They discovered 247 SNPs among analysed genotypes which present useful markers for genetic analysis in grapevine.

Troggio et al. [81] also successfully used SSCP methodology and mini-sequencing for the development of SNP markers in grapevines, showing this to be an affordable mid-throughput methodology, which could be used for medium sized marker assisted selection projects.

Dong et al. [85] developed 21 primer pairs from grapevine EST sequences, generating 144 sequences by PCR amplification which revealed 154 SNPs. A phylogenetic tree was constructed from these data, which discriminated well among the analyzed 16 cultivars (11 Eurasian and 5 Euramerica cultivars), proving SNPs to be effective for grapevine genotype identification.

Lijavetzky et al. [86] employed high throughput SNP discovery approach for analysing 230 gene fragments of eleven genotypes. The approach enabled the discovery of 1573 SNPs of which 96 were submitted to high throughput genotyping technology for marker development. 80 SNPs were successfully genotyped in 360 grapevine genotypes, with a success rate of 93.5% within a sample.

At the start of large-scale development of SNP markers, low and mid throughput methods were available for SNP detection and identification of grapevines. Pindo et al. [87] provided a high throughput SNP genotyping method (SNPlex genotyping system), which correlated with the completion of the sequencing of the heterozygous genome of Pinot Noir [83]. About 950 candidates SNP from non-repetitive contigs of the assembled genome of Pinot Noir, were tested on 90 progeny of a Syrah × Pinot Noir cross. They obtained 563 new eSNPs and mapped them according to their quality values. This methodology was shown to be accurate and reproducible, and the high level of throughput enabled analysis of several hundred SNP in a hundred samples at the same time. Myles et al. [88] identified 469,470 SNPs from reduced representation libraries from 17 grapevine samples (10 *V. vinifera* L. cultivars and 7 wild species), which were sequenced using sequencing-by-synthesis technology. A subset consisting of 8898 SNPs were validated which are referred to as a Vitis9KSNP genotyping array. This 9K array demonstrated the power to distinguish between *V. vinifera* L. cultivars, hybrids and wild species, resolving the genetic relationships among diverse cultivars.

Cultivar identification is one of the many applications of the various marker systems. In relation to the greatly used microsatellites, it has been proved that six SSR loci are enough for genetic identification of most cultivars, with a cumulative probability of identity of 4.3 × 10-9 [51]. Lijavetzky et al. [86] found that SNP markers generated a lower PIC than microsatellites, thus requiring a higher numbers of markers to achieve similar PI values. It has been estimated that 20 SNPs with a minor allele frequency above 0.30 are needed to achieve a similar PI as when six SSR loci are used. The advantage of SNPs is reflected in their bi-allelic nature, since there are still frequent problems of microsatellite allele identification among different labs using different techniques for allele separation.

A set of 48 SNPs was proposed as a standard set for grapevine genotyping [89]. For successful genotyping, these 48 SNPs were chosen from an initial set of 332 SNPs, and are showing high information content, small minor allele frequency and are equally distributed across 17 chromosomes of grapevine (2-3 SNPs per chromosome). They have similar discrimination power to a set of 15 microsatellite markers.

SNPs markers have been shown to be efficient in parentage/offspring and kinship analysis. Zinelabidine et al. [90] used SNP markers to assess the role of the cultivar Cayetana Blanca in terms of its genetic relationships with other Iberian and Mediterranean cultivars. A total of 427 cultivars were analyzed as possible parent candidates, using 243 SNPs. It was discovered that Cayetana Blanca is a putative parent of several other Iberian varieties. Cayetana

Blanca and Alfrocheiro Preto gave rise to 5 cultivars used in Portugal and found in this study to be sibling cultivars. Cayetana Blanaca parents remain unknown but the analysis indicated that this cultivar is the progenitor of several cultivars that are grown on the Iberian Peninsula, thus also being of Iberian origin.

SNP markers are useful in genetic mapping studies particulary in search of trait-linked markers. SNP markers highly associated with berry weight variability in grapevines have been identified. While searching for SNP markers linked to the fleshless berry mutation, 554 SNPs were identified along the *flb* region (assumed to comprise four genes involved in berry weight variation). This nucleotide diversity demonstrated by the discovered SNPs could be further used for developing a genotyping chip useful for fine mapping of the *flb* gene and analysis of genetic diversity [91]. Emanuelli et al. [92] confirmed the role of the candidate gene *VvDXS* in determining the muscat flavour in grapevines. This study revealed three SNPs that are significantly associated with muscat flavoured varieties, while an SNP in the coding region of *VvDXS* has been suggested as the causal gain of function mutation. Polymorphisms in the nucleotide sequence of *VvDXS* could be applied in marker assisted selection for rapid screening of seedlings for their potential to express muscat flavour.

Single nucleotide polymorphisms represent a new generation marker system that is nowadays compared favourably to the greatly used microsatellite markers in grapevines. The major advantage of SNPs is their higher abundance within a genome, and they are more present in coding regions with a high possibility of being trait linked in genome mapping. Since the assessment of the grapevine genome sequence of a highly homozygous genotype [82] and heterozygous clone of Pinot Noir [83], high throughput methodologies for SNP detection and identification have become available, with the results easily transferable between different laboratories. This transferability is also reflected in the bi-allelic nature of SNPs as opposed to the allele bining related to microsatellites, and no use of reference cultivars is needed. The allele bining issue in microsatellites has been partially overcome with the discovery of 3 to 5 core repeats and microsatellites still remain markers with higher PIC values than SNPs.

Author details

Lidija Tomić[1,2*], Nataša Štajner[2] and Branka Javornik[2]

*Address all correspondence to: lidija.tomic@agrofabl.org

1 University of Banjaluka Faculty of Agriculture, Bosnia and Herzegovina

2 University of Ljubljana Biotechnical Faculty, Slovenia

References

[1] Pelsy, F. (2010). Molecular and cellular mechanisms of diversity within grapevine varieties. *Heredity*, 104(4), 331-340.

[2] Cipriani, G., Frazza, G., Peterlunger, E., & Testolin, R. (1994). Grapevine fingerprinting using microsatellite repeats. *Vitis*, 33(4), 211-215.

[3] Sefc, K. M., Lefort, F., Grando, MS, Scott, K., Steinkellner, H., & Thomas, M. (2001). Microsatellite markers for grapevine: A state of the art. *Amsterdam: Kluwer Publishers*.

[4] Bachmann, O. (1994). Peroxidase isoenzyme patterns in Vitaceae. *Vitis*, 33(3), 151-153.

[5] Royo, J. B., Cabello, F., Miranda, S., Gogorcena, Y., Gonzalez, J., Moreno, S., et al. (1997). The use of isoenzymes in characterization of grapevines (Vitis vinifera L). Influence of the environment and time of sampling. *Scientia Horticulturae*, 69(3-4), 145-155.

[6] Sanchez-Escribano, E., Ortiz, J. M., & Cenis, J. L. (1998). Identification of table grape cultivars (Vitis vinifera L.) by the isoenzymes from the woody stems. *Genetic Resources and Crop Evolution*, 45(2), 173-179.

[7] Jahnke, G., Korbuly, J., Majer, J., & Molnar, J. G. (2007). Discrimination of the grapevine cultivars'Picolit' and'Keknyelu' with molecular markers. *Scientia Horticulturae*, 114(1), 71-73.

[8] Bourquin, J. C., Tournier, P., Otten, L., & Walter, B. (1992). Identification of sixteen grapevine rootstocks by RFLP and RFLP analysis of nuclear DNA extracted from the wood. *Vitis*, 31(3), 157-162.

[9] Bourquin, J. C., Otten, L., & Walter, B. (1991). Identification of grapevine root-stocks by RFLP. *Comptes Rendus De L Academie Des Sciences Serie Iii-Sciences De La Vie-Life Sciences*, 312(12), 593-598.

[10] Bourquin, J. C., Sonko, A., Otten, L., & Walter, B. (1993). Restriction fragment length polymorphism and molecular taxonomy in Vitis vinifera L. *Theoretical and Applied Genetics*, 87(4), 431-438.

[11] Bourquin, J. C., Otten, L., & Walter, B. (1995). PCR-RFLP analysis of Vitis, Ampelopsis and Parthenocissus and its application to the identification of rootstocks. *Vitis*, 34(2), 103-108.

[12] Weising, K., Nybom, H., Wolff, K., & Kahl, G. (2005). DNA fingerprinting in plants: principles, methods and applications:. *CRC Press Taylor & Francis Group*.

[13] Collins, G., & Symons, R. (1993). Polymorphisms in grapevine DNA detected by the RAPD PCR technique. *Plant Molecular Biology Reporter*, 11(2), 105-112.

[14] Jeanjaques, I., Defontaine, A., & Hallet, J. N. (1993). Characterization of Vitis vinifera cultivars by random amplified polymorphic DNA markers. *Vitis*, 32(3), 189-190.

[15] Grando, M. S., Demicheli, L., Biasetto, L., & Scienza, A. (1995). RAPD markers in wild and cultivated Vitis vinifera. *Vitis*, 34(1), 37-39.

[16] Stavrakakis, M. N., Biniari, K., & Hatzopoulos, P. (1997). Identification and discrimination of eight Greek grape cultivars (Vitis vinifera L.) by random amplified polymorphic DNA markers. *Vitis*, 36(4), 175-178.

[17] Kocsis, M., Jaromi, L., Putnoky, P., Kozma, P., & Borhidi, A. (2005). Genetic diversity among twelve grape cultivars indigenous to the Carpathian Basin revealed by RAPD markers. *Vitis*, 44(2), 87-91.

[18] Ergul, A., Marasali, B., & Agaoglu, Y. S. (2002). Molecular discrimination and identification of some Turkish grape cultivars (Vitis vinifera L.) by RAPD markers. *Vitis*, 41(3), 159-160.

[19] Shubhada, A. T., Patil, S. G., & Rao, V. S. (2001). Assessment of the genetic diversity of some important grape genotypes in India using RAPD markers. *Vitis*, 40(3), 157-161.

[20] This, P., Cuisset, C., & Boursiquot, J. M. (1997). Development of stable RAPD markers for the identification of grapevine rootstocks and the analysis of genetic relationships. *American Journal of Enology and Viticulture*, 48(4), 492-501.

[21] Lodhi, M. A., Daly, M. J., Ye, G. N., Weeden, N. F., & Reisch, B. I. (1995). A molecular marker based linkage map of Vitis. *Genome*, 38(4), 786-794.

[22] Striem, M. J., Ben-Hayyim, G., & Spiegel-Roy, P. (1996). Identifying molecular genetic markers associated with seedlessness in grape. *Journal of the American Society for Horticultural Science*, 121(5), 758-763.

[23] Luo, S. L., He, P. C., Zhou, P., & Zheng, X. Q. (2001). Identification of molecular genetic markers tightly linked to downy mildew resistant genes in grape. *Acta genetica sinica*, 28(1), 77-82.

[24] Merdinoglu, D., Wiedeman-Merdinoglu, S., Coste, P., Dumas, V., Haetty, S., Butterlin, G., et al. (2003). Genetic analysis of downy mildew resistance derived from Muscadinia rotundifolia. *Acta Horticulturae*, 603, 451-456.

[25] Fanizza, G., Chaabane, R., Lamaj, F., Ricciardi, L., & Resta, P. (2003). AFLP analysis of genetic relationships among aromatic grapevines (Vitis vinifera). *Theoretical and Applied Genetics*, 107(6), 1043-1047.

[26] Cervera, M. T., Cabezas, J. A., Sanchez-Escribano, E., Cenis, J. L., & Martinez-Zapater, J. M. (2000). Characterization of genetic variation within table grape varieties (Vitis vinifera L.) based on AFLP markers. *Vitis*, 39(3), 109-114.

[27] Scott, K. D., Eggler, P., Seaton, G., Rossetto, M., Ablett, E. M., Lee, L. S., et al. (2000). Analysis of SSRs derived from grape ESTs. *Theoretical and Applied Genetics*, 100(5), 723-726.

[28] Cervera, M. T., Cabezas, J. A., Rodriguez-Torres, I., Chavez, J., Cabello, F., & Martinez-Zapater, J. M. (2002). Varietal diversity within grapevine accessions of cv. *Tempranillo. Vitis,* 41(1), 33-36.

[29] Tomić, L. (2009). Genetic characterization of the grapevine variety Žilavka (Vitis vinifera L.) with DNA markers. *MSc thesis. University of Ljubljana.*

[30] Moncada, X., & Hinrichsen, P. (2007). Limited genetic diversity among clones of red wine cultivar'Carmenere' as revealed by microsatellite and AFLP markers. *Vitis,* 46(4), 174-180.

[31] Fanizza, G., Lamaj, F., Resta, P., Ricciardi, L., & Savino, V. (2005). Grapevine cvs Primitivo, Zinfandel and Crljenak kastelanski: Molecular analysis by AFLP. *Vitis,* 44(3), 147-148.

[32] Fossati, T., Labra, M., Castiglione, S., Failla, O., Scienza, A., & Sala, F. (2001). The use of AFLP and SSR molecular markers to decipher homonyms and synonyms in grapevine cultivars: the case of the varietal group known as "Schiave". *Theoretical and Applied Genetics,* 102(2-3), 200-205.

[33] Labra, M., Winfield, M., Ghiani, A., Grassi, F., Sala, F., Scienza, A., et al. (2001). Genetic studies on Trebbiano and morphologically related varieties by SSR and AFLP markers. *Vitis,* 40(4), 187-190.

[34] Upadhyay, A., Saboji, M. D., Reddy, S., Deokar, K., & Karibasappa, G. S. (2007). AFLP and SSR marker analysis of grape rootstocks in Indian grape germplasm. *Scientia Horticulturae,* 112(2), 176-183.

[35] Dalbo, M. A., Ye, G. N., Weeden, N. F., Steinkellner, H., Sefc, K. M., & Reisch, B. I. (2000). A gene controlling sex in grapevines placed on a molecular marker-based genetic map. *Genome,* 43(2), 333-340.

[36] Doligez, A., Bouquet, A., Danglot, Y., Lahogue, F., Riaz, S., Meredith, C. P., et al. (2002). Genetic mapping of grapevine (Vitis vinifera L.) applied to the detection of QTLs for seedlessness and berry weight. *Theoretical and Applied Genetics,* 105(5), 780-795.

[37] Fischer, B. M., Salakhutdinov, I., Akkurt, M., Eibach, R., Edwards, K. J., Topfer, R., et al. (2004). Quantitative trait locus analysis of fungal disease resistance factors on a molecular map of grapevine. *Theoretical and Applied Genetics,* 108(3), 501-515.

[38] Doucleff, M., Jin, Y., Gao, F., Riaz, S., Krivanek, A. F., & Walker, M. A. (2004). A genetic linkage map of grape, utilizing Vitis rupestris and Vitis arizonica. *Theoretical and Applied Genetics,* 109(6), 1178-1187.

[39] Mejia, N., Gebauer, M., Munoz, L., Hewstone, N., Munoz, C., & Hinrichsen, P. (2007). Identification of QTLs for seedlessness, berry size, and ripening date in a. seedless x seedless table grape progeny. *American Journal of Enology and Viticulture,* 58(4), 499-507.

[40] Costantini, L., Battilana, J., Lamaj, F., Fanizza, G., & Grando, MS. (2008). Berry and phenology-related traits in grapevine (Vitis vinifera L.): From Quantitative Trait Loci to underlying genes. *BMC Plant Biology*, 8.

[41] Pauquet, J., Bouquet, A., This, P., & Adam-Blondon, A. F. (2001). Establishment of a local map of AFLP markers around the powdery mildew resistance gene Run1 in grapevine and assessment of their usefulness for marker assisted selection. *Theoretical and Applied Genetics*, 103(8), 1201-1210.

[42] Thomas, M. R., & Scott, N. S. (1993). Microsatellite repeats in grapevine reveal DNA polymorphisms when analyzed as sequence-tagged sites (STSs). *Theoretical and Applied Genetics*, 86(8), 985-990.

[43] Thomas, M. R., Cain, P., & Scott, N. S. (1994). DNA typing of grapevines: A universal methodology and database for describing cultivars and evaluating genetic relatedness. *Plant Molecular Biology*, 25(6), 939-949.

[44] Bowers, J. E., Dangl, G. S., Vignani, R., & Meredith, CP. (1996). Isolation and characterization of new polymorphic simple sequence repeat loci in grape (Vitis vinifera L.). *Genome*, 39(4), 628-633.

[45] Bowers, J., Boursiquot, J. M., This, P., Chu, K., Johansson, H., & Meredith, C. (1999). Historical genetics: The parentage of Chardonnay, Gamay, and other wine grapes of northeastern France. *Science*, 285(5433), 1562-1565.

[46] Sefc, K. M., Regner, F., Turetschek, E., Glossl, J., & Steinkellner, H. (1999). Identification of microsatellite sequences in Vitis riparia and their applicability for genotyping of different Vitis species. *Genome*, 42(3), 367-373.

[47] Arroyo-Garcia, R., Ruiz-Garcia, L., Bolling, L., Ocete, R., Lopez, M. A., Arnold, C., et al. (2006). Multiple origins of cultivated grapevine (Vitis vinifera L. ssp sativa) based on chloroplast DNA polymorphisms. *Molecular Ecology*, 15(12), 3707-3714.

[48] Di Gaspero, G., Cipriani, G., Marrazzo, M. T., Andreetta, D., Castro, M. J. P., Peterlunger, E., et al. (2005). Isolation of (AC)n-microsatellites in Vitis vinifera L. and analysis of genetic background in grapevines under marker assisted selection. *Molecular Breeding*, 15(1), 11-20.

[49] Goto-Yamamoto, N., Mouri, H., Azumi, M., & Edwards, K. J. (2006). Development of grape microsatellite markers and microsatellite analysis including oriental cultivars. *American Journal of Enology and Viticulture*, 57(1), 105-108.

[50] Sefc, K. M., Regner, F., Glossl, J., & Steinkellner, H. (1998). Genotyping of grapevine and rootstock cultivars using microsatellite markers. *Vitis*, 37(1), 15-20.

[51] This, P., Jung, A., Boccacci, P., Borrego, J., Botta, R., Costantini, L., et al. (2004). Development of a standard set of microsatellite reference alleles for identification of grape cultivars. *Theoretical and Applied Genetics*, 109(7), 1448-1458.

[52] Lefort, F., Kyvelos, C. J., Zervou, M., Edwards, K. J., & Roubelakis-Angelakis, K. A. (2002). Characterization of new microsatellite loci from Vitis vinifera and their conservation in some Vitis species and hybrids. *Molecular Ecology Notes*, 2(1), 20-21.

[53] Lopes, M. S., Sefc, K. M., Dias, E. E., Steinkellner, H., Machado, M. L. D., & Machado, A. D. (1999). The use of microsatellites for germplasm management in a Portuguese grapevine collection. *Theoretical and Applied Genetics*, 99(3-4), 733-739.

[54] Lefort, F., & Roubelakis-Angelakis, K. K. A. (2001). Genetic comparison of Greek cultivars of Vitis vinifera L. by nuclear microsatellite profiling. *American Journal of Enology and Viticulture*, 52(2), 101-108.

[55] Gonzalez-Andres, F., Martin, J. P., Yuste, J., Rubio, J. A., Arranz, C., & Ortiz, J. M. (2007). Identification and molecular biodiversity of autochthonous grapevine cultivars in the'Comarca del Bierzo', Leon, Spain. *Vitis*, 46(2), 71-76.

[56] Ladoukakis, E. D., Lefort, F., Sotiri, P., Bacu, A., Kongjika, E., & Roubelakis-Angelakis, K. A. (2005). Genetic characterization of Albanian grapevine cultivars by microsatellite markers. *Journal International des Sciences de la Vigne et du Vin*, 39(3), 109-119.

[57] Karatas, H., Degirmenci, D., Velasco, R., Vezzulli, S., Bodur, C., & Agaoglu, Y. S. (2007). Microsatellite fingerprinting of homonymous grapevine (Vitis vinifera L.) varieties in neighboring regions of South-East Turkey. *Scientia Horticulturae*, 114(3), 164-169.

[58] Štajner, N., Korošec-Koruza, Z., Rusjan, D., & Javornik, B. (2008). Microsatellite genotyping of old Slovenian grapevine varieties (Vitis vinifera L.) of the Primorje (coastal) winegrowing region. *Vitis*, 47(4), 201-204.

[59] Štajner, N., Rusjan, D., Korošec-Koruza, Z., & Javornik, B. (2011). Genetic characterization of old slovenian grapevine varieties of Vitis vinifera L. by microsatellite genotyping. *American Journal of Enology and Viticulture*, 62(2), 250-255.

[60] Štajner, N., Angelova, E., Božinović, Z., Petkov, M., & Javornik, B. (2009). Microsatellite marker analysis of Macedonian grapevines (Vitis vinifera L.) compared to the Bulgarian and Greek cultivars. *Journal International des Sciences de la Vigne et du Vin*, 43(1), 29-34.

[61] Laiadi, Z., Bentchikou, M. M., Bravo, G., Cabello, F., & Martinez-Zapater, J. M. (2009). Molecular identification and genetic relationships of Algerian grapevine cultivars maintained at the germplasm collection of Skikda (Algeria). *Vitis*, 48(1), 25-32.

[62] Dzhambazova, T., Tsvetkov, I., Atanassov, I., Rusanov, K., Zapater, J. M. M., Atanassov, A., et al. (2009). Genetic diversity in native Bulgarian grapevine germplasm (Vitis vinifera L.) based on nuclear and chloroplast microsatellite polymorphisms. *Vitis*, 48(3), 115-121.

[63] Gheorghe, R. N., Popescu, C. F., Pamfil, D., Ciocirlan, C. N., & Sestras, R. (2010). Genetic diversity of some Romanian grapevine cultivars as revealed by microsatellite markers. *Romanian Biotechnological Letters*, 15(2), 26-31.

[64] Tomić, L., Štajner, N., Jovanović-Cvetković, T., Cvetković, M., & Javornik, B. (2012). Identity and genetic relatedness of Bosnia and Herzegovina grapevine germplasm. *Scientia Horticulturae*, 143, 122-126.

[65] Crespan, M., Meneghetti, S., & Cancellier, S. (2009). Identification and genetic relationship of the principal rootstocks cultivated in Italy. *American Journal of Enology and Viticulture*, 60(3), 349-356.

[66] Sefc, K. M., Steinkellner, H., Glossl, J., Kampfer, S., & Regner, F. (1999). Reconstruction of a grapevine pedigree by microsatellite analysis. *Theoretical and Applied Genetics*, 97(1-2), 227-231.

[67] Piljac, J., Maletic, E., Kontic, J. K., Dangl, G. S., Pejic, I., Mirosevic, N., et al. (2002). The parentage of Posip bijeli, a major white wine cultivar of Croatia. *Vitis*, 41(2), 83-87.

[68] Crespan, M. (2003). The parentage of Muscat of Hamburg. *Vitis*, 42(4), 193-197.

[69] Crespan, M., Cabello, F., Giannietto, S., Ibanez, J., Kontic, J. K., Maletic, E., et al. (2006). Malvasia delle Lipari, Malvasia di Sardegna, Greco di Gerace, Malvasia de Sitges and Malvasia dubrovačka- synonyms of an old and famous grape cultivar. *Vitis*, 45(2), 69-73.

[70] Vouillamoz, J. F., Monaco, A., Costantini, L., Stefanini, M., Scienza, A., & Grando, MS. (2007). The parentage of'Sangiovese', the most important Italian wine grape. *Vitis*, 46(1), 19-22.

[71] Akkak, A., Boccacci, P., & Botta, R. (2007). Cardinal' grape parentage: a case of a breeding mistake. *Genome*, 50(3), 325-328.

[72] Cipriani, G., Spadotto, A., Jurman, I., Di Gaspero, G., Crespan, M., Meneghetti, S., et al. (2010). The SSR-based molecular profile of 1005 grapevine (Vitis vinifera L.) accessions uncovers new synonymy and parentages, and reveals a large admixture amongst varieties of different geographic origin. *Theoretical and Applied Genetics*, 121(8), 1569-1585.

[73] Tomić, L. (2012). Molecular characterization and analysis of the genetic relatedness of old grapevine (Vitis vinifera L.) cultivars from the Western Balkan. *PhD thesis. University of Ljubljana.*

[74] Laucou, V., Boursiquot, J. M., Lacombe, T., Bordenave, L., Decroocq, S., & Ollat, N. (2008). Parentage of grapevine rootstock'Fercal' finally elucidated. *Vitis*, 47(3), 163-167.

[75] Riaz, S., Dangl, G. S., Edwards, K. J., & Meredith, CP. (2004). A microsatellite marker based framework linkage map of Vitis vinifera L. *Theoretical and Applied Genetics*, 108(5), 864-872.

[76] Adam-Blondon, A. F., Roux, C., Claux, D., Butterlin, G., Merdinoglu, D., & This, P. (2004). Mapping 245 SSR markers on the Vitis vinifera genome: a tool for grape genetics. *Theoretical and Applied Genetics*, 109(5), 1017-1027.

[77] Barker, C. L., Donald, T., Pauquet, J., Ratnaparkhe, M. B., Bouquet, A., Adam-Blondon, A. F., et al. (2005). Genetic and physical mapping of the grapevine powdery mildew resistance gene, Run1, using a bacterial artificial chromosome library. *Theoretical and Applied Genetics*, 111(2), 370-377.

[78] Hoffmann, S., Di Gaspero, G., Kovacs, L., Howard, S., Kiss, E., Galbacs, Z., et al. (2008). Resistance to Erysiphe necator in the grapevine'Kishmish vatkana' is controlled by a single locus through restriction of hyphal growth. *Theoretical and Applied Genetics*, 116(3), 427-438.

[79] Fernandez, L., Doligez, A., Lopez, G., Thomas, M. R., Bouquet, A., & Torregrosa, L. (2006). Somatic chimerism, genetic inheritance, and mapping of the fleshless berry (flb) mutation in grapevine (Vitis vinifera L.). *Genome*, 49(7), 721-728.

[80] Mandl, K., Santiago, J. L., Hack, R., Fardossi, A., & Regner, F. (2006). A genetic map of Welschriesling x Sirius for the identification of magnesium-deficiency by QTL analysis. *Euphytica*, 149(1-2), 133-144.

[81] Troggio, M., Malacarne, G., Vezzulli, S., Faes, G., Salmaso, M., & Velasco, R. (2008). Comparison of different methods for SNP detection in grapevine. *Vitis*, 47(1), 21-30.

[82] Jaillon, O., Aury, J. M., Noel, B., Policriti, A., Clepet, C., Casagrande, A., et al. (2007). The grapevine genome sequence suggests ancestral hexaploidization in major angiosperm phyla. *Nature*, 449(7161), 463-468.

[83] Velasco, R., Zharkikh, A., Troggio, M., Cartwright, D. A., Cestaro, A., Pruss, D., et al. (2007). A high quality draft consensus sequence of the genome of a heterozygous grapevine variety. *PLOS ONE*, 2(12), e1326.

[84] Salmaso, M., Faes, G., Segala, C., Stefanini, M., Salakhutdinov, L., Zyprian, E., et al. (2004). Genome diversity and gene haplotypes in the grapevine (Vitis vinifera L.), as revealed by single nucleotide polymorphisms. *Molecular Breeding*, 14(4), 385-395.

[85] Dong, Q. H., Cao, X., Yang, G. A., Yu, H. P., Nicholas, K. K., Wang, C., et al. (2010). Discovery and characterization of SNPs in Vitis vinifera and genetic assessment of some grapevine cultivars. *Scientia Horticulturae*, 125(3), 233-238.

[86] Lijavetzky, D., Cabezas, J. A., Ibanez, A., Rodriguez, V., & Martinez-Zapater, J. M. (2007). High throughput SNP discovery and genotyping in grapevine (Vitis vinifera L.) by combining a re-sequencing approach and SNPlex technology. *BMC Genomics*, 8, 424.

[87] Pindo, M., Vezzulli, S., Coppola, G., Cartwright, D. A., Zharkikh, A., Velasco, R., et al. (2008). SNP high-throughput screening in grapevine using the SNPlex™ genotyping system. *BMC Plant Biology*, 8, 12.

[88] Myles, S., Chia, J. M., Hurwitz, B., Simon, C., Zhong, G. Y., Buckler, E., et al. (2010). Rapid genomic characterization of the genus Vitis. *PLOS ONE*, 5(1), e8219.

[89] Cabezas, J. A., Ibanez, J., Lijavetzky, D., Velez, D., Bravo, G., Rodriguez, V., et al. (2011). A 48 SNP set for grapevine cultivar identification. *Bmc Plant Biology*, 11.

[90] Zinelabidine, L. H., Haddioui, A., Rodriguez, V., Cabello, F., Eiras-Dias, J. E., Zapater, J. M. M., et al. (2012). Identification by SNP analysis of a major role for Cayetana Blanca in the genetic network of Iberian Peninsula grapevine varieties. *American Journal of Enology and Viticulture*, 63(1), 121-126.

[91] Houel, C., Bounon, R., Chaib, J., Guichard, C., Peros, J. P., Bacilieri, R., et al. (2010). Patterns of sequence polymorphism in the fleshless berry locus in cultivated and wild Vitis vinifera accessions. *BMC Plant Biology*, 10, 284.

[92] Emanuelli, F., Battilana, J., Costantini, L., Le Cunff, L., Boursiquot, J. M., This, P., et al. (2010). A candidate gene association study on muscat flavor in grapevine (Vitis vinifera L.). *BMC Plant Biology*, 10(1), 241.

Chapter 2

Application of Microsatellite Markers in Grapevine and Olives

Jernej Jakše, Nataša Štajner, Lidija Tomić and Branka Javornik

Additional information is available at the end of the chapter

http://dx.doi.org/10.5772/ 53411

1. Microsatellite markers

Since their discovery in the 80s, microsatellites have become a popular molecular marker for studying plant genomes and are still the marker system of choice for various applications, such as genetic diversity and genetic structure studies, fingerprinting of individuals, parentage analyses and mapping studies. Although they have been used as a PCR marker system for more than 20 years now [1, 2], the numerous recent publications on their use confirm their durability and relevance. This is mainly due to their intrinsic properties (associated high polymorphisms) and a constant evolution of the technical methodology in terms of high throughput, ease of use and price. The starting methodology was based on radioactive labelled amplified microsatellite alleles separated on polyacrilamide gels. Nowadays, highly multiplexed fluorescently labelled microsatellites are commonly genotyped in capillary based automatic systems.

1.1. Microsatellite specifications, nomenclature and definitions

Microsatellites are part of tandemly repeated sequences of the genome, where a specific core motif is repeated several times. The term microsatellite is coined from the term "satellite", which originates from DNA buoyant density gradient centrifugation experiments, in which DNA fragments with different base composition were separated from the main genomic DNA and formed a so-called "satellite" band. It was found that these satellite bands contain tandem arrays of repetitive sequences [3]. Based on the length of the core repeat unit, the repetitive DNA is classified as satellite, minisatellite or microsatellite DNA. While the repeat units in satellite and minisatellite DNA can be from 100 kb to over Mb and from 10 to 80 bp long, respectively, the core repeat unit of microsatellites is the shortest and in a range from 2 to 8 bp [4]. Some researchers also consider mononucleotide tracts (e.g., $(A)_n$) to be part of the microsatellite DNA [5], although they are less suitable for marker development and genotyping purposes, due to their properties. In some classifications, only repeats up to 5 bp are considered to be part of microsatellite DNA [6]. Nevertheless, the commonest targets for marker development are di-, tri- and tetranucleotide microsatellites.

In addition to microsatellites, several synonymous terms are used to describe the smallest class of tandem repeats. The term microsatellite was initially used to describe the most frequent human dinucleotide repeat $(CA)_n/(GT)_n$ [2] and various terms were used for other types. Synonyms are also often used for describing microsatellite sequences, such as "simple sequence repeats" (SSR), "short tandem repeats" (STR), and "variable number of tandem repeat" (VNTR). The VNTR term is particularly suitable for describing both microsatellite and minisatellite sequences and for bridging the gap between these two types [7]. Hancock [8] proposed that only the term microsatellite should be used, to avoid confusion. Based on the repeat type and its composition, the following nomenclature and classes of microsatellites have been proposed [7]:

a. a pure or perfect microsatellite consists of only one type of microsatellite repeat, e.g., $(AG)_{14}$, $(ACA)_9$,

b. a compound microsatellite consists of at least two different types of microsatellite repeats, e.g., $(CT)_{10}(AT)_{12}$,

c. an interrupted microsatellite (often also listed as imperfect) has a core sequence repeat interrupted by a short insertion of bases not following the repeat type, e.g., $(AG)_8CCC(AG)_{10}$; they can be of pure or compound type,

d. authors also use the term complex microsatellite, in which short arrays of repeats are interrupted by sequences that are themselves short repeats.

Another phrase that describes microsatellite-like sequences and is useful for proper annotation of such sequence arrays is cryptic simplicity [8]. Such regions resemble microsatellite repeats but are interrupted many times with irregularities. The authors suggested that these sequences are an intermediate stage during the birth or death of the microsatellite.

1.2. Microsatellite frequencies and distributions in plant genomes

Numerous publications deal with analysis of the frequencies and distributions of microsatellites. Citing all of them is beyond the scope of this chapter. We will highlight the first published papers related to database searches of plant sequences, and data on two model plants - rice and *Arabidopsis* - as representatives of monocot and dicot kingdoms. In grapevine, the genome sequence is available and positions of microsatellite sequences are known. In olive, however, the amount of sequence data is still scarce. Microsatellites were at first considered to be part of the "junk" part of the genome but there is planty of evidence today that they are also abundant in genes as part of promoters, UTRs, introns or even coding sequences.

The first surveys of publicly available sequence data of higher plants for the presence, abundance and ubiquity of di- and trinucleotide repeats were conducted in 1993 [9, 10]. They found that the most frequent dinucleotide repeats were $(AT)_n$ tracts with 74%, followed by $(AG)_n$/$(TC)_n$ with 24% and $(AC)_n$/$(TG)_n$ with 1%. These were the first publications to indicate the different frequencies of microsatellite repeats in plants compared to animals and humans, in which $(AC)_n$/$(TG)_n$ repeats are by far the most frequent class and the $(AT)_n$ type quite rare. The most abundant trinucleotide repeats were $(TAT)_n$ and $(TCT)_n$ microsatellites, accounting for 27.5% and 25%, respectively. Based on the volume of data they searched, they estimated that the average distance between microsatellites would be about 50 kb. With respect to the coding sequences, they found that 22% of dinucleotide types of repeats can be associated with the 5′ or 3′ UTR regions and introns, whereas trinucleotides can also be found in coding sequences. This is because the change in the repeat length of trinucleotide microsatellites does not disrupt the reading frame. A study by Lagercrantz et al. [9] augmented database search with Southern blot analyses of the microsatellite repeats. A study by Wang et al. [11] searched for microsatellite presence in organellar (1.2 Mb) and genomic (3 Mb) plant DNA sequences. They found a low frequency of organelle specific microsatellites, while in general confirming data found by Morgante and Olivieri [10]. Numerous publications followed, analyzing ever larger volumes of plant sequences or even whole genome data. The results mostly narrowed down the average distance between microsatellite loci, correcting the frequency distributions of specific repeats and highlighting species specific details.

A species specific search was conducted on a large set of rice sequences, with an emphasis on express sequence tags (ESTs) to develop markers for mapping [12]. The most abundant dinucleotides were $(GA)_n$ repeats, while among trinucleotides, GC rich repeats of $(CGG)_n$ and $(GAG)_n$ types were most common. The latter may be due to the higher GC content of Poales genomes [13] or the specific poly amino acid tracts present in certain coding sequences. The next rice study searched over 58 Mb of rice DNA sequences [14], which confirmed GC rich trinucleotides to be the most abundant microsatellites in the rice genome. The authors also noted the association of $(AT)_n$ microsatellites with miniature inverted-repeat transposable elements, which make them unusable for marker development. With the availability of whole genome sequences of rice [15], a complete genome survey of rice microsatellites was possible and a list was published of 18,828 perfect microsatellite repeats in a length > 20 bp, which behave as hypervariable loci. The whole genome scan confirmed previous reports that $(AT)_n$ and $(CCG)_n$ repeats are the most common ones in rice (> 35% and ~ 10%).

A study by Cardle et al. [16] investigated the expanding quantity of sequencing data in public databases and compared *Arabidopsis* genomic DNA sequences > 10 kbp and EST data searches with data of certain other plants. The results showed a lower frequency of microsatellites in EST data, with an average distance between microsatellite loci in genomic data being 6.04 kb and 14 kb for ESTs. In genomic data, the frequency of di- and trinucleotides was comparable, while in EST data trinucleotides were more than 2 times more abundant than dinucleotide repeats. Although the amount of genomic sequences from other plants was lower than with *Arabidopsis*, the average distance between microsatellite loci was comparable with *Arabidopsis*, being 7.4 kb in barley and 6.4 kb in potato. Finally, the *Arabidopsis* genome was the first sequenced plant genome to become available, at the end of 2000 [17]. A study by Morgante et al. [18], in which genome and EST sequences of *Arabidopsis* and 4 major crops were used to estimate microsatellite densities, showed that overall microsatellite frequency is related to the investigated genome size and the amount of its repetitive DNA, but the proportion of microsatellite sequences in the transcribed part of the genome remained constant. The authors concluded that plant microsatellites reside in the low-copy part of the genome, which predates known expansions that have occurred in many species.

Due to its economic and cultural importance and relatively small genome size, the genome sequence of the grapevine (highly selfed Pinot Noir and Pinot Noir) is available [19, 20] and the microsatellite content and distribution has been analyzed [20]. The authors reported on 73,853 microsatellite loci (2-8 bp core repeat unit length) totalling up to 1.8 Mb of the grapevine genome.

Olive is a rather neglected species in terms of the availability of sequences compared to other crops or fruit species. The largest available set of olive EST data was obtained by next generation sequencing methodology (454), by which several thousand microsatellites were detected in raw sequencing data [21]. The analyzed data are accessible through WWW available Olea EST db in which 13,636 unique sequences contain microsatellites (including mononucleotide tracts), representing 5.2% of total sequences.

1.3. Searching for microsatellites

Due to their high polymorphism, which is reflected in multi-allelic patterns at a particular locus, microsatellites are ideal targets for the development of molecular markers. Several strategies have been developed for this purpose, the most ideal of which is locus specific amplification of a microsatellite site by PCR [10]. For this purpose, the DNA sequences surrounding the microsatellite need to be known, so sequence data is required as the first step. Where species specific sequence information is not available, therefore, genomic libraries need to be developed and screened for the presence of microsatellites. These isolation methods can be classified as traditional and specific ones, implementing enrichment strategies and, recently, also next generation sequencing (NGS) approaches.

The traditional microsatellite isolation method makes use of a classical genomic library and Southern screening of such a library with a microsatellite sequence [22]. A problem of such an approach is screening several thousand bacterial clones to obtain only a few microsatellite sequences, due to the low frequency of microsatellite containing clones. This approach was used in the first studies of isolating grapevine microsatellites of VVS and VVMD sets, in which reports on 5 [23] and 4 [24] developed markers was published. The authors reported 0.5% and 1.2% of colonies being positive for two different dinucleotide microsatellites [24] and 0.6% of positive ones for one type of dinucleotide repeat [23]. The first microsatellite markers published for olive of ssrOeUA set were also isolated using the classical approach [25].

Because the traditional approach was very labour intensive, various enrichment strategies were adopted to increase the number of microsatellites in genomic libraries. Such strategies were based on different approaches, e.g., using a *dut/ung* bacterial selection [26] or hybridization capture using either biotylinated microsatellite probes and magnetic particles [27, 28] or microsatellite probes attached to small pieces of nylon membrane [4, 29, 30]. These procedures substantially increased the proportion of microsatellite sequences in libraries up to 95%, which in some cases enabled skipping the tedious Southern screening of the library. Such approaches were used in the discovery and development of additional microsatellite markers for olive [31-33] and *Vitis* species [34].

The emergence of NGS enabled a quantum leap in microsatellite discovery, since massive sequencing enabled the production of a huge amount of sequencing data for several species at the same time [35, 36]. The Southern screening step is no longer needed with the NGS approach.

Where larger amounts of species specific DNA sequences are available, they can be mined for microsatellite repeats using devoted software tools, omitting the costly step of library development. A comprehensive overview of mining tools with specific characteristics and their limitations is available [37]. The database mining approach has been used extensively for mining new microsatellite markers in grapevine, for which public DNA sequences were already available [38, 39].

1.4. Genotyping methodology

Several advances in genotyping methodology enable studies partially to automate the process, populate data in real time and to compare and store the genotyping data easily and efficiently. Inter-laboratory comparison of the genotyping data has become easy. All advances have sought to achieve two goals to make genotyping faster and cheaper. Microsatellite genotyping has basically followed the advances of Sanger sequencing, since the same equipment and methodology is used – separating the fragment within a resolution of 1 bp.

Initially, thin denaturating polyacrilamide gels were used and fragments visualized either by means of radioactive nucleotides [2] or radioactively labelled primers [1] or, in laboratories without "hot rooms", silver staining procedures were adopted [40].

Automated laser induced fluorescence sequencing revolutionized DNA sequencing and the first fluorescent dyes were introduced, which were also successfully adopted for genotyping purposes. Equipment still relied on polyacrilamide gel electrophoresis but was able to acquire the data in real-time and no post gel handling was required. Gel based systems were later replaced by capillary ones, whereby a substantial breakthrough in automated sample handling was achieved. These systems are nowadays widely used in microsatellite genotyping applications.

Another achievement that can speed up analysis and reduce the costs is multiplexing – a procedure by which several microsatellite loci are co-amplified together in a single tube. The procedure relies on non-overlapping allele sizes of the loci used and on using different fluorescent fluorophores. Up to five different fluorescent dyes can nowadays be used simultaneously in genotyping applications. Multiplexing requires careful development of primers and precise determination of optimal reaction conditions to achieve co-amplification of several loci, since interactions during PCR are more likely to occur when several loci are amplified together. A multiplexing approach has been developed for grapevine [41]. An easier approach that is often used is post-PCR multiplexing, in which single loci amplifications are pooled together after PCR and separated in a single lane [42]. A problem associated with the use of fluorescently labelled primers is the high price of the dye. An economic labelling method, based on the elongation of one primer for a common sequence and using a third labelled primer in a PCR reaction, has been developed [43] and is now widely used, especially when a new set of markers is in the developing and optimisation phase.

2. Application of microsatellite markers in grapevine

2.1. Microsatellite marker development

Methods that enable analysis at the level of cultivar genotype have been developed because identification of grapevine cultivars based on morphological differences between plants may be incorrect due to the influence of ecological factors. In the last twenty years, various techniques for the characterization of cultivars at the level of DNA (RFLP, RAPD, AFLP, SCAR and SSR markers) and isoenzymes have been established, of which the most appropriate for genotyping are those using microsatellite markers. Microsatellites, in addition to some basic applications, allow the identification and determination of genetic relationships and the origin of varieties and grapevines preserved in collections or found only in vineyards, where they are usually grown only to a minor extent. Many grapevine varieties have several synonyms, meaning that they have different names, although they carry an identical genotype, which can be proved by analysis of microsatellite loci. In some cases, there are also groups or pairs of varieties that have the same or a very similar name but a different genetic background; such varieties are called homonyms.

Microsatellites or simple sequence repeats (SSRs) have proved to be the most effective markers for grapevine genotyping [24, 44-50]. Many microsatellites are highly variable both within and between species. The polymorphism between individuals is mainly accounted for changes in the number of repetitions of the basic motif [51]. The great variability of microsatellites is associated with the fact that from 10^4 to 10^5 microsatellite loci are randomized in the genome of eukaryotes, which means a large number of polymorphic sites that can be used for genetic markers. Because of the high mutation rate of microsatellite sequences, they are highly informative molecular markers, with a maximum value of polymorphism information and as such have been established for the identification of grapevine cultivars.

Thomas et al. [24] first used microsatellites for the identification of grapevine cultivars and demonstrated that microsatellite sequences are often represented in the grapevine genome and are very informative for the identification of *V. vinifera* cultivars. Detection of microsatellite polymorphism by the PCR technique is fast, easy and efficient, even with a very low quantity of DNA, which means that in the case of grapevine, products such as must and wine can be used for DNA analysis instead of plant tissue [52, 53]. Because of these characteristics, microsatellites have proved to be very effective as molecular markers for genotyping, identification studies, for solving dilemmas of synonyms, homonyms or the origin of varieties, relatedness studies, for population genetic studies, for the identification of clones and for marker assisted selection.

2.2. Comparison of developed markers

Microsatellites are known to have different mutation rates between loci [54] and there are several potential factors that contribute to the diverse dynamics of the development of microsatellite sequences: the number of repetitions, type of repeat sequence motif, the length of repeat units, interruptions in microsatellite, flanking regions, recombination rate etc.

Hundreds of microsatellite markers for grapevines have been developed and most of them are publicly available [23, 38, 41, 55-60], large set also by the Vitis Microsatellite Consortium by the company Agrogene (France). The extraordinary potential of some of them and their usefulness in determining grapevine cultivars and rootstocks has been demonstrated in many studies and they have been used for identification in most European winegrowing regions. A set of six (VVS2, VVMD5, VVMD7, VVMD27, VrZag62, VrZAG79) or nine (+ VVMD32, VVMD36, VVMD25) microsatellite markers has mostly been used in grapevine gentyping studies, which are highly polymorphic and most appropriate for determining genetic variability among European grapevine cultivars [61, 62]. Microsatellite markers are evaluated on the basis of various parameters of variability: observed heterozygosity (Ho) is the proportion of heterozygous individuals in the analyzed sample; expected heterozygosity (He) or genetic diversity shows the percentage of the population that would be heterozygous if an accidental cross occurs between individuals; the polymorphic information content (PIC) includes both the number of alleles detected at each locus, as well as the frequency of each allele and is the rate at which a marker unambiguously determines the genetic identity of an individual; the probability of identity (PI) is the likelihood of two randomly chosen individuals having two identical alleles at any locus; the power of discrimination (PD) is the probability that two randomly sampled accessions in the studied population can be differentiated by their allelic profile at a given locus. Higher PI values or lower PD values show a low discrimination power of the locus, which is usually the consequence of a small number of alleles or the high frequency of one allele.

On average, the number of amplified alleles per locus has been similar among different studies [46, 57, 63, 64] but the variability mostly depends on the size and heterogeneity of the sample. In contrast, the discriminative power of loci can vary significantly; for example,, in Slovenian grapevines SsrVrZAG79 proved to be the most informative locus, with a PD value of 0.928 [65] but in Portuguese grape varieties [63], this locus was considered to be least informative. The comparison confirmed the findings of Sefc et al. [46] that the discrimination power of each marker depends on the set of analyzed samples, which is related to the fact that different alleles are dominant in different regions the vines are growing.

Locus VVMD5 also proved to have high discriminative power in analysis of Slovenian grapevines (0.925) [65], Castilian – Spain grapevines (0.934) [48] and also in the analysis of grapevines collected in Balkan countries (0.932) [66]. In the last study, the maximum power associated with high PD values (0.96, 0.94) was evidenced separately for loci VVMD28 and Vchr8b. Locus Vchr8b is one of the 'new' microsatellite markers, containing tri-, tetra- and penta-nucleotide repeats selected from a total of 26,962 perfect microsatellites in the genome sequence of grapevine PN40024 [38]. In the study by Cipriani et al. [49], based on the genotyping of 1005 grapevine accessions with a 'new' set of 34 SSR markers with a long core repeat optimized for grape genotyping [38], the loci with the highest power of discrimination were Vchr3a and Vchr8b. However, from later results it can be concluded that locus Vchr8b is highly discriminative but also shows a high estimated frequency of null alleles (>0.20), which may indicate an excess of homozygotes, expected to some extent in grape or a mutation at the priming site of the locus. The presence of null alleles for the loci, as for example Vchr8b and VVMD36 was observed in different studies [49, 65, 67, 68] and usually loci with null alleles resulted in no PCR amplification for samples representing the homozygous genotypes and lead to greater number of missing data in the study.

The comprehensive ranking of 'new' and 'old' SSR markers was facilitated in the study of Tomić [68] where all potentially good markers were evaluated together and according to their power of discrimination (only for loci with PD>0.9) ranked as follows: VVMD28, VChr8b, VVMD5, VrZAG79, VVMD32, VChr3a.

Based on high values for power of discrimination (PD), it can be said that alleles are uniformly distributed among the analyzed samples and that loci are very informative. A low PD value despite a large number of amplified alleles at a specific locus is sometimes due to the uneven distribution of allele frequencies in the analyzed sample, as for example at locus VVMD7 [65], where the frequencies of three out of ten alleles added up to 85%. Locus Vchr8b amplified 21 alleles in two studies [49, 68] but only 6 alleles were shown to be effective and two alleles prevailed, with frequencies over 20% [49].

A study by Laucou et al. [50] comprises the largest analysis of genetic diversity in grape ever, with an estimate of the usefulness of 20 SSR markers scattered throughout the genome in a set of 4,370 accessions [3,727 Vitis vinifera subsp. sativa accessions, 80 Vitis vinifera subsp. sylvestris individuals, 364 interspecific Vitis hybrid accessions used for fruit production and 199 Vitis rootstocks). Of these markers, 11 were from previous studies [61] and 9 from a genetic map [59], chosen according to their position and ease of genotyping. When arranged according to PI, a set of eight markers (VVIp31, VVMD28, VVMD5, VVS2, VVIv37, VMC1b11, VVMD27 and VVMD32) was determined as sufficient for identification of all the cultivars. The highest observed PD calculated from 2,739 single accessions was obtained for VVIp31 and VVMD28 markers [0.982 and 0.981, respectively) and five out of the eight most discriminative markers belong to a previously reported set of 'old' markers. Based on criteria such as multiplexing and easy-scoring, Laucou et al. [50] defined another minimum set of nine SSR markers (VVMD5, VVMD27, VVMD7, VVMD25, VVIh54, VVIp60, VVIn16, VVIb01, VVIq52) and proposed them for the routine analysis of European grapevines.

However, there are some limitations even with SSR markers, such as when the PCR amplification gives instead of one or two expected fragments (alleles), a group of fragments that differ by only 2 bp. Additional fragments, also called secondary fragments (stutter bands), are usually caused by slippage during amplification with *Taq* polymerase and the determination of allele lengths can therefore be difficult, especially if the two alleles differ only by two bp and it is necessary to distinguish homo-and heterozygous form. In reviewing for stutter bands the set of nine di-nucleotide markers currently in use, locus VVS2 has by far the strongest stutter bands,VVMD32 has two or three stutters, but not distracting because the "main" peak is well established, VVMD5, VVMD7, VVMD27 and ZAG62 all have one stutter and VVMD25, ZAG79, VVMD28 have no stutter bands. Tri-, tetra- and penta-nucleotide SSR markers are less prone to stuttering and the space between adjacent alleles is larger than in di-nucleotide SSRs, which enable a clear distinction between true alleles and stutter bands and minimize miscalling of the true allele. To overcome these limitations, Cipriani et al. [38] developed 'new' tri-, tetra- and penta-nucleotide repeated markers, which have proved to be very efficient [68].

2.3. Effectiveness of microsatellite markers in different applications

2.3.1. Chimerism

Microsatellite markers have often been used to differentiate grapes at a cultivar level and have been less interesting and less effective for the study of clonal variation [46]. Many cases have been recently described in which clones of grape varieties can be distinguished with microsatellite markers, such as 'Pinot Noir' 'Pinot gris', 'Pinot blanc' [69], 'Pinot Meunier' [70] 'Chardonnay' [71], synonyms of variety 'Black Currant' and 'Mavri Corinthiaki' [72], 'Pikolit' [73], etc. Laucou et al. [50] tested whether SSR markers could easily identify cultivars and clones when applied to a very large set of grape samples. Five percent of differentiated clones revealed between 1 and 3 differences (and only one mutant with four differences). Differences were sometimes of a homozygote versus heterozygote type or size shifts in 1 allele. It was demonstrated that cultivars showed at least four allelic differences, while clones showed fewer than four allelic differences but can also be distinguished. Studies of microsatellites have also demonstrated that the main type of mutation that leads to clonal variation is the development of chimeric growing tips. A chimera is a specific type of genetic mosaic, which is usually the result of mutation in one cell of the shoot apical meristem, spread by replication and cell division. The presence of a third allele suggests that the plant is a periclinal chimera, in which a mutant allele is present only in the L1 layer, as described by Riaz et al. [71]. Most chimeric cultivars do not exhibit special phenotypes, although some of them do so, such as Pinot Meunier (trichomes) and pinot gris (berry colour). Chimerism is usually detected at various loci, such as in studies by Tomić [68] and Stenkamp et al. [70], in which it was detected at five (VVMD7, VVMD32, VChr8a, VChr8b and VChr9a) and four (VMC9a3.1, VVS5, VVMD7, and VrZag79) different loci, respectively. Triallelic profiles at loci VVS2 and VVS5 were detected for Pinot Meunier clone [70], and for Primitivo di Gioia at locus VVS19 [74]. In the review of research, we found that a three-allelic profile has appeared several times at locus VVMD7 [68, 69, 75] and, in the last study [68] VVMD7 was shown to carry three alleles in 12 different cultivars out of 16 cultivars showing chimerism.

2.3.2. Cultivar identity/synonyms detection

There are around 10,000 grapevine cultivars held in germplasm collections worldwide [48] but, based on DNA analyses, the number of grapevine varieties is estimated at approx. 5,000 [76]. This proves the need for identifying synonyms and homonyms in collections to remove redundant accessions and improve management.

Local winegrowers in the past cultivated primarily less known varieties, but with the intensive renovation of vineyards these have almost disappeared and been replaced by new varieties grown elsewhere in Europe. Most winegrowing countries have initiated a global campaign of collecting, preserving and evaluating old cultivars and clones, as well as organizing collections. Some of these indigenous or local varieties are particularly promising in terms of high quality but, in the past, for various reasons, have not been adequately exploited. Wine produced today from native varieties provides a new niche in the competitive market. The descriptions of some of these varieties and associated data available are incomplete and it is necessary to identify them or resolve their description. In addition, the populations of *Vitis vinifera* L. are often very heterogeneous and vines of each clone can be very different, which also hampers identification at the morphological level [50]. Diverse historical development and multilingual areas have contributed to differences in the naming of local varieties, which have resulted in the high number of synonyms and homonyms [76].

Laucou et al. [50] recently published data that, among 4,370 accessions maintained in the INRA germplasm collection, 1,050 cases of questionable synonyms were discovered or confirmed. Santana et al. [48] found 300 synonymic samples among 421 Spanish grapevines and Cipriani et al. [49] 260 out of 1005 international, national and local grapevine accessions. Tomić [68] reported 58 synonyms out of 196 samples included in SSR analysis, discovering 20 groups of synonyms and 12 groups of homonyms associated with the wrong description due to local denominations. In the latest study [68] cultivar identification was performed also by comparing the set of 138 unique profiles (without redundant genotypes) with approximately 2000 other grape genotypes grown in Europe (personal communication with Vouillamoz, Jose) and 15 groups of synonyms and 3 groups of homonyms were found. Comparison of Slovenian genotypes [65] with 161 European varieties described by Sefc et al. [46] helped to identify 3 new pairs of synonyms: Volovnik = Vela Pergolla (Croatia), Pregarc = Garnache Tintorera (Spanish) and Kanarjola = Trebbiano Toscano (Italian).

Microsatellite similarity analysis of Slovenian genotypes [65, 75] confirmed some suspicions of identical varieties made on the basis of morphological characters; such as the variety Ferjanščkova, which was shown to be a synonym for Merlot and Grganc a synonym for Rebula, which means that the ancient names have apparently been preserved in some areas in Slovenian Istria. A group of five varieties (Glera = Prosecco = Briška Glera = Števerjana = Beli teran) is among synonyms that were *de novo* obtained by analysis of microsatellites; some of them have been previously described based on morphological similarity, while, for example, Števerjana is a new synonym of these varieties. A high diversity of microsatellite loci was detected between the varieties Briška Glera and White Glera, which could be explained by the fact that the Glera name was often used in the past for a variety of white grapevine varieties grown in the sub-Mediterranean part of Slovenia. Another group of homonyms represent varieties called Ribolla (Rebula, Old Rebula and Rebula-100 years) also revealing high polymorphism among them. A comparison of genotypes of Slovenian [65, 75] and Croatian varieties [77], performed on the basis of 7 microsatellite loci, also revealed synonyms between Muscat Ruža Porečki (Croatia) and Cipro (Slovenia) and between Ranfol bijeli (Croatia) and Belina Pleterje (Slovenia). Homonymy was detected between the Croatian variety Plavina described by Calo et al. [78] and the Slovenian variety with the same name, although their similarity based on SSR analysis was only 20% [65]. Varieties called Pagadebiti from Slovenia [65], Croatia [78] and Italy [79] also revealed very different SSR-allelic profiles.

2.3.3. Genetic relatedness, structure and parentage

Additional important applications of genotyping are analyses of genetic variability, genetic structure and parentage. When the data are comparable between different studies or within a larger group of cultivars, it is possible to identify the origin and relationships of cultivars. For example, a comparison of Slovenian genotypes with 161 genotypes [46] from eight European winegrowing regions showed that they are more related to Croatian and Greek varieties than to those from the adjacent Italian peninsula, but most genetically distant from French varieties, which may be a result of maritime trade across the Mediterranean Sea or along commercial routes through the Balkans [65].

Genetic clustering of varieties from the Castilian Plateau of Spain revealed three differentiated grups: Muscat-type accessions and interspecific Vitis hybrids, accessions from France and the western Castilian Plateau, and accessions from the central Castilian Plateau together with local table grapes. The close relatedness of accessions from the western plateau among each other and to French varieties suggested the introduction of the latter along the pilgrimage route to Santiago de Compostela [48].

Analysis of genetic relatedness of Balkan genotypes [68] showed that genotypes from Serbia, Bosnia and Slovenia are genetically fairly similar to each other, while genotypes from Macedonia and Montenegro are genetically more distant from the rest.

Microsatellite analysis and grouping of 1005 international, national and local grapevine accessions resulted in a weak correlation with their geographical origin and/or current area of cultivation, showing a large admixture of local varieties with those most widely cultivated, as a result of ancient commerce and population flows [49].

2.4. *Vitis* microsatellite databases

The main purpose of assembling data in databases with open access is to enlarge the number of varieties available for comparison and to facilitate the identification of genotypes. The largest international Vitis Microsatellite Collection is currently available within the European Vitis Database, which was constructed within the context of the European projects Genres081, GrapeGen06 and maintains SSR-marker data of 4364 accessions evaluated at 9 SSR loci [80]. High priority in these projects was given to the trueness-to-type of valuable and unique genotypes and a prerequisite for true-to-type identification is analysis of identity based on microsatellites. SSR-marker data within this database can be retrieved in two ways; search by cultivars or search by allele lengths. The database also includes SSR-marker data of 46 reference varieties, which enables comparison of data from different laboratories. The database has open access to partners providing SSR-marker data.

Some minor databases also exist, such as the publicly available Swiss microsatellite database (SVMD) [81], which includes 170 domestic and foreign genotypes growing in the given area and their SSR data for six microsatellite loci (VVMD5, VVMD7, VVMD27, VVS2, VrZAG62 and VrZAG79). A Greek collection (Greek Vitis Microsatellite Database) includes all possible information about grapevines that grow in Greece and is a combination of two older ampelographic databases, supplemented by microsatellite data (298 varieties and rootstocks) [82]. The Italian database (GMC - Grape Microsatellite Collection) provides a complete overview of microsatellite analysis of grapevine performed in different laboratories/countries and also includes information on authors and methods of work [83].

The reference varieties presented in the database are prerequisite for the comparison of data revealed from different systems/laboratories. Due to different electrophoresis systems, a difference between the lengths of alleles (shift of relative allele length) can be detected and data needs to be standardized. The length of alleles can be changed or standardized, so that analysis of genotyping includes some of the reference samples on which to compare 'unknown' samples and their allele lengths can be adjusted. Differences in allele lengths are the same within each locus, so reference samples included in the analysis can be used as a base to standardize all 'unknown' samples [61]. Information on allele lengths obtained in different laboratories can thus be compared and combined into a common database. An alternative for grapevine genotyping, where the complete genome sequence is available, is identification of thousands of single nucleotide polymorphisms (SNP), which can be very useful for genotyping purposes, since they can be multiplexed and need no standardization of results with additional reference cultivars. Because SNP markers are bi-allelic, genotypes obtained with different equipment and by different laboratories are always fully comparable [84].

3. Application of microsatellite markers in olives

Microsatellite markers or SSR (simple sequence repeats) have found wide applications in genetic studies of olives, including cultivar identification, assessment of genetic diversity in different sets of genotypes, evaluation of relationships among olive cultivars and among cultivated and wild olives, designation of geographic origin, genetic mapping, construction of core collections and similar studies.

This contribution presents a short review of SSR marker application in olives.

3.1. Microsatellite marker development

In view of their characteristics (high abundance and random in genome, high polymorphism, co-dominant inheritance, locus specific) SSR are desirable markers in plant genetic studies, although considerable input is required for initial marker development. The main features of SSR marker development in olive is summarized in Table1. The first SSR markers in olives were developed in 2000 by two groups. Sefc et al. [25] constructed a genomic library using the DNA of three Portuguese olive cultivars for the identification of SSR loci. The genomic library was probed by $(GA)_n$ and $(CA)_n$ repeats and 28 microsatellite containing sequences suitable for primer development were found and 15 SSR loci gave specific amplifications under optimized PCR conditions. These markers were designated ssrOeUA-DCA, followed by a two digit number, in short, a DCA series. Markers were tested on 48

Iberian and Italian olive trees for the number of amplified alleles (on average 8.3 alleles per primer pair), and observed and expected heterozygosity (Ho and He) showed the characteristics of each SSR marker (Table 1). The second group [33] developed 5 SSR markers out of 13 microsatellite loci obtained from a GA-enriched genomic library. The 5 SSR markers were tested on a set of 46 olive cultivars for their characteristics, giving an average of 5.2 alleles per marker. They were were designated IAS-olio, followed by a two digit number. Three new series of SSR markers for olives followed in 2002. Carriero et al. [31] developed 20 SSR markers out of a highly $(GA)_n$ enriched genomic library and 10 markers were further characterized on twenty olive cultivars, amplifying 5.7 alleles per marker. These markers are designated GAPU, followed by a three digit number. Six SSR markers (EMO, followed by two digits) derived from a $(GA)_n$ and $(CA)_n$ enriched genomic library and one marker (EMOL) developed from a gene sequence containing a $(GA)_n$ microsatellite motif, were tested on 23 olive cultivars, giving an amplification of 6.1 alleles per primer pair [32]. Three of these markers also amplified microsatellite alleles in other species of Oleaceae, showing their transferability. Cipriani et al. [85] published 30 SSR markers designated UDO99-, followed by three digits but they are usually designated UDO-two digits. These markers were tested on a small set of 12 olive cultivars, amplifying 1-7 (average 3.6) alleles per primer pair and five markers gave an allelic profile of duplicated loci. A Spanish group undertook the development of a second set of IAS-olio SSR markers [86]. Primer pairs were designed for 24 microsatellite containing sequences, of which 12 loci gave an amplification product of expected profile; 10 markers gave a single locus amplification and 2 markers duplicated loci, confirmed by segregation analysis. Markers were characterized on a set of 51 olive cultivars, giving on average 5.6 alleles per locus. The most recent 12 SSR markers for olives were developed by Gil et al. [87], which were characterized on 33 olive cultivars giving an average of 6.75 alleles per locus. These SSR markers were designated ssrOeIGP, followed by digits.

3.2. Comparison of developed markers

The developed olive SSR markers, particularly those from 2000 and 2002, have been used by various research groups working on olives. The choice of markers from the literature was mainly based on the researchers' selection based on their own experimental results, usually testing the SSR markers on a small set of genotypes and then selecting the markers with the best performance in terms of single locus amplification, stutter of bands, weak amplification of longer alleles, stability of repeats and number of alleles per marker [88] or by the SSR marker characteristics (number of alleles, Ho and He, polymorphic information content (PIC) or discrimination power (DP) provided in the literature. The citation index (Table 1) gives an idea of the most frequently used SSR markers in olives.

However, comparison of the allelic profiles of olive cultivars across different studies has been hindered by the use of different sets of markers and experimental conditions, resulting in discrepancies in allele size assignment. Bandelj et al. [89] carried out one of the first identifications of 19 olive cultivars by SSR markers, using a sequencing gel for allele separation and silver staining. The allele sizes were determined by 10 bp size ladder and sequencing reaction. Sarri et al. [90] used the same nine SSR markers in an analysis of 118 olive cultivars, separating the alleles with sequencing apparatus and sizing them with computer software. A comparison of allele sizes at the same loci between those two works shows discrepancies of 1-2bp per allele, making it difficult to decide whether an allele is 238 bp, 239 bp or 240 bp long. Allele size discrepancy is also reflected in the genotypes of a particular cultivar analysed in different laboratories. For example, the cultivar Arbequina was genotyped at eight common SSR loci by Bandelj et al. [89] and Doveri et al. [91] but showed no match at any loci.

A first attempt to provide some common SSR markers for olive cultivar identification and discrimination was reported by Doveri et al. [91]. Four partner laboratories tested eight SSR markers from the DCA series, on seventeen selected cultivars using ABI and LICOR systems for fragment analysis and allele sizing. The allele sizes of each marker from the different laboratories were harmonized by comparison and by the use of three cultivars with standard alleles for each loci. Markers DCA3, DCA8, DCA11, DCA13, DCA14 and DCA15 were assessed as the most reproducible among the four laboratories, stressing that reproducibility depends on the use of the same source of plant material, the same reference cultivars and standardization of analytical conditions. Baldoni et al. [92] later published the most comprehensive evaluation of available SSR markers and produced a consensus list of 11 SSR markers for olive genotyping. Thirty-seven SSR markers were tested for reproducibility (low stutter, strong peak signal, single loci amplification and no null alleles) on a set of 21 cultivars, among four laboratories, three using a capillary sequencer (two labs MegaBACE 1000, one lab ABI3130) and one a 2100 Bioanalyzer Agilent. Up to 5 bp discrepancies in allele size were observed among the labs, mainly due to the use of different sequencers and internal allele references. They selected 11 SSR markers for which an allelic ladder at each locus is provided. Alleles were further sequenced to estimate the true size and to characterize the repeat motifs and mapped such that only unlinked loci were selected. The selected markers, ranked by their information value UDO-043, DCA9, GAPU103A, DCA18, DCA16, GAPU101, DCA3, GAPU71B, DCA5, DCA14 and EMO-90, were further tested on a larger set of 77 cultivars to calculate their genetic parameters. This consensus list of SSR markers, together with allelic references, provides a solid platform for olive genotyping by different labs, enabling inter-lab comparison and the construction of an SSR database of olive genotypes, which would be of great help for true-to-type cultivar identification and management of olive germplasm banks.

3.3. Application of SSR markers

Olive trees have been grown for oil and table olive production in the Mediterranean basin since ancient times. The genetic diversity of cultivated olives is abundant and is characterized by a numerous local cultivars vegatatively propagated by farmers. Bartolini et al. [93] collected information on more than 1,208 cultivars from 52 countries, conserved in 94 collections. The number of cultivars is probably much higher, bearing in mind the lack of information on minor cultivars in different olive growing regions. Cultivar surveys have been initiated in many olive growing countries in order to describe existing cultivars, thus obtaining information for germplasm preservation, description of cultivars of specific growing regions and for breeding purposes. For the description and management of the existing genetic diversity in olives, molecular markers have been found to be particularly valuable because of such characteristics as high genetic informativeness, environmental independence, relatively easy use and the possibility of accumulating a large amount of data. SSR markers, in particular, have been extensively used in olives for cultivar identification, assessment of genetic diversity and other genetic studies.

Cultivar identification in olives is important to confirm true-to-type denominated cultivars, solve problems relating to synonyms, homonyms and mislabelled planting material. One of the first cultivar identifications by SSR markers was done on a small set of Slovene and Italian [19] cultivars using the DCA series [89]. The work was extended to practical application of SSR markers for confirmation of true-to type denomination of 13 olive samples from nursery using two DCA markers [94]. Comparison of the genotyped samples with the genotypes of reference cultivars obtained from three collections enabled confirmation of the correct denomination of six samples, 5 samples were mislabelled and no reference cultivar was available for two samples. Thirty-five Spanish and Italian olive cultivars of commercial interest were then genotyped by UDO series [95]. Olive cultivars were further genotyped for identification purposes or for assessment of genetic diversity on international (world germplasm collections), national (Spain, Italy, Tunis, Morocco, Turkey, Greece, Croatia, Slovenia, Portugal, Lebanon, Alger) and regional scales (olive growing region with characteristic variety structure). There have been numerous publications from these studies and we present here only a few examples. Sarri et al. [90] genotyped 118 olive cultivars from several Mediterranean countries by use of twelve SSR markers (10 DCA series, GAPU89 and UDO12) showing high discrimination power. A combination of only three markers distinguished almost all analysed cultivars and a selection of six markers was sufficient to assign cultivars to their geographic origin, divided into eastern, central and western Mediterranean. Geographic structuring of diversity was also found in a set of 211 autochthonous cultivars in six southern Italian olive growing regions [96]. The cultivars were analysed by 11 SSR loci (DCA, GAPU and UDO), which discriminated 199 unique genotypes and identified ten pairs of synonyms, four cases of homonyms and a possible parent-offspring relationship. Poljuha et al. [97] analysed 27 olive accessions from an olive growing region in Croatia and Slovenia (Istria), using 12 SSR markers (DCA) and finding a distinction between native and introduced cultivars, as well as some cases of synonyms and homonyms.

Khadari et al. [98] analysed 215 olive trees sampled in all Moroccan traditional growing regions. Using 15 SSR (4 DCA, 3 GAPU and 8 UDO) they, identified 60 SSR profiles among which 52 genotypes belonged to cultivated trees with no denomination, demonstrating high genetic diversity in Moroccan olive germplasm. However, a single Moroccan cultivar, belonging to a different gene pool to local cultivars, which were probably derived from local domestication, was predominant in all growing regions. Local olive domestication in two out of three sampled olive growing regions in Spain was also suggested by Belaj et al. [99] in a study of the relationship between wild and cultivated olives using eight SSR markers (4 DCA, 3 UDO and EMO). A low level of local olive domestication was found in a study of Sardinian wild (21), local (22) and ancient cultivars (35) [100], using 6 DCA, 4 UDO and 3 GAPU SSR markers, however most of the Sardinian local cultivars were also very closely related to ancient cultivars analysed. The relationship between ancient olive trees and cultivars in Southern Spain is slightly different, since only 9.6% of 106 ancient trees matched olive cultivars, as revealed by analysis using 14 SSR markers (7 DCA, 2 GAPU and 5 UDO) [101].

Several cases of synonyms, homonyms and mislabelled samples, as well as high diversity were revealed in a survey of 84 accessions from a Tunisian germplasm collection, using eight SSR markers of series DCA (5), GAPU(2) and UDO. On the basis of the SSR analysis, an improved classification of accessions was proposed for better management of the germplasm collection [102].

Proper management of germplasm collections in terms of evaluation, documentation, regeneration and effective use of available genetic diversity present in a collection is hindered by the large sizes of collections, redundancy and lack of accession information. In order to overcome these problems, core collections have been established that contain a limited number of accessions, capturing maximum allelic diversity. There are two world olive germplasm collections, one in Cordoba, Spain (C1) and the other in Marrakech (M), Morocco, which have in common 153 accessions and both core collections have been established using SSR markers for measuring genetic diversity [103, 104]. In the Marrakech collection, 561 accessions were analysed by 12 SSR markers (8 DCA, 2 GAPU, 1 UDO, 1 EMO) and the estimated core collection comprises 67 accessions; a slightly lower number [56] of accessions to represent the total allelic diversity was estimated in the Cordoba collection on the basis of analysing 378 accessions with 14 SSR markers (6 DCA, 4 GAPU, 4 UDO, 1 EMO). The Cordoba (C2) collection of 361 accessions was additionally assessed with 23 SSR markers (5 DCA, 6GAPU, 8 UDO, 1 EMO, 3 GP) as well as DaRT, SNP and morphological markers and their estimate for a core collection adequate for conservation of genetic diversity was 68 accessions [105]. Enormous work was carried out in genotyping all these accessions. However, the set of SSR markers used were unfortunately selected arbitrarily. Seven SSR markers were the same in M and C1 collections but only three and one SSR markers were in common with the C2 collection, respectively. In comparison with the Baldoni et al. [92] recommended list of SSR markers, the C1 collection had in common 8 markers, the M collection 6 and the C2 collection only 2 SSR markers. The advantages of SSR markers, which enable inter-laboratory comparison, in these cases not really fully exploited, since not only the same markers but also harmonized protocols are needed for reliable comparison of analysed genotypes.

In conclusion, SSR markers have been proven through numerous applications to be a very powerful tool in studies of olive genetic structure, domestication processes, genetic relationships among different cultivars, wild and cultivated olives, in the management of germplasm collection etc.

Some sort of agreement on the use of SSR markers and protocols should be reached in the future, which would allow inter-laboratory comparisons and, most importantly, the establishment of an international olive microsatellite database.

SSRs series	Source	Screening	No. SSR loci	No. cvs. tested	No. alleles	No. alleles/ locus	H_o range	H_e range	Reference	Cited (SCI)
DAC	genomic library; DNA cvs: Porto Martins, Terceira, Açores	(GA)n, (CA)n	15	47	124	8.3	0.28-0.98	0.36-0.86	[25]	131
IAS-oli	Enriched genomic library, DNA cv: Arbequina	(GA)n	5 (13)*	46	26	5.2		0.46-0.71	[33]	120
GAPU	Enriched genomic library, DNA cvs: 6 different cultivars	(GA)n	10(20)*	20	57	5.7			[31]	104
EMO	Enriched genomic library, DNA cv: Picual	(GA)n, (CA)n	7	23	43	6.1	0.39-0.91	0.62-0.81	[32]	71
UDO	Enriched genomic library, DNA cv: Frantoio	(AC)n, (AG)n	29 (30)*	12	103	3.6		0.44-0.77	[85]	139
IAS-oli	Enriched genomic library, DNA cv: Arbequina	(GA)n, (GT)n, (ACT)n	10(2)*	51	68	5.6	1-0.82	1-0.94	[86]	26
ssrOeIGP	Enriched genomic library, DNA cv: Lezzo	?	12 (19)*	33	60	6.7	0.42-0.89	0.19-0.81	[87]	12

Table 1. SSR markers development in olives * single locus amplification (monomorphic, two or multiple loci amplification)

Author details

Jernej Jakše[1*], Nataša Štajner[1], Lidija Tomić[2] and Branka Javornik[2*]

*Address all correspondence to: branka.javornik@bf.uni-lj.si

1 University of Ljubljana, Biotechnical Faculty, Slovenia

2 University of Banja Luka Faculty of Agriculture, Bosnia and Herzegovina

References

[1] Litt, M., & Luty, J. A. (1989). A hypervariable microsatellite revealed by invitro amplification of a dinucleotide repeat within the cardiac-muscle actin gene. *American Journal of Human Genetics.*, Mar, 44(3), 397-401.

[2] Weber, J. L., & May, P. E. (1989). Abundant class of human dna polymorphisms which can be typed using the polymerase chain-reaction. *American Journal of Human Genetics.*, Mar, 44(3), 388-96.

[3] Miklos, G. L., & John, B. (1979). Heterochromatin and satellite DNA in man: properties and prospects. *Am J Hum Genet.*, May, 31(3), 264-80.

[4] Armour, J. A. L., Neumann, R., Gobert, S., & Jeffreys, A. J. (1994). Isolation of human simple repeat loci by hybridization selection. *Human Molecular Genetics*, Apr, 3(4), 599-605.

[5] Goldstein, D. B., & Pollock, D. D. (1997). Launching microsatellites: A review of mutation processes and methods of phylogenetic inference. *Journal of Heredity.*, Sep-Oct, 88(5), 335-42.

[6] Schlotterer, C., & Tautz, D. (1992). Slippage synthesis of simple sequence dna. *Nucleic Acids Research*, Jan, 20(2), 211-5.

[7] Chambers, G. K., & Mac, Avoy. E. S. (2000). Microsatellites: consensus and controversy. *Comparative Biochemistry and Physiology B-Biochemistry & Molecular Biology.*, Aug, 126(4), 455-76.

[8] Hancock, J. M., Goldstein, D. B., & Schlötterer, C. (1999). Microsatellites and other simple sequences: genomic context and mutational mechanisms. Microsatellites: Evolution and Applications. *Oxford: Oxford University Press*, 1-9.

[9] Lagercrantz, U., Ellegren, H., & Andersson, L. (1993). The abundance of various polymorphic microsatellite motifs differs between plants and vertebrates. *Nucleic Acids Research*, Mar, 21(5), 1111-5.

[10] Morgante, M., & Olivieri, A. M. (1993). PCR-amplified microsatellites as markers in plant genetics. *The Plant journal : for cell and molecular biology*, Jan, 3(1), 175-82.

[11] Wang, Z., Weber, J. L., Zhong, G., & Tanksley, S. D. (1994). Survey of plant short tandem dna repeats. *Theoretical and Applied Genetics*, Apr, 88(1), 1-6.

[12] Akagi, H., Yokozeki, Y., Inagaki, A., & Fujimura, T. (1996). Microsatellite DNA markers for rice chromosomes. *Theoretical and Applied Genetics*, Nov, 93(7), 1071-7.

[13] Kuhl, J. C., Cheung, F., Yuan, Q. P., Martin, W., Zewdie, Y., Mc Callum, J., et al. (2004). A unique set of 11,008 onion expressed sequence tags reveals expressed sequence and genomic differences between the monocot orders Asparagales and Poales. *Plant Cell.*, Jan, 16(1), 114-25.

[14] Temnykh, S., De Clerck, G., Lukashova, A., Lipovich, L., Cartinhour, S., & Mc Couch, S. (2001). Computational and experimental analysis of microsatellites in rice (Oryza sativa L.): Frequency, length variation, transposon associations, and genetic marker potential. *Genome Research.*, Aug, 11(8), 1441-52.

[15] Matsumoto, T., Wu, J. Z., Kanamori, H., Katayose, Y., Fujisawa, M., Namiki, N., et al. (2005). The map-based sequence of the rice genome. *Nature.*, Aug, 436(7052), 793-800.

[16] Cardle, L., Ramsay, L., Milbourne, D., Macaulay, M., Marshall, D., & Waugh, R. (2000). Computational and experimental characterization of physically clustered simple sequence repeats in plants. *Genetics.*, Oct, 156(2), 847-54.

[17] Kaul, S., Koo, H. L., Jenkins, J., Rizzo, M., Rooney, T., Tallon, L. J., et al. (2000). Analysis of the genome sequence of the flowering plant Arabidopsis thaliana. *Nature.*, Dec, 408(6814), 796-815.

[18] Morgante, M., Hanafey, M., & Powell, W. (2002). Microsatellites are preferentially associated with nonrepetitive DNA in plant genomes. *Nature Genetics*, Feb, 30(2), 194-200.

[19] Jaillon, O., Aury, J. M., Noel, B., Policriti, A., Clepet, C., Casagrande, A., et al. (2007). The grapevine genome sequence suggests ancestral hexaploidization in major angiosperm phyla. *Nature.*, Sep, 449(7161), 463-U5.

[20] Velasco, R., Zharkikh, A., Troggio, M., Cartwright, D. A., Cestaro, A., Pruss, D., et al. (2007). A High Quality Draft Consensus Sequence of the Genome of a Heterozygous Grapevine Variety. *Plos One.*, Dec, 2(12).

[21] Alagna, F., D'Agostino, N., Torchia, L., Servili, M., Rao, R., Pietrella, M., et al. (2009). Comparative 454 pyrosequencing of transcripts from two olive genotypes during fruit development. *Bmc Genomics.*, Aug, 10.

[22] Maniatis, T., Sambrook, J., Fritsch, E. F., Cold, Spring., & Harbor, L. (1982). Molecular cloning : a laboratory manual / T. Maniatis, E.F. Fritsch, J. Sambrook. *Cold Spring Harbor, N.Y. : Cold Spring Harbor Laboratory*.

[23] Bowers, J. E., Dangl, G. S., Vignani, R., & Meredith, CP. (1996). Isolation and characterization of new polymorphic simple sequence repeat loci in grape (Vitis vinifera L). *Genome.*, Aug, 39(4), 628-33.

[24] Thomas, M. R., & Scott, N. S. (1993). Microsatellite repeats in grapevine reveal dna polymorphisms when analyzed as sequence-tagged sites (stss). *Theoretical and Applied Genetics.*, Sep, 86(8), 985-90.

[25] Sefc, K. M., Lopes, S., Mendonca, D., Dos, Santos. M. R., Machado, M. L. D., & Machado, A. D. (2000). Identification of microsatellite loci in olive (Olea europaea) and their characterization in Italian and Iberian olive trees. *Molecular Ecology*, Aug, 9(8), 1171-3.

[26] Ostrander, E. A., Jong, P. M., Rine, J., & Duyk, G. (1992). Construction of small-insert genomic dna libraries highly enriched for microsatellite repeat sequences. *Proceedings of the National Academy of Sciences of the United States of America*, Apr, 89(8), 3419-23.

[27] Hamilton, M. B., Pincus, E. L., Di Fiore, A., & Fleischer, R. C. (1999). Universal linker and ligation procedures for construction of genomic DNA libraries enriched for microsatellites. *Biotechniques*, Sep, 27(3), 500-507.

[28] Kijas, J. M. H., Fowler, J. C. S., Garbett, C. A., & Thomas, M. R. (1994). Enrichment of microsatellites from the citrus genome using biotinylated oligonucleotide sequences bound to streptavidin-coated magnetic particles. *Biotechniques.*, Apr, 16(4), 656-662.

[29] Edwards, K. J., Barker, J. H. A., Daly, A., Jones, C., & Karp, A. (1996). Microsatellite libraries enriched for several microsatellite sequences in plants. *Biotechniques*, May, 20(5), 758-760.

[30] Karagyozov, L., Kalcheva, I. D., & Chapman, V. M. (1993). Construction of random small-insert genomic libraries highly enriched for simple sequence repeats. *Nucleic Acids Research*, Aug, 21(16), 3911-2.

[31] Carriero, F., Fontanazza, G., Cellini, F., & Giorio, G. (2002). Identification of simple sequence repeats (SSRs) in olive (Olea europaea L.). *Theoretical and Applied Genetics*, Feb, 104(2-3), 301-307.

[32] De La Rosa, R., James, C. M., & Tobutt, K. R. (2002). Isolation and characterization of polymorphic microsatellites in olive (Olea europaea L.) and their transferability to other genera in the Oleaceae. *Molecular Ecology Notes*, Sep, 2(3), 265-7.

[33] Rallo, P., Dorado, G., & Martin, A. (2000). Development of simple sequence repeats (SSRs) in olive tree (Olea europaea L.). *Theoretical and Applied Genetics*, Oct, 101(5-6), 984-989.

[34] Lefort, F., Kyvelos, C. J., Zervou, M., Edwards, K. J., & Roubelakis-Angelakis, K. A. (2002). Characterization of new microsatellite loci from Vitis vinifera and their conservation in some Vitis species and hybrids. *Molecular Ecology Notes*, Mar, 2(1), 20-1.

[35] Malausa, T., Gilles, A., Meglecz, E., Blanquart, H., Duthoy, S., Costedoat, C., et al. (2011). High-throughput microsatellite isolation through 454 GS-FLX Titanium pyrosequencing of enriched DNA libraries. *Molecular Ecology Resources*, Jul, 11(4), 638-44.

[36] Santana, Q. C., Coetzee, M. P. A., Steenkamp, E. T., Mlonyeni, O. X., Hammond, G. N. A., Wingfield, M. J., et al. (2009). Microsatellite discovery by deep sequencing of enriched genomic libraries. *Biotechniques.*, Mar, 46(3), 217-23.

[37] Sharma, P. C., Grover, A., & Kahl, G. (2007). Mining microsatellites in eukaryotic genomes. *Trends in Biotechnology*, Nov, 25(11), 490-8.

[38] Cipriani, G., Marrazzo, M. T., Di Gaspero, G., Pfeiffer, A., Morgante, M., & Testolin, R. (2008). A set of microsatellite markers with long core repeat optimized for grape (Vitis spp.) genotyping. *Bmc Plant Biology.*, Dec, 8.

[39] Scott, K. D., Eggler, P., Seaton, G., Rossetto, M., Ablett, E. M., Lee, L. S., et al. (2000). Analysis of SSRs derived from grape ESTs. *Theoretical and Applied Genetics.*, Mar, 100(5), 723-6.

[40] Echt, C. S., May, Marquardt. P., Hseih, M., & Zahorchak, R. (1996). Characterization of microsatellite markers in eastern white pine. *Genome.*, Dec, 39(6), 1102-8.

[41] Merdinoglu, D., Butterlin, G., Bevilacqua, L., Chiquet, V., Adam-Blondon, A. F., & Decroocq, S. (2005). Development and characterization of a large set of microsatellite markers in grapevine (Vitis vinifera L.) suitable for multiplex PCR. *Molecular Breeding*, May, 15(4), 349-66.

[42] Hall, J. M., Le Duc, C. A., Watson, A. R., & Roter, A. H. (1996). An approach to high-throughput genotyping. *Genome Research.*, Sep, 6(9), 781-90.

[43] Schuelke, M. (2000). An economic method for the fluorescent labeling of PCR fragments. *Nature Biotechnology*, Feb, 18(2), 233-4.

[44] Cipriani, G., Frazza, G., Peterlunger, E., & Testolin, R. (1994). Grapevine fingerprinting using microsatellite repeats. *Vitis.*, Dec, 33(4), 211-5.

[45] Sefc, K. M., Regner, F., Glossl, J., & Steinkellner, H. (1998). Genotyping of grapevine and rootstock cultivars using microsatellite markers. *Vitis.*, Mar, 37(1), 15-20.

[46] Sefc, K. M., Lopes, M. S., Lefort, F., Botta, R., Roubelakis-Angelakis, K. A., Ibanez, J., et al. (2000). Microsatellite variability in grapevine cultivars from different European regions and evaluation of assignment testing to assess the geographic origin of cultivars. *Theoretical and Applied Genetics.*, Feb, 100(3-4), 498-505.

[47] Sanchez-Escribano, E. M., Martin, J. R., Carreno, J., & Cenis, J. L. (1999). Use of sequence-tagged microsatellite site markers for characterizing table grape cultivars. *Genome*, Feb, 42(1), 87-93.

[48] Santana, J. C., Heuertz, M., Arranz, C., Rubio, J. A., Martinez-Zapater, J. M., & Hidalgo, E. (2010). Genetic Structure, Origins, and Relationships of Grapevine Cultivars

from the Castilian Plateau of Spain. *American Journal of Enology and Viticulture*, 61(2), 214-24.

[49] Cipriani, G., Spadotto, A., Jurman, I., Di Gaspero, G., Crespan, M., Meneghetti, S., et al. (2010). The SSR-based molecular profile of 1005 grapevine (Vitis vinifera L.) accessions uncovers new synonymy and parentages, and reveals a large admixture amongst varieties of different geographic origin. Nov. *Theoretical and Applied Genetics*, 121(8), 1569-85.

[50] Laucou, V., Lacombe, T., Dechesne, F., Siret, R., Bruno, J. P., Dessup, M., et al. (2011). High throughput analysis of grape genetic diversity as a tool for germplasm collection management. *Theoretical and Applied Genetics*, Apr, 122(6), 1233-45.

[51] Eisen, J. A. (1999). Mechanistic basis of microsatellite instability. *Goldstein DB, Schlotterer C, editors. Microsatellites: Evolution and Applications. Oxford: Oxford University Press.*

[52] Faria, M. A., Magalhaes, R., Ferreira, M. A., Meredith, C. P., & Monteiro, F. F. (2000). Vitis vinifera must varietal authentication using microsatellite DNA analysis (SSR). *Journal of Agricultural and Food Chemistry.*, Apr, 48(4), 1096-100.

[53] Zulini, L., Russo, M., & Peterlunger, E. (2002). Genotyping wine and table grape cultivars from Apulia (Southern Italy) using microsatellite markers. *Vitis*, 41(4), 183-7.

[54] Di Rienzo, A., Donnelly, P., Toomajian, C., Sisk, B., Hill, A., Petzl-Erler, M. L., et al. (1998). Heterogeneity of microsatellite mutations within and between loci, and implications for human demographic histories. *Genetics.*, Mar, 148(3), 1269-84.

[55] Bowers, J. E., Dangl, G. S., & Meredith, C. P. (1999). Development and characterization of additional microsatellite DNA markers for grape. *American Journal of Enology and Viticulture*, 50(3), 243-6.

[56] Sefc, K. M., Regner, F., Turetschek, E., Glossl, J., & Steinkellner, H. (1999). Identification of microsatellite sequences in Vitis riparia and their applicability for genotyping of different Vitis species. *Genome.*, Jun, 42(3), 367-73.

[57] Lefort, F., & Roubelakis-Angelakis, K. K. A. (2001). Genetic comparison of Greek cultivars of Vitis vinifera L. by nuclear microsatellite profiling. *American Journal of Enology and Viticulture*, 52(2), 101-8.

[58] Arroyo-Garcia, R., & Martinez-Zapater, J. M. (2004). Development and characterization of new microsatellite markers for grape. *Vitis*, 43(4), 175-8.

[59] Adam-Blondon, A. F., Roux, C., Claux, D., Butterlin, G., Merdinoglu, D., & This, P. (2004). Mapping 245 SSR markers on the Vitis vinifera genome: a tool for grape genetics. *Theoretical and Applied Genetics.*, Sep, 109(5), 1017-27.

[60] Di Gaspero, G., Cipriani, G., Marrazzo, M. T., Andreetta, D., Castro, M. J. P., Peterlunger, E., et al. (2005). Isolation of (AC)n-microsatellites in Vitis vinifera L. and anal-

ysis of genetic background in grapevines under marker assisted selection. *Molecular Breeding.*, Jan, 15(1), 11-20.

[61] This, P., Jung, A., Boccacci, P., Borrego, J., Botta, R., Costantini, L., et al. (2004). Development of a standard set of microsatellite reference alleles for identification of grape cultivars. *Theoretical and Applied Genetics.*, Nov, 109(7), 1448-58.

[62] Sefc, K. M., Lefort, F., Grando, Scott. K., Steinkellner, H., & Thomas, M. (2001). Microsatellite markers for grapevine: A state of the art. *Roubelakis-Angelakis KA, editor. Amsterdam: Kluwer Publishers,* 407-438.

[63] Lopes, M. S., Sefc, K. M., Dias, E. E., Steinkellner, H., Machado, M. L. D., & Machado, A. D. (1999). The use of microsatellites for germplasm management in a Portuguese grapevine collection. *Theoretical and Applied Genetics.*, Aug, 99(3-4), 733-9.

[64] Ibanez, J., de Andres, M. T., Molino, A., & Borrego, J. (2003). Genetic study of key Spanish grapevine varieties using microsatellite analysis. *American Journal of Enology and Viticulture*, 54(1), 22-30.

[65] Stajner, N., Rusjan, D., Korosec-Koruza, Z., & Javornik, B. (2011). Genetic Characterization of Old Slovenian Grapevine Varieties of Vitis vinifera L. by Microsatellite Genotyping. *American Journal of Enology and Viticulture*, 62(2), 250-5.

[66] Tomić, L. (2012). Molecular characterization and analysis of the genetic relatedness of old grapevine (Vitis vinifera L.) cultivars from the Western Balkan. *Ljubljana.*

[67] Vouillamoz, J. F., Maigre, D., & Meredith, CP. (2004). Identity and parentage of two alpine grape cultivars from Switzerland (Vitis vinifera L. Lafnetscha and Himbertscha). *Vitis*, 43(2), 81-7.

[68] Tomić, L. (2012). Molecular characterization and analysis of the genetic relatedness of old grapevine (Vitis vinifera L.) cultivars from the Western Balkan [doctoral dissertation]. *Ljubljana: University of Ljubljana.*

[69] Hocquigny, S., Pelsy, F., Dumas, V., Kindt, S., Heloir, M. C., & Merdinoglu, D. (2004). Diversification within grapevine cultivars goes through chimeric states. *Genome.*, Jun, 47(3), 579-89.

[70] Stenkamp, S. H. G., Becker, M. S., Hill, B. H. E., Blaich, R., & Forneck, A. (2009). Clonal variation and stability assay of chimeric Pinot Meunier (Vitis vinifera L.) and descending sports. *Euphytica.*, Jan, 165(1), 197-209.

[71] Riaz, S., Garrison, K. E., Dangl, G. S., Boursiquot, J. M., & Meredith, CP. (2002). Genetic divergence and chimerism within ancient asexually propagated winegrape cultivars. *Journal of the American Society for Horticultural Science*, Jul, 127(4), 508-14.

[72] Ibanez, J., de Andres, M. T., & Borrego, J. (2000). Allelic variation observed at one microsatellite locus between the two synonym grape cultivars Black Currant and Mavri Corinthiaki. *Vitis*, Dec, 39(4), 173-4.

[73] Zulini, L., Fabro, E., & Peterlunger, E. (2005). Characterisation of the grapevine cultivar Picolit by means of morphological descriptors and molecular markers. *Vitis*, 44(1), 35-8.

[74] Franks, T., Botta, R., & Thomas, M. R. (2002). Chimerism in grapevines: implications for cultivar identity, ancestry and genetic improvement. *Theoretical and Applied Genetics.*, Feb, 104(2-3), 192-9.

[75] Stajner, N., Korosec-Koruza, Z., Rusian, D., & Javornik, B. (2008). Microsatellite genotyping of old Slovenian grapevine varieties (Vitis vinifera L.) of the Primorje (coastal) winegrowing region. *Vitis*, 47(4), 201-4.

[76] This, P., Lacombe, T., & Thomas, M. R. (2006). Historical origins and genetic diversity of wine grapes. *Trends in Genetics*, Sep, 22(9), 511-9.

[77] Maletic, E., Sefc, K. M., Steinkellner, H., Kontic, J. K., & Pejic, I. (1999). Genetic characterization of Croatian grapevine cultivars and detection of synonymous cultivars in neighboring regions. *Vitis.*, Jun, 38(2), 79-83.

[78] Calo, A., Costacurta, A., Maras, V., Meneghetti, S., & Crespan, M. (2008). Molecular correlation of Zinfandel (Primitivo) with Austrian, Croatian, and Hungarian cultivars and Kratosija, an additional synonym. *American Journal of Enology and Viticulture*, 59(2), 205-9.

[79] Muganu, M., Dangl, G., Aradhya, M., Frediani, M., Scossa, A., & Stover, E. (2009). Ampelographic and DNA Characterization of Local Grapevine Accessions of the Tuscia Area (Latium, Italy). *American Journal of Enology and Viticulture*, 60(1), 110-5.

[80] The European Vitis Database [Internet]. Available from:, http://www.eu-vitis.de/index.php.

[81] Swiss Vitis microsatellite database [Internet]. Available from:, http://www1.unine.ch/svmd/.

[82] Greek Vitis Database [Internet]. Available from:, http://gvd.biology.uoc.gr/gvd/contents/databases/index.htm.

[83] Grape Microsatellite Collection [Internet]. Available from:, http://meteo.iasma.it/genetica/gmc.html.

[84] Cabezas, J. A., Ibanez, J., Lijavetzky, D., Velez, D., Bravo, G., Rodriguez, V., et al. (2011). A 48 SNP set for grapevine cultivar identification. *Bmc Plant Biology*, Nov, 11.

[85] Cipriani, G., Marrazzo, M. T., Marconi, R., Cimato, A., & Testolin, R. (2002). Microsatellite markers isolated in olive (Olea europaea L.) are suitable for individual fingerprinting and reveal polymorphism within ancient cultivars. *Theoretical and Applied Genetics.*, Feb, 104(2-3), 223-8.

[86] Diaz, A., De la Rosa, R., Martin, A., & Rallo, P. (2006). Development, characterization and inheritance of new microsatellites in olive (Olea europaea L.) and evaluation of

their usefulness in cultivar identification and genetic relationship studies. *Tree Genetics & Genomes.*, Jul, 2(3), 165-75.

[87] Gil, F. S., Busconi, M., Machado, A. D., & Fogher, C. (2006). Development and characterization of microsatellite loci from Olea europaea. *Molecular Ecology Notes*, Dec, 6(4), 1275-7.

[88] Bandelj, D., Jakse, J., & Javornik, B. (2004). Assessment of genetic variability of olive varieties by microsatellite and AFLP markers. *Euphytica*, 136(1), 93-102.

[89] Bandelj, D., Jakse, J., & Javornik, B. (2002). DNA fingerprinting of olive varieties by microsatellite markers. *Food Technology and Biotechnology.*, Jul-Sep, 40(3), 185-90.

[90] Sarri, V., Baldoni, L., Porceddu, A., Cultrera, N. G. M., Contento, A., Frediani, M., et al. (2006). Microsatellite markers are powerful tools for discriminating among olive cultivars and assigning them to geographically defined populations. *Genome.*, Dec, 49(12), 1606-15.

[91] Doveri, S., Gil, F. S., Diaz, A., Reale, S., Busconi, M., Machado, A. D., et al. (2008). Standardization of a set of microsatellite markers for use in cultivar identification studies in olive (Olea europaea L.). *Scientia Horticulturae*, May, 116(4), 367-73.

[92] Baldoni, L., Cultrera, N. G., Mariotti, R., Ricciolini, C., Arcioni, S., Vendramin, G. G., et al. (2009). A consensus list of microsatellite markers for olive genotyping. *Molecular Breeding.*, Oct, 24(3), 213-31.

[93] Bartolini, G., Prevost, G., Messeri, C., & Carignani, C. (2005). Olive germplasm: cultivars and world-wide collections Rome: FAO/Plant Production and Protection. [cited 2012, 19 Aug 2012]. Available from:, www.oleadb.it.

[94] Bandelj, D., & Javornik, B. (2007). Microsatellites as a powerfull tool for identification of olive (Olea europaea L.) planting material in nurseries. *Annales, Series Historia Naturalis*, 17, 133-8.

[95] Belaj, A., Cipriani, G., Testolin, R., Rallo, L., & Trujillo, I. (2004). Characterization and identification of the main Spanish and Italian olive cultivars by simple-sequence-repeat markers. *Hortscience.*, Dec, 39(7), 1557-61.

[96] Muzzalupo, I., Stefanizzi, F., & Perri, E. (2009). Evaluation of Olives Cultivated in Southern Italy by Simple Sequence Repeat Markers. *Hortscience*, Jun, 44(3), 582-8.

[97] Poljuha, D., Sladonja, B., Setic, E., Milotic, A., Bandelj, D., Jakse, J., et al. (2008). DNA fingerprinting of olive varieties in Istria (Croatia) by microsatellite markers. *Scientia Horticulturae*, Feb, 115(3), 223-30.

[98] Khadari, B., Charafi, J., Moukhli, A., & Ater, M. (2008). Substantial genetic diversity in cultivated Moroccan olive despite a single major cultivar: a paradoxical situation evidenced by the use of SSR loci. *Tree Genetics & Genomes*, Apr, 4(2), 213-21.

[99] Belaj, A., Munoz-Diez, C., Baldoni, L., Satovic, Z., & Barranco, D. (2010). Genetic diversity and relationships of wild and cultivated olives at regional level in Spain. *Scientia Horticulturae,* Apr, 124(3), 323-30.

[100] Erre, P., Chessa, I., Munoz-Diez, C., Belaj, A., Rallo, L., & Trujillo, I. (2010). Genetic diversity and relationships between wild and cultivated olives (Olea europaea L.) in Sardinia as assessed by SSR markers. *Genetic Resources and Crop Evolution,* Jan, 57(1), 41-54.

[101] Diez, C. M., Trujillo, I., Barrio, E., Belaj, A., Barranco, D., & Rallo, L. (2011). Centennial olive trees as a reservoir of genetic diversity. *Annals of Botany,* Oct, 108(5), 797-807.

[102] Fendri, M., Trujillo, I., Trigui, A., Rodriguez-Garcia, M. I., & Ramirez, J. D. A. (2010). Simple Sequence Repeat Identification and Endocarp Characterization of Olive Tree Accessions in a Tunisian Germplasm Collection. *Hortscience.,* Oct, 45(10), 1429-36.

[103] Haouane, H., El Bakkali, A., Moukhli, A., Tollon, C., Santoni, S., Oukabli, A., et al. (2011). Genetic structure and core collection of the World Olive Germplasm Bank of Marrakech: towards the optimised management and use of Mediterranean olive genetic resources. *Genetica.,* Sep, 139(9), 1083-94.

[104] Diez, C. M., Imperato, A., Rallo, L., Barranco, D., & Trujillo, I. (2012). Worldwide Core Collection of Olive Cultivars Based on Simple Sequence Repeat and Morphological Markers. *Crop Science,* Jan, 52(1), 211-21.

[105] Belaj, A., Dominguez-Garcia, M. D., Atienza, S. G., Urdiroz, N. M., De la Rosa, R., Satovic, Z., et al. (2012). Developing a core collection of olive (Olea europaea L.) based on molecular markers (DArTs, SSRs, SNPs) and agronomic traits. *Tree Genetics & Genomes.,* Apr, 8(2), 365-78.

Chapter 3

The Current Status of Wild Grapevine Populations (*Vitis vinifera* ssp *sylvestris*) in the Mediterranean Basin

Rosa A. Arroyo García and Eugenio Revilla

Additional information is available at the end of the chapter

http://dx.doi.org/10.5772/52933

1. Introduction

The Eurasian grapevine (*Vitis vinifera* L) is the most widely cultivated and economically important fruit crop in the world (Mattia *et al.* 2008). *Vitis vinifera* L includes the cultivated form *V. vinifera* ssp *vinifera* and the wild form *V. vinifera* ssp *sylvestris*, considered as two subspecies based on morphological differences. However, it can be argued that those differences are the result of the domestication process (This *et al.* 2006). The wild form, considered the putative ancestor of the cultivated form, represents the only endemic taxon of the *Vitaceae* in Europe and the Maghreb (Heywood and Zohary 1991). Grapevine domestication has been linked to the discovery of wine (McGovern 2004). Although wild grapevines were spread over Southern Europe and Western and Central Asia during the Neolithic period, archeological and historical evidence suggest that primo domestication events would had occurred in the Near-East (McGovern *et al.* 1996). In addition, several studies have shown evidence supporting the existence of secondary domestication events along the Mediterranean basin (Aradhya *et al.* 2003; Grassi *et al.* 2003, Arroyo-García *et al.* 2006; Lopes *et al.* 2009; Andres *et al.*, 2012). Recent genetic analyses using a large SNP platform provided genetic evidence supporting the Eastern origin of most cultivated germplasm as well as the existence of introgression from wild germplasm in Western regions, likely as the consequence of those predicted secondary domestication events (Myles *et al.* 2010). Distinction between wild and cultivated forms of *Vitis vinifera* L is mainly based on morphological traits. The most conspicuous differential trait is plant sex: wild grapevines are dioecious (male and female plants), while cultivated forms are mostly hermaphrodite plants, with self fertile hermaphrodite flowers (This *et al.* 2006).

Wild grapevines can still be found in Eastern and Western Europe (Arnold et al. 1998). The South Caucasus (Azerbaijan, Armenia and Georgia), together with eastern Anatolia, has been considered for a longtime as the birth place for viticulture with the earliest examples of wine-making (This *et al.* 2006, McGovern 2003, Zohary 1995, Olmo 1995, Levadoux 1956, Negrul 1938). A 1998 census (Arnold et al. 1998) showed that wild grapevine were present in Spain, Italy, Switzerland, Romania, Bulgaria, Hungary, Austria, and in the countries of former Yugoslavia (Figure 1). Apparently, Spain and Italy harbor the highest number of recorded populations and they were proposed to work as shelters for *V. vinifera* during the last glaciation (about 12,000 years ago) as well as putative sources of postglacial colonization and diversification (Levadoux 1956). Wild vines were abundant in their indigenous range in Europe until the middle of the 19th century, when the arrival of North American pests (Phylloxera) and pathogens (downy and powdery mildews) and the destruction of their habitats drove European wild vines close to extinction (IUCN 1997). The solution to generate resistance to Phylloxera was the use of American species and hybrids as rootstocks and many varieties of rootstocks were developed by breeders (Arraigo and Arnold, 2007).

Currently, vines found in natural habitats are considered to be a mixture of wild forms, naturalized cultivated forms and rootstocks escaped from vineyards as well as hybrids derived from spontaneous hybridizations among those species and forms (Laguna 2003, Lacombe *et al.* 2003, This *et al.* 2006). Recently, Arrigo and Arnold (2007) compared ecological features and genetic diversity among populations of naturalized rootstocks and native wild grapevines and did not detect the existence of genetic flux between them. The genetic analysis of wild grapevine populations from France and Spain (Di Vecchi *et al* 2009; Andres *et al* 2012) detected the existence of gene flow between cultivated and wild grapevine, estimating up to 3% of pollen migration between the cultivated fields and closely located wild grape. These pollen fluxes may have a significant effect on the evolution of those populations. Currently, wild grapevine is endangered throughout all its distribution range, (Di Vecchi *et al.* 2009) and conservation efforts are required to maintain the genetic integrity and survival of the remnant populations. Within this context, information on the amount and distribution of wild grapevine genetic diversity is crucial for the development of conservation strategies.

Figure 1. Localization of wild grapevine population in the Mediterranean basin. (Heywood and Zohary, 1991).

The principal key ideas of this chapter is a better understanding of the exact status of the remaining wild grape populations and their relationships with existing varieties using the molecular markers and genetic analysis approaches that it has been published about some wild grapevine populations around the Mediterranean basin.

2. Chlorotype variation and distribution in *V. vinifera* ssp. sylvestris around de Mediterranean basin

The chlorotype variation is based on specific features of the chloroplast genome as well as its conserved gene order and coding sequences in different species and its general lack of heteroplasmy and recombination. Furthermore, chloroplasts are uniparentally transmitted in most species (usually maternal in angiosperms and paternal in gymnosperms). The low mutations rates observed in the chloroplast genome represent a drawback to their wide application in the study of population history and dynamics within a given species. However, this problem has been overcome by the identification of variable intergenic regions and introns flanked by conserved sequences in many species as well as by the identification of chloroplast microsatellites which consist of mononucleotide repeats. Chloroplast microsatellites have been found in all plant species analyzed and they frequently are highly polymorphic (Provan *et al.* 2001). One problem associated with chloroplast microsatellites is their high homoplasy due to the recurrent generation of alleles of the same length that creates alleles which being identical by state are not identical by descent. High levels of homoplasy can confound estimates of population differentiation and the recurrent generation of alleles could mimic gene flow (Goldstein and Pollock 1997). The risk is however reduced in intraspecific analysis (Arnold *et al.* 2002).

As in other angiosperms, grapevine chloroplasts are maternally inherited (Arroyo-García *et al.* 2002) and therefore transmitted through seeds and cuttings. The chloroplast genome of grape is 160,928 bp in length and its gene content and gene order are identical to many other unarranged angiosperm chloroplast genomes (Jansen *et al.* 2006). Genetic diversity at the grape chloroplast has so far only been analyzed at the level of chloroplast microsatellite loci. Polymorphisms were searched by Arroyo-García *et al.* (2006) with 54 chloroplast microsatellite markers corresponding to 34 different loci in sample sets of four Vitis species (*Vitis berlandieri* Planchon, *V. riparia* Mich., *V. rupestris* Scheele and *V. vinifera* L.), using primer pairs developed for tobacco (Bryan et al. 1999; Weising and Gardner 1999; Chung and Staub, 2003) and Arabidopsis (Provan 2000). Nine loci were initially found polymorphic due to differences in the number of mononucleotide repeats in poly T/A stretches (Arroyo-García *et al.* 2006), which after comparison with the complete chloroplast genome sequence (Jansen *et al.* 2006) corresponded to five different loci: cpSSR3 (equivalent to NTCP-8), cpSSR5 (equivalent to NTCP-12 and ccSSR5), cpSSR10 (equivalent to ccSSR14), ccSSR9 and ccSSR23. These loci were genotyped in a sample of more than 1,200 genotypes of *V. vinifera* which uncovered the presence of two to three alleles per polymorphic locus and a total of eight chlorotypes. Among them, only four (A, B, C and D) had global frequencies greater than 5%. Chlorotype diversity is moderate in grapevine with diversity values (H) reaching 0.44 in the most diverse populations or cultivars groups that contrast with average H values of 0.55 reported in Arabidopsis (Picó *et al.* 2008) or H values higher than 0.95 observed in *Pinus sylvestris* (Provan *et al.*, 1998).

Very small and isolated populations of *V. vinifera* ssp. *sylvestris* can still be found in European temperate regions along deep river banks. Among them, Arroyo-Garcia et al, (2006) have performed an exhaustive screening of Iberian and Anatolian populations in the two ends of the Mediterranean basin and have included additional populations representative of other regions; they considered that all the natural populations were grouped in eight population groups following a geographic criterion. No clear-cut geographic structure was found among the seven *sylvestris* population groups considered. However, the most frequent chlorotypes displayed a different geographic distribution. As seen in Fig. 2, chlorotype A is very prevalent in West European *sylvestris* populations (IBP, CEU), but was not found in the Near East (NEA, MEA). In contrast, chlorotypes C, D and G are frequent in Near Eastern populations (NEA, MEA), but were not found farther west (e.g. IBP and CEU).

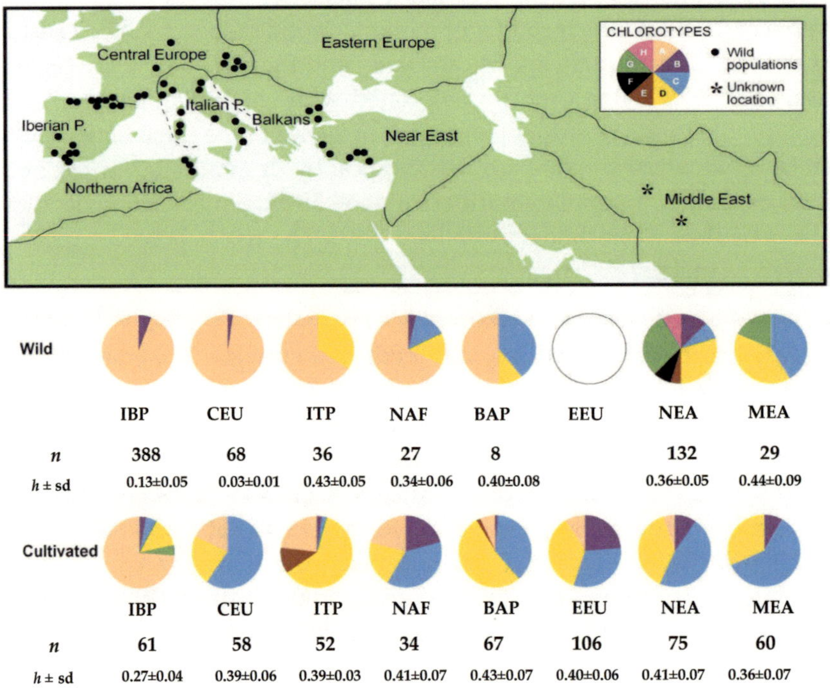

Figure 2. Chlorotype distribution in *sylvestris* and *sativa* population groups. Geographic areas considered are separated by lines when needed. Black periods do not mark specific *sylvestris* populations but river valleys where wild genotypes were collected at several locations. Asterisks indicate that specific locations of collection in the area are unknown. *Sativa* and *sylvestris* genotypes are grouped in eight population groups. From west to east: Iberian Peninsula (IBP), Central Europe (CEU), Northern Africa (NAF), Italian Peninsula (ITP), Balkan Peninsula (BAP), Eastern Europe (EEU), Near East (NEA) and Middle East (MEA). The figure also shows the values of unbiased chlorotype diversity and the number of genotypes considered within each population group. Chlorotype colour codes are as in Figure.

3. Multiple origins for cultivated grapevine

The chlorotype distributions observed among *sylvestris* populations allow for testing the two basic hypotheses on the origin of cultivated grapevine, proposed above, since they lead to different predictions regarding the amount and distribution of chloroplast genetic variation (Arroyo-García *et al.* 2006). The restricted origin hypothesis predicts that the chlorotype diversity of cultivated Eurasian grape should be limited to a few founder chlorotypes. In contrast, a multiple-origin hypothesis would predict greater diversity in cultivated grapevine groups than in *sylvestris* population groups. As shown in Fig. 2, unbiased chlorotype diversity is very similar in all the cultivated groups (from 0.36 to 0.43 with the exception of a lower value for IBP) and in most cases cultivated diversity values are higher than diversity values observed in *sylvestris* population groups. These results are also consistent with the existence of higher genetic differentiation (*GST*) among population groups of *sylvestris* (0.353 ± 0.10) than *sativa* (0.169 ± 0.07) grapevines. Interestingly, the geographic distribution observed for some chlorotypes in *sylvestris* groups can still be observed in cultivated groups (Fig. 2). In this way, cultivars with chlorotype A are highly abundant in Western Europe while they were not observed in Near and Middle East samples. Similarly, chlorotypes C and D, which are very common among NEA and MEA cultivars, are less frequent among IBP cultivars. To test further the origin hypotheses, they analyzed the genetic relationships among *sylvestris* and *sativa* population groups, since single- or multiple-origin hypotheses would predict different patterns of genetic relationships. All analyses grouped the cultivated population groups in two major clusters (Fig. 3). One cluster with high bootstrap values related the IBP cultivated group with the western, IBP, CEU, and Northern Africa, NAF *sylvestris*, population groups. The second main cluster showed that all the other cultivated groups considered are highly related to eastern *sylvestris* groups NEA and MEA. BAP and ITP *sylvestris* population groups appeared more related to the NEA/MEA cluster than to the western *sylvestris* cluster. These inferences were independent of the genetic model assumed, as the same partitioning was supported by all analyzed models. The statistical analysis was also robust for different clustering methods, including agglomerative and *K*-means, the latter indicating two as the optimum number of clusters. In summary, these results support the existence of a relevant genetic contribution of eastern and western *sylvestris* population groups to the genetic make-up of current grapevine cultivars and could suggest the existence of at least two origins of *sativa* cultivars: (i) an eastern origin related to NEA and MEA *sylvestris* population groups and characterized by chlorotypes C and D, and (ii) a western origin related to IBP, CEU and NAF *sylvestris* population groups and characterized by chlorotype A. Whether this second origin represents independent domestication events or developed as a consequence of the east to west transmission of the 'wine culture' will require further archaeological research. One palaeobotanical study (Hopf 1991) of grape pollen and seeds suggests that the Eurasian grapevine was exploited by Neolithic populations of the Iberian Peninsula before contact with Eastern cultures took place. This implies that grapevine could have been independently domesticated in Eastern and Western Europe. The putative existence of western and eastern domestication events is consistent with the morphotype classification of cultivated grapes proposed by Negrul (1938), who distinguished an *occidentalis* group, characterized by the small berry grapes of Western Europe, an *orientalis* group comprised of the large berry cultivars of Central Asia, and a *pontica* group including the intermediate types from the Black Sea basin and Eastern Europe. The results show by Arroyo-García *et al*. (2006) do not exclude the existence of additional genetic contributions of local *sylvestris* wild germplasm or even domestication events in other regions of the species distribution. However, sample size and the limited chloroplast genetic variation found in the Eurasian grape do not provide enough resolution to detect them. In fact, putative genetic relationships between cultivated varieties and local *sylvestris* populations have been proposed in other regions (Grassi *et al*. 2003; Di Vecchi *et al*., 2009).

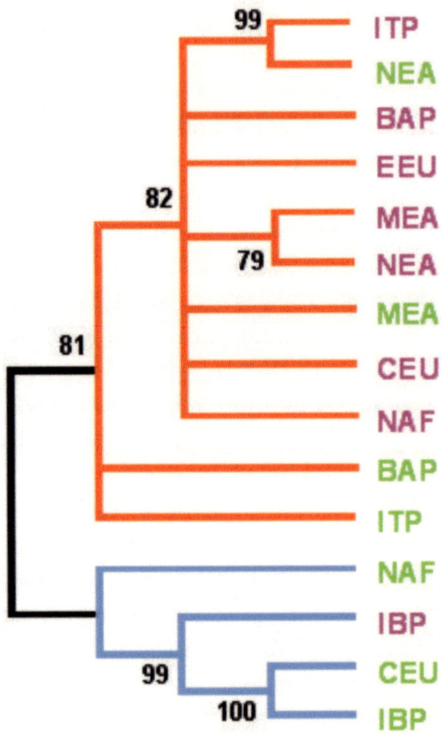

Figure 3. Genetic relationships among *sylvestris* and *sativa* grapevine population groups. The tree was constructed using the neighbor joining method on the Dmyu distance matrix calculated for all pairwise combinations of population groups. Bootstrap support values exceeding 70 are indicated. Branches with low bootstrap support were collapsed. Major clusters are depicted with red and blue colours. *Sylvestris* population groups are depicted in green and *sativa* population groups in magenta. Population codes are as Fig 2.

4. Nuclear diversity in cultivated and wild grapevine

The characterization of the genetic diversity and its distribution throughout the species range is important for our understanding about the adaptation and survival of wild species to ensure that genetic resources are available for use in research and breeding programs (This *et al.,* 2006). Microsatellite markers, being abundant, multi-allelic and polymorphic, provide a means of detecting genetic polymorphism. Due to their co-dominant structure this marker system enables studies on population genetic analysis, assessment of genetic structures and differentiation in germplasm collections and natural populations. The cultivated grapevine (*Vitis vinifera* L.) is very diverse, with 6,000–10,000 cultivars believed to exist in the world (Galet 2000), and many grape collections (http://www.vitaceae.org/index.php/Grape_Germplasm_Resources). This large diversity is mostly due to the long history of grapevine cultivation (McGovern 2003), and vegetative propagation, which has enabled the conservation of cultivars over centuries. There is also a large diversity of complex Vitis hybrids and rootstocks (Galet 2000).

Cipriani et al., (2010) have analyzed a collection of 1005 grapevine accessions; they were genotyped at 34 microsatellite loci with the aim of analyzing genetic diversity and exploring parentages. This study constitutes the largest analysis of genetic diversity in cultivated grape and confirms previous analyses suggesting that grape is a very diverse species (Martinez *et al.* 2006; Ibanez *et al.* 2009). The genetic diversity on average is quite high for *V. vinifera* ssp vinifera (0.769) and even higher for rootstocks and hybrids. It is as diverse as poplar (Smulders *et al.* 2008), rose (Esselink *et al.* 2003), wild populations of rice (Gao *et al.* 2006), and much more diverse than tomato (Ranc *et al.* 2008). High genetic distance is a good indication that grape has been widely exchanged and crossed in order to increase its diversity level (This *et al.* 2006). The analysis of kinship uncovered 74 complete pedigrees, with both parents identified. Many of these parentages were not previously known and are of considerable historical interest. Grouping the accessions by profile resulted in a weak correlation with their geographical origin and current area of cultivation, revealing a large admixture of local varieties with those most widely cultivated, as a result of ancient commerce and population flow.

Several studies have described successfully used microsatellite markers to genotype *V. vinifera* ssp. sylvestris and *V. vinifera* ssp. vinifera (e.g., Aradhya *et al.* 2003; Dangl *et al.* 2001; Imazio *et al.* 2003; Lacombe *et al.* 2003; Regner *et al.* 2000; Lopes *et al.,* 2009; ; Laucou *et al.,* 2011; Andrés *et al.,* 2012). However, *V. vinifera* ssp sylvestris was found to be less diverse than Hybrids or Rootstocks, in accordance with previous observations (De Andres et al. 2007). In general, *V. vinifera* ssp sylvestris is less diverse than the domesticated forms, which could be due to the scarcity of the endangered wild form, small natural populations and the small number of samples available in the collections. The distribution of the wild grapevine has dramatically been reduced over the last 150 years, with the spread of pathogens from North America (phylloxera, oidium, mildew). Most of them died, except in floodplain forests as the root–host homoptera phylloxera was sensitive to flooding (Ocete *et al.,* 2004). Massive death also occurred in vineyards. In France, most vineyards were destroyed and replanted afterwards using American rootstock. Phylloxera did not disappear and continued to infect populations of wild grapevines surviving in the floodplain forests in zones where the water table sank. Intensive river management, starting in the middle of the 19th century, enhanced this process. Two other human impacts also contributed to the destruction of populations of wild Vitis. Shortly after river management, most of the floodplain forests were fragmented and replaced by arable crops or meadows. In remnant forests, the intensification of forest management led to the removal of the vines, considered detrimental to tree growth. Fragmentation of wild grapevine habitats had an enormous impact on gene exchanges between populations, leading to a bottleneck, especially in gyno-dioicious plants. This also reduced the adaptability of the plant to habitat changes.

The total genetic diversity values found in wild grape individuals from Anatolia region are higher than of wild type accessions from other regions such as those described for the Mediterranean basin (Andrés *et al.*, 2012; Di Vecchi *et al.*, 2009; Lopes *et al.*, 2009; Zinelabidine *et al.* 2010). In general, these values are similar for outcrossing vegetative propagated perennial species (Bejaj *et al.* 2007). The observed heterozygosity (Ho) is not significantly different ($P \leq 0.01$) than expected heterozygosity (He) in the wild group, indicating a random mating population. However, reduction in observed heterozygosity has been observed in wild grapevine populations analyzed in Spain, Portugal, France or Italy (Andrés *et al* 2012; Lopes *et al.* 2009; Di Vecchi *et al.* 2006; Grassi *et al.* 2003), most likely due to the reduction of these populations by human action. The comparison of the genetic diversity values with the autoctonous grape cultivars from Anatolia region indicated that diversity is greater in the wild grapes than in the cultivated ones. Similar results have been found in other studies (Lopes *et al.*, 2009; Riani *et al.*, 2010). The wild grapevine population from the both ends of the Mediterranean basin showed a higher genetic variability in Anatolian wild grape populations than in Spanish populations (Ergul *et al.*, 2011). This result is in agreement with the comparison of the number of alleles at the 15 shared SSR loci between Spanish and Anatolian populations. Of 229 total alleles detected at these loci, 189 were observed only in Spanish while 237 were observed only in Turkish populations. The number of unique alleles in Anatolian populations was also much higher than in Spanish populations. This result was expected as Anatolian populations are located at the primary center of diversity and thus are more diverse than in the peripheral populations. At the same time, the Iberian wild grape populations are small, showed lower genetic diversity values and suffered from inbreeding depression (Andres *et al.*, 2012).

In conclusion, the present study suggests that there is no immediate reason for concern about any demographic bottlenecks facing the wild grape populations of the Anatolian region, and the presence of high number of rare alleles in populations investigated here is clear evidence for this finding. At the same time, the wild population from the western and central Europe pointed out that they are suffering inbreeding depression due to the low level of genetic diversity. For the future, in situ conservation of wild grapevine populations around the Mediterranean basin should be advanced by a dynamic approach to keep the level and composition of genetic diversity as high as possible for safeguarding these precious genetic resources for crop improvement.

5. Genetic relationship: Cultivated versus wild compartment of grape

The picture arising today is of a low but clear genetic differentiation of cultivars and wild grape based either on chloroplast markers (Arroyo-Garcia *et al.* 2006; Grassi *et al.* 2006), nuclear microsatellites (Snoussi *et al.* 2004; Grassi *et al.* 2003; Lopes *et al.*, 2009; Ergul *et al.*, 2011; Andres *et al.*, 2012) or both (Grassi *et al.* 2003; Sefc *et al.* 2003). The wild individuals also cluster according to their populations (Grassi *et al.* 2008). The positive Fis values observed in the wild grapevine accessions suggest a high level of genetic relationship among the individuals of the same wild populations. In fact, the detection of potential parent-progeny relationships within wild populations supports that possibility (Andres *et al.*, 2012). At the same time, the detection of gene flow between both compartments (Di Vecchi *et al.* 2009; Andres *et al.*, 2012) could have in the future strong consequences. Therefore, the histories of both compartments are also different and as a consequence linkage disequilibrium is more important in cultivated grape (Barnaud *et al.* 2006) than in wild individuals (Barnaud *et al.* 2010).

Until now a systematic genetic and morphological characterization of the individual accessions had been done with some wild grapevine population in order to confirm whether they could correspond to bona fide ssp. sylvestris individuals, naturalized grapevine cultivars, rootstocks, or spontaneous hybrids derived from wild and cultivated forms as previously described (Di Vecchi *et al.* 2009; Zecca *et al.*, 2011; Andres *et al.*, 2012). The results of the genotypic and phenotypic analyses of wild grapevine accessions from Spain allowed classifying approximately 19% of the samples as naturalized cultivated forms (Andres *et al.*, 2012). These samples could have "escaped" from old abandoned vineyards. As expected for an outcrossing dioecius subspecies they have observed the existence of spontaneous hybrids (4% of the collected samples) between wild and cultivated forms (Andres *et al.*, 2012). The existence of cross hybridization between wild and cultivated forms has been shown to be a widespread phenomenon in many species (Arnold 1998; Papa and Gepts, 2003; Di Vecchi *et al.* 2009). The detection of spontaneous hybrids in grapevine wild populations is in agreement with the previous detection of pollen flow between vineyards and wild plants reported by Di Vecchi *et al.* (2009). This level of gene flow between wild and cultivated forms taking place during many generations might have consequences, as introgression, pollution of the gene pool and genetic loss, on the evolution of these small wild populations (Grassi *et al.* 2006). In addition, these results showed no evidence of hybridization between rootstocks and wild individuals (Andres *et al.* 2012). This could be due to the existence of genetic barriers between both taxa such as the phenological mismatches suggested by Arrigo and Arnold (2007).

Different studies suggest genetic exchange between cultivated and wild grapevines (Cunha *et al.*, 2009; Di Vecchi *et al.*, 2009; Grassi *et al.*, 2003). The genetic relationship between cultivated varieties and wild grapevine populations from Spain suggests a genetic contribution of Southern wild populations in the autochthonous grapevine cultivars varieties (Andres *et al.*, 2012). Therefore, it seems that in opposition to the established dominant theory on the origin of the domestication of grapevine, many of the varieties of the Iberian Peninsula and from other European countries could have local origins.

The genetic analysis of wild grapevine from Spain and cultivars from European countries showed the partition in wild and cultivated forms from that region. The STRUCTURE analysis identifies two genetic groups (clusters C1 and C2) which included all the wild accessions from Spain and correspond to Northern and Southern populations and two other (C3 and C4) including the majority of the analyzed cultivars (Figure 4). The existence of two genetic groups within the wild accessions suggests some level of isolation among those genetic lineages. One possible scenario to generate such structure is that it derives from the isolation created by the last Pleistocene glaciations. As reviewed by Gomez and Lunt (2006), the fragmented nature of the Iberian Peninsula habitat favored the occurrence of multiple glacial refugees isolated from each other. Phylogeographic studies of different European species such as olive trees have shown the existence of strong genetic differentiation within the Iberian Peninsula (Belaj *et al.* 2007). Alternatively, these two genetic groups could represent different colonization events of the Iberian Peninsula by the species *Vitis vinifera* L. what could have taken place following Northern (the Pyrynees) and Southern pathways (Gibraltar). The common chlorotype A identified both in Western Europe and Northern Africa (Arroyo-Garcia *et al.* 2006) seems to suggest a single common origin for all the ancestral populations favoring the first hypothesis. Alternatively, we cannot discard that part of the moderate genetic differentiation observed between the two genetic groups could result from their different history of relationship with the cultivated forms. In fact, we have found a high number of wild genotypes from Southern group showing high ancestry values of clusters C3 and C4 that mainly group cultivated forms of grapevine. In the same direction, we found higher genetic differentiation (Fst = 0.13) between cluster 1 (Northern group) and the analyzed cultivars than between cluster 2 (Southern group; Fst =0.07) and the analyzed cultivars. On the other hand, genetic differentiation between clusters 1 and 2 would be reduced by the existence of gene flow between both genetic groups, what seems to be suggested by the presence of some genotypes showing high ancestry values from both genetic clusters.

Two different genetic clusters could also be detected within the analyzed cultivars although showing very low genetic differentiation (Fst=0.0048). This low genetic differentiation would result from the high level of gene flow between grapevine cultivars. Myles *et al.*, (2011) have proposed that the genetic structure of the vinifera cultivars represents a large complex pedigree resulting from a number of spontaneous and inter-generation crosses between cultivars that have been vegetatively propagated for centuries. Still within this complex pedigree structure, it could be possible to distinguish different groups of more strongly related cultivars that would vary depending on the set of cultivars analyzed. In this case, an analysis of cluster 3 and 4 identified mainly Iberian cultivars as having higher ancestry in genetic cluster 3 and central European cultivars and Northern Iberian cultivars as having higher ancestry in genetic cluster 4.

Interestingly, the analyses of the ancestry values showed by analyzed cultivars identify some of them with a high ancestry value of cluster 1 and cluster 2. These grapevine cultivars correspond to the Spanish cultivars; Allarén, Benedicto, Listan Negro, Malvasia de Lanzarote and Malvasia Blanca and the European cultivars Cabernet Franc, Petit Verdot, Pinot Meunier and Sangiovese. These cultivars have been described as more closely related to wild accessions (This *et al.*, 2006) or are considered autochthonous cultivars. Therefore, these results support the existence of introgression from Western wild forms of *Vitis vinifera* in the pedigree of some of the current Western European cultivars. Finally, the genetic differentiation observed between wild and cultivated forms of grapevine in the Iberian Peninsula point out the interest to characterize and conserved that the existent Western populations as a source of novel alleles for the future understanding and improvement of the genetics of grapevine cultivated forms.

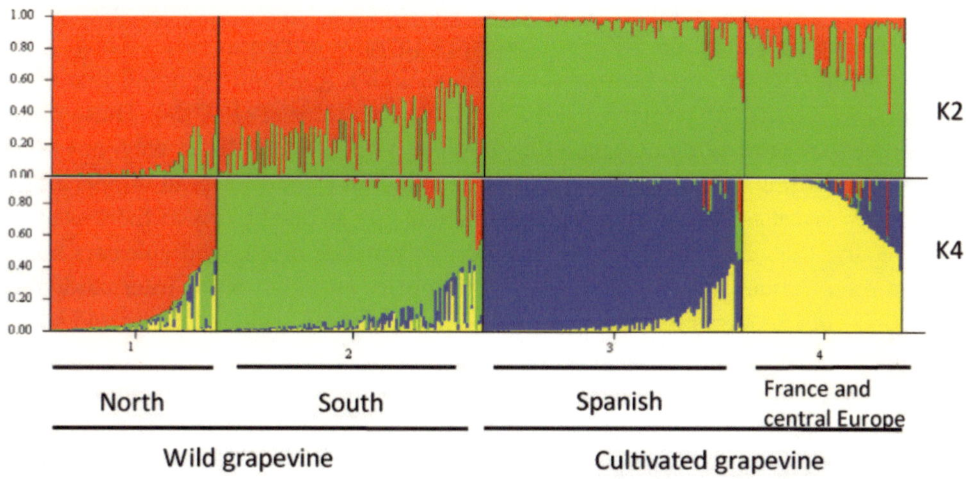

C1	C2	C3	C4	Clusters
73	116	113	71	Individuals (N)
0,37	0,54	0,05	0,04	wild (193)
0,03	0,02	0,58	0,37	cultivated (180)

Figure 4. Graphical representation of ancestry membership coefficients of all individuals analyzed (Cultivated and wild grapevine from Spain). Each individual is shown as a vertical line divided into segments representing the estimated membership proportions in the two and four ancestral genetic clusters inferred with STRUCTURE. Individuals within each cluster are arranged according to estimated cluster membership proportions. (Bottom) Number of individuals and the mean membership fractions in the four genetic clusters.

In conclusion, molecular marker analysis have shown clear divergence between wild and cultivated grapes and low level of introgression (Grassi *et al.* 2003, Ergul *et al.* 2011, Andres et al. 2012), but they are still connected through gene-flow (Regner *et al.* 2000, Lopes *et al.* 2009). Some studies (Grassi *et al.* 2003, Arroyo-García *et al.* 2006, Lopes *et al.* 2009; Andres *et al.*, 2012) have reported the possibility of multiple domestication events in different geographic locations in the origin of cultivated grape. The several geographic sources of wild and cultivated grapes, supports at least two separate domestication events that gave raise to cultivated grape; one derived from the wild grape from Transcaucasia, and another from the wild grape of southern European and North African origin. Probably, with wider representation of wild grape, one may be able to demonstrate the multiple domestication events supporting diffused center of domestication of cultivated grape.

6. Wild grapes as phytogenetic resource

Genetic erosion was perceived as global scale problem in the middle of the twentieth century. It was found out that the introduction of new grapevine cultivars had rapidly displaced the varieties traditionally cultivated resulting in great uniformity of cultivated crops. Therefore, the genetic diversity of those species became alarmingly scarce. This situation led to the implementation of measures for the conservation of plant genetic resources. In the vine, as in other crops, genetic erosion or loss of variability is occurring. That is, it is reducing dangerously agrobiodiversity, the genetic base on which natural selection acts, increasing dramatically the vulnerability of different cultivars to new environmental changes or the appearance of new pests and diseases (Ocete *et al.*, 2007). It should be noted that the wild forms contain diversity for ongoing feedback to relatives (This *et al.*, 2006). These plant genetic resources are generally not a material that is exploitable in a direct way, but it can be used in plant breeding, because wild populations still conserves an overall important genetic diversity (Grassi *et al.*, 2003). This rich genetic pool can be used to avoid the loss of biodiversity affecting the current viticulture. Indeed, the number of allowed cultivars has been reduced to the detriment of several traditional minority varieties. Some international cultivars, like Cabernet Sauvignon, Merlot, Shyrah, Chardonnay, Sauvignon Blanc and so on are being planted in vineyards of all over the world. At the same time, only few numbers of clones from each cultivar are available (Ocete *et al.*, 2004). These facts contribute to a great extent to speed up the problem of genetic erosion in modern viticulture and mainly lead to increase a risk of rapid propagation of new devastating pests and diseases. Some interesting characteristics of wild plants can be transferred throughout the breeding to cultivars suitable of wine making, table grapes and also rootstocks.

Genetic resources in *V. vinifera* are likely limited to only several thousand genotypes in germplasm stock centers or in endangered wild populations. Inter-fertility between species of the genus *Vitis* opens the genetic variation available for breeding across the whole genus. Considering the relevance of genetic resources for the future of the crop and their current scarcity, major efforts should be dedicated to the collection and characterization of the existing resources in the species and the genus. Genomic tools and information can help to rapidly generate genotypic information; however, collection of phenotypic data requires more careful characterization at morphological, biochemical, physiological or pathological and environmental response levels. Open databases with these phenotypic and genotypic data are required as well as more efficient ways to store and exchange biological materials representing all the available genetic diversity.

Together with the genetic variation characterized in the population screened in European countries could be interesting to generate a collection of genotypes that can still represent part of the existent natural genetic variation of the species. This collection could be phenotype in different environments and these genetic tools could be the basis for further studies to establish the relationship between phenotypic variation and nucleotide diversity in grapevine. Understanding grapevine natural genetic variation will help the improvement and breeding of grapevine cultivars.

7. Phenotypic characterization of wild grapevine populations

The analysis of large sets of genetic resources at the morphological level has not been intensive. One of the reasons might be the complexity of the methods available so far or the fact that phenotyping grape is expensive, time consuming and requires a lot of space. Most of the work in the past years has been devoted to the development of methods for many traits from composition of berries to disease resistance and abiotic stresses tolerance but development of rapid methods and non-destructives ones should still be a priority in order to speed up the analysis of genetic resources.

7.1. Enological characterization of wild grapevine populations from Spain

The anthocyanin composition of female grape accessions, mostly Spanish, preserved at El Encin Germoplasm Bank (Madrid, Spain) was analysed during several years. After the extraction from grape skins, total anthocyans were determined by spectrophotometry, and the anthocyanin fingerprint of grapes by HPLC, considering the relative amount of 15 anthocyanins (Revilla *et al.*, 2010). Some typical chromatograms are shown in Figure 5.

The anthocyanin fingerprint of grapes revealed the presence of three types of accessions (Revilla *et al.*, 2010; Revilla *et al.*, 2011). In the first group (23 accessions), grapes did not contain acylated anthocyanins (Revilla *et al.*, 2012). This character is unusual in cultivated grapevines, occurring primarily in Pinot noir and its mutants (Wenzel *et al.*, 1987, Mattivi *et al.*, 2007) and in some grey and rosé cultivars that may be mutants of red grapes (e.g., Pinot gris) or white grapes (e.g., Muscat Rouge de Madere). To our knowledge, this type of anthocyanin fingerprint has not been described in grape cultivars usually considered of Spanish origin (García-Beneytez *et al.*, 2002, Pomar *et al.*, 2005, Gómez-Alonso *et al.*, 2007). In the second group (17 accessions), grapes contained acylated anthocyanins and a high proportion of cyanidin-derived monoglucosides. This character is rare in cultivated grapevines, although it has been reported and was observed in 12 cultivars among the 64 studied (Mattivi *et al.*, 2007). Most were grey or rosé cultivars, or even mutants of white cultivars (e.g., Gewürztraminer). To our knowledge, this anthocyanin fingerprint is rare in grape cultivars usually considered of Spanish origin, with Brancellao as the most remarkable exception (Pomar *et al.*, 2005). In the third group (86 accessions), grapes contained acylated anthocyanins and a large proportion of delphinidin-derived monoglucosides, as do most grapevine cultivars (Wenzel *et al.*, 1987, García-Beneytez *et al.*, 2002, Pomar *et al.*, 2005, Mattivi *et al.*, 2007). In most of these accessions (53), *p*-coumarylated derivatives were more abundant than acetylated derivatives. This character is quite common in red cultivars usually considered as Spanish (e.g., Garnacha and Tempranillo), as described previously (García-Beneytez *et al.*, 2002). On the other hand, acetylated anthocyanins were more abundant than *p*-coumarylated derivatives in 33 accessions. This character is well documented in several French cultivars (e.g., Cabernet Sauvignon and Merlot), but is rare in Spanish cultivars. Among the Spanish cultivars commonly grown, only Mencía presents this type of fingerprint (García-Beneytez *et al.* 2002, Pomar *et al.* 2005).

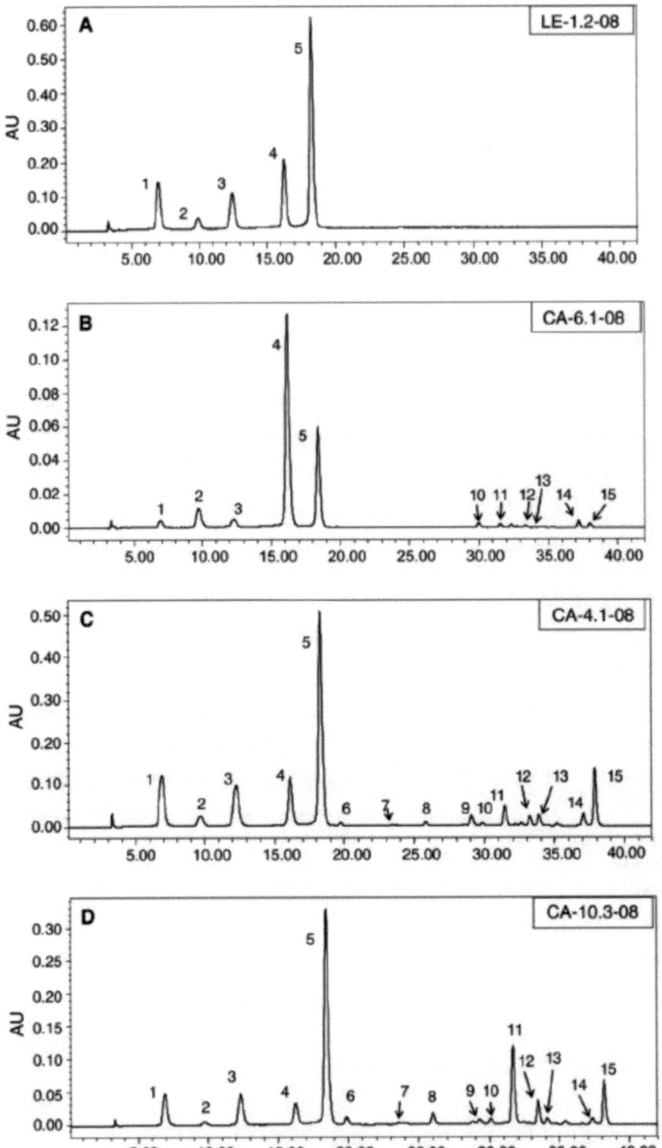

Figure 1 Chromatograms at 520 nm for grape skins extracts of four different wild grapevine accessions: LE-1.2.08, CA-6.1.08, CA-4.1.08, and CA-10.3.08. Peak 1, DpGl; 2, CyGl; 3, PtGl; 4, PnGl; 5, MvGl; 6, DpGlAc; 7, CyGlAc; 8, PtGlAc; 9, DpGlCm; 10, PnGlAc; 11, MvGlAc; 12, MvGlCf; 13, PtGlCm; 14, PnGlCm; 15, MvGlCm.

Figure 5. Chromatograms at 520 nm for grape skins extracts of four different wild grapevine accessions: LE-1.2.08, CA-6.1.08, CA-4.1.08, and CA-10.3.08. Peak 1, DpGl; 2, CyGl; 3, PtGl; 4, PnGl; 5, MvGl; 6, DpGlAc; 7, CyGlAc; 8, PtGlAc; 9, DpGlCm; 10, PnGlAc; 11, MvGlAc; 12, MvGlCf; 13, PtGlCm; 14, PnGlCm; 15, MvGlCm.

Nevertheless, the intensity of acylation is quite variable in this group of accessions, and in about 30% of them the proportion of acylated derivatives is <15%, revealing that the expression of genes involved in the acylation of anthocyanins is quite variable among the accessions.

Results obtained by two-factor ANOVA (accession and year) of the 15 variables used to describe the anthocyanin fingerprint of grapes, using a group of 21 accessions sampled during three consecutive years, suggest that variations in the anthocyanin profile among wild grape accessions were more important than differences among years for a given accession (Revilla *et al.*, 2010). Weather conditions affect to some extent the relative proportion of primitive anthocyanins (DpGl and CyGl) and of some acylated derivatives. Similar results were obtained previously in studies with cultivated varieties (Ryan and Revilla 2003, Revilla *et al.*, 2009). Variance component analysis confirmed that the factor *accession* contributed to variance more than the factor *year*, except for MvGlCf. Moreover, the factor *year* is relevant for primitive anthocyanins (DpGl and CyGl) and MvGl. The influence of year may be related to data on Tempranillo wines, which show different amounts of DpGl when grapes were grown in different environments, but collected at similar stages of ripening and made into wine with the same technology (Revilla *et al.*, 2005).

In conclusion, the maintenance of genetic variability and the phenotypic characterization within wild grape populations has become a priority primarily due to the concurrent risks of increased human impact on flood-plain areas and the spread of new pests. Fragmentation of species habitat will reduce both the number and size of the population, and decrease the genetic variation within populations. So the existence of different genetic pools within this population is remarkable and the conservation of this germplasm becoming more interesting. This population, as the rest situated in Spain, has not a specific preservation statute. It is necessary to take into account that Spain is the country with the largest area of vineyards all over the world, and it is affected by a heavy process of genetic erosion (Ocete *et al.* 2007). In consequence, there is an urgent need to bring this material that could be propagated to nurseries for use in the restoration of riparian forests and undertake breeding programs of cultivars and rootstocks. Particularly, the low incidence of pests and diseases is remarkable, the high acidity of the wines and their high intensity of color total, interesting characteristics can be transferred by crossing with cultivars from Mediterranean areas. On the other hand, the immersion tolerance, absence of rot root and symptoms caused by nematodes could be interesting for obtaining new rootstocks, hybriding with traditional rootstocks, especially when many vineyards have fertirrigation or are planted on clayey soils under a rainy climate, as it was indicated by Ocete *et al* (2010). These phenotypic data will be used to incorporate the wild populations found to the European Vitis Data Base, according to the postulates of the COST Action FA-1003 of Viticulture (EU).

Acknowledgements

Part of the work described in this chapter was funded by the grant number RTA2008-00032-C02-01 and the grant number RTA2011-00029-C02-01. We are grateful to Rafael Ocete and Osvaldo Failla for their collaboration in the COST Action FA1003 and for sharing their expertise in wild grapevine.

Author details

Rosa A. Arroyo García[1] and Eugenio Revilla[2]

*Address all correspondence to: rarroyo@inia.es

1 CBGP-INIA Campus de Montegancedo. Autovía Pozuelo de Alarcón, Madrid, Spain

2 Departamento de Química Agrícola, Facultad de Ciencias, Universidad Autónoma de Madrid, Madrid, Spain

References

[1] Andrés, M. T., Benito, A., Perez-Rivera, G., Ocete, R., Lopez, Gaforio. L., Muñoz, G., Cabello, F., Martínez-Zapater, J. M., & Arroyo-García, R. (2012). Genetic diversity of wild grapevine populations in Spain and their genetic relationship with cultivated grapevines. Molecular Ecology 21; , 800-816.

[2] Aradhya MK, Dang GS, Prins BH, Boursiquot JM, Walker MA, Meredith CP, Simon CJ(2003). Genetic structure and differentiation in cultivated grape Vitis vinifera L. Genetic Resources, , 81, 179-192.

[3] Arnold, C., Gillet, F., & Gobat, J. M. ((1998).) Situation de la vigne sauvage Vitis vinifera ssp silvestris en Europe. Vitis, ., 41, 159-170.

[4] Arnold, C., Rossetto, M., Mcnally, J., & Henry, R. J. (2002). The application of SSRs characterized for grape (V. vinifera) to conservation studies in Vitaceae. American Journal of Botany , 89, 22-28.

[5] Arroyo-García, R., Lefort, F., de Andrés, M. T., et al. (2002). Chloroplast microsatellites polymorphisms in Vitis species. Genome, , 45, 1142-1149.

[6] Arroyo-García, R., Ruiz-García, L., Bolling, L., Ocete, R., Lopez, et., & al, . (2006). Multiple origins of cultivated grapevine (Vitis vinifera L ssp sativa) based on chloroplats DNA polymorphims. Molecular Ecology, , 15, 3707-3714.

[7] Arrigo, N., & Arnold, C. (2007). Naturalised Vitis Rootstocks in Europe and Consequences to Native wild Grapevine. PLos One, 2, (6) 521e.

[8] Barnaud, A., Lacombe, T., & Doligez, A. (2006). Linkage disequilibrium in cultivated grapevine, Vitis vinifera L. Theoretical Applied Genetics , 112, 708-716.

[9] Barnaud, A., Laucou, V., This, P., Lacombe, T., & Doligez, A. (2010). Linkage disequilibrium in wild European grapevine, Vitis vinifera L. ssp. silvestris. Heredity , 104, 431-437.

[10] Belaj, A., Muñoz-Diez, C., Baldoni, L., Procedí, A., Barranco, D., & Satovic, Z. (2007). Genetic Diversity and Populations Structure of Wild Olives from the North-western Mediterranean Assessed by SSR Marker. s. Annals of Botany, , 100, 449-458.

[11] Bryan, G. J., Mc Nicoll, J., Ramsay, G., Meyers, R. C., & De Jong, W. S. (1999). Polymorphic simple sequence repeat markers in chloroplastgenomes of Solanaceous plants. Theoretical and Applied Genetics,, 99, 859-867.

[12] Chung SM, Staub JE(2003). The development and evaluation of consensus chloroplast primer pairs that possess highly variable sequence regions in a diverse array of plant taxa. Theoretical and Applied Genetics, , 107, 757-767.

[13] Cipriani, G., Spadotto, A., Jurman, I., Di Gaspero, G., Crespan, M., Meneghetti, S., Frare, E., Vignani, R., Cresti, M., Morgante, M., Pezzotti, M., Pe, E., Policriti, A., & Testolin, R. (2010). The SSR-based molecular profile of 1005 grapevine (Vitis vinifera L.) accessions uncovers new synonymy and parentages, and reveals a large admixture amongst varieties of different geographic origin. Theoretical Applied Genetics , 121, 1569-1585.

[14] Cunha, J., Balerias-Couto, M., Cunha, J. P., Banza, J., Soveral, A., Carneiro, L. C., & Eiras-Dias, J. E. (2007). Characterization of Portuguese populations of Vitis vinifera L ssp sylvestris (Gmelin) Hegi. *Genetic Resources and Crop Evolution*, 981 EOF-988 EOF.

[15] Dangl GS, Mendum ML, Prins BH, Walker MA, Meredith CP, Simon CJ. (2001). Simple sequence repeat analysis of a clonally propagated species: a tool for managing a grape germplasm collection. *Genome*, 44, 432-438.

[16] Di Vecchi, M., Lucou, V., Bruno, G., Lamcombe, T., Gerber, S., Bourse, T., Boselli, M., & This, P. (2009). Low level of Pollen-mediated gene flow from cultivated to wild grapevine: Consequences for the evolution of the endangered subspecies Vitis vinifera L. ssp silvestris. *Journal of Heredity*, 66 EOF-75 EOF.

[17] De Andres, M. T., Cabezas, J. A., Cerveza, M. T., Borrego, J., Martinez-Zapater, J. M., & Jouve, N. (2007). Molecular characterization of grapevine rootstocks maintained in germplasm collections. American Journal of Enology and Viticulture , 58, 75-86.

[18] Ergul, A., Perez-Rivera, G., Soybelezoglu, G., Kazan, K., Arroyo-Garcia, R., & 201, . (2011). Genetic diversity in Anatolian wild grapes (Vitis vinifera subsp sylvestris) estimated by SSR markers. Plant Genetic Resources , 9(3), 375-383.

[19] Esselink, G. D., Smulders, M. J. M., & Vosman, B. (2003). Identification of cutrose (Rosa hybrida) and rootstock varieties using robust sequence tagged microsatellite site markers. Theoretical Applied Genetics , 106, 277-286.

[20] Galet, P. (2000). Dictionnaire encyclopédique des cépages. Hachette, Paris, France.

[21] Gao LZ, Zhang CH, Li DY, Pan DJ, Jia JZ, Dong YS(2006). Genetic diversity within Oryza rufipogon germplasms preserved in Chinese fi eld gene banks of wild rice as revealed by microsatellite markers. Biodiver Conserv , 15, 4059-4077.

[22] García-Beneytez, E., Revilla, E., & Cabello, F. (2002). Anthocyanin pattern of several red grape cultivars and wines made with them. European Food Research and Technology, , 215, 32-37.

[23] Grassi, F., Labra, M., Imazio, S., Spada, A., Sgorbati, S., Scienza, A., & Sala, F. (2003). Evidence of a secondary grapevine domestication centre detected by SSR analysis. Theoretical and Applied Genetics, , 107, 1315-1320.

[24] Grassi, F., Labra, M., Imazio, S., Ocete, Rubio. R., Failla, O., Scienza, A., & Sala, F. (2006). Phylogeographical structure and conservation genetics of wild grapevine. *Conservation Genetics*, 7, 837-845.

[25] Grassi, F., De Mattia, F., Zecca, G., Sala, F., & Labra, M. (2008). Historical isolation and Quaternary range expansion of divergent lineages in wild grapevine. Biological Journal of the Linnea Society, , 95, 611-619.

[26] Gomez, A., & Lunt, D. (2006). Refugia within refugia: patterns of phylogeographic concordance in the Iberian Peninsula. Edited by S Weiss and N Ferrand. Phylogeography of Southern European Refugia. Springer Dordrecht, The Netherlands., 155-188.

[27] Goldstein DB, Linares AR, Cavalli-Sforza LL, Feldman MW(1995). An evaluation of genetic distances for use with microsatellitesloci. Genetics, , 139, 463-471.

[28] Goldstein DG, Pollock DD(1997). Launching microsatellites: a review of mutation processes and methods of phylogenetic inference. Journal Heredity , 88, 335-342.

[29] Gómez-Alonso, S., Férnandez-González, M., Mena, A., Martínez, J., & García-Romero, E. (2007). Anthocyanin profile of Spanish Vitis vinifera L. red grape varieties in danger of extinction. *Australian Journal of Grape and Wine Research*, 13, 150-156.

[30] Heywood, V., & Zohary, D. ((1991).) A catalogue of wild relatives of cultivated plants native to Europe. Flora Mediterranea ., 5, 375-415.

[31] Hopf, M. (1991). In: Die funde der Südostspanishchen Bronzezeit aus der Sammlung Siret (eds Schubart H, Ulreich H), Philipp von Zabern, Mains., 397-413.

[32] Ibanez, J., Velez de, Andres. M. T., & Borrego, J. (2009). Molecular markers for establishing distinctness in vegetatively propagated crops: a case study in grapevine. Theoretical Applied Genetics , 119, 1213-1222.

[33] Imazio, S., Labra, M., Grassi, F., Winfield, M., Bardini, M., & Scienza, A. (2002). Molecular tools for clone identification: the case of the grapevine cultivar 'traminer''. Plant Bre. ed , 121, 531-535.

[34] IUCN ((1997).) A Global Overview of Forest Protected Areas on The World Heritage List. Jim Thorsell and Todd Sigaty. (Eds.) IUCN.

[35] Jansen, R. K., Kaittanis, C., Saski, C., Lee, S. B., Tomkins, J., Alverson, A. J., & Daniell, H. (2006). Phylogenetic analyses of Vitis (Vitaceae) based on complete chloroplast genome sequences: effects of taxon sampling and phylogenetic methods on resolving relationships among rosids. BMC Evol Biol 6: 32.

[36] Lacombe, T., Laucou, V., Di Vecchi, M., Bordenave, L., Bourse, T., Siret, R., David, J., Boursiquot, J. M., Bronner, A., Merdinoglu, D., & This, P. (2003). Inventory and characterization of Vitis viniferassp. silvestris in France. Acta Horticulturae , 553-557.

[37] Laguna, A. (2003). Sobre las formas naturalizadas de Vitis en la Comunidad Valenciana I. Las especies. Flora Mon. tiberica, , 23, 46-82.

[38] Laucou, V., Lacombe, T., Dechesne, F., Siret, R., Bruno, J. P., Dessup, M., Dessup, T., Ortigosa, P., Parra, P., Roux, C., Santoni, S., Vare`s, D., Pe´ros, J. P., Boursiquot, J. M., & This, P. (2011). High throughput analysis of grape genetic diversity as a tool for germplasm collection management. Theor Appl Gene , 122, 1233-1245.

[39] Levadoux, L. (1956). Les populations sauvages et cultivées de Vitis vinifera L. A. nnales d´Amelloration del Plante, , 1, 59-118.

[40] Lopes, Mendoça. D., Rodrigues, Santos. J. E., Eiras-Dias, J. E., da, Camara., & Machado, A. (2009). New insights on the genetic basis of Portuguese grapevine and on grapevine domestication. *Genome*, 52, 790-800.

[41] Mattia, F., Imazio, S., Grassi, F., Doulati, H., Scienza, A., & Labra, M. (2008). Study of genetic relationships between wild and domesticated grapevine distributed from middle east regions to European countries. *Rendiconti Lincei*, 19, 223-240.

[42] Mattivi, F., Guzzon, R., Vrhovsek, U., Stefanini, M., & Velasco, R. (2007). Metabolite profiling of grape: flavonols and anthocyanins. *Journal of Agricultural and Food Chemistry*, 54, 7692-7702.

[43] Martinez, L. E., Cavagnaro, P. F., Masuelli, R. W., & Zuniga, M. (2006). SSR-based assessment of geneticdiversity in South American V. vinifera varieties. Plant Science , 170, 1036-1044.

[44] Mc Govern, P. E., Glusker, D. L., Exener, L. J., & Voigt, . (1996). Neolithic resin wine. Nature, 381(6528), 480-481.

[45] McGovern PE(2003). Ancient wine. The search for the origins of viniculture. Princeton University Press, Princeton, NJ.

[46] Myles, S., Boyko, A. R., Owens, C., Brown, P., Grassi, F., Aradhya, M. K., Prins, B., Reynolds, A., Chia, J. M., Ware, D., Bustamante, C. D., & Buckler, E. (2011). Genetic structure and domestication history of the grape. Proceedings National Academic of Science USA, 108, (9), 3530-5.

[47] Negrul AM(1938). Evolucija kuljturnyx form vinograda. Doklady Akademii nauk SSSR, , 8, 585-585.

[48] Ocete, R., Lopez, Gallardo. A., Perez, Troncoso. A., Cantos, M., Arnold, C., & Perez, F. (2004). Las poblaciones anadaluzas de vid silvestre, Vitis vinífera L subspecie sylvestris (Gmelin) Hegi: estudio ecológico, ampelográfico, sanitario, y estrategias de conservación. Ed Consejeria de Medio Ambiente, Junta de Andalucia, Sevilla (Spain).

[49] Ocete, R., Cantos, M., López, Gallardo. A., Pérez, Troncoso. A., Lara, M., Failla, O., Ferragut, F. J., & Liñán, J. (2007). Caracterización y conservación del recurso fitogenético vid silvestre en Andalucía. Ed. Falcor. Sevilla (Spain).

[50] Ocete, R., Arroyo-Garcia, R., Morales, M. L., Cantos, M., Gallardo, A., Perez, Gomez. I., & Lopez, (2011. (2011). Characterization of Vitis vinifera L. subspecies sylvestris (Gmelin) Hegi in the Ebro river Basin (Spain). Vitis , 50(1), 11-16.

[51] Olmo HP(1995). The origin, domestication of the vinifera grape In: PE Mc Govern, SJ Fleming,SH Katz (eds) The Origins and Ancient History of Wine. Gordon and Breach Publishers, Philadelphia, USA, , 23-30.

[52] Papa, R., & Gepts, P. (2003). Asymetry of gene flow and differential geographical structure of molecular diversity in wild and domesticated common bean (Phaseolus vulgaris) from Mesopotamia. Theoretical Applied Genetics, , 106, 239-250.

[53] Pico, F. X., Mendez-Vigo, B., Martinez-Zapater, J. M., & Alonso-Blanco, C. (2008). Natural genetic variation of Arabidopsis thaliana is geographically structured in the Iberian Peninsula. *Genetics*, 180, 1009-1021.

[54] Pomar, F., Novo, M., & Masa, A. (2005). Varietal differences among the anthocyanin profiles of 50 red table grape cultivars studied by high performance liquid chromatography. Journal of Chromatography , 1094, 34-41.

[55] Provan, J., Soranzo, N., Wilson, N. J., Mc Nicol, J. W., Forrest, G. I., Cottrell, J., & Powell, W. (1998). Genepool variation in Caledonian and European Scots pine (Pinus sylvestris L.) revealed by chloroplast simple sequence repeats. Proc Roy Soc London B Biol Sci , 265, 1697-1705.

[56] Provan, J., Soranzo, N., Wilson, N. J., Goldstein, D. B., & Powell, W. (1999). A low mutation rate for chloroplasts microsatellites. Genetics, , 153, 943-947.

[57] Provan, J. (2000). Novel chloroplast microsatellites reveal cytoplasmic variation in Arabidopsis thaliana. *Molecular Ecology*, 9, 2183-2185.

[58] Provan, J., Powell, W., & Hollingsworth, P. M. (2001). Chloroplast microsatellites: new tools for studies in plant ecology and evolution. Trends Ecology Evolution , 16, 142-147.

[59] Revilla, E., González-Reig, Garcinuño. P., & García-Beneytez, E. (2005). Role of anthocyanins in the differentiation of Tempranillo wines. In Food Flavor and Chemistry: Exploration into the 21st Century. A.M. Spanier et al. (Eds.), Royal Society of Chemistry, Cambridge., 72-81.

[60] Revilla, E., García-Beneytez, E., & Cabello, F. (2009). Anthocyanin fingerprint of clones of Tempranillo grapes and wines made with them. *Australian Journal of Grape and Wine Research*, 15, 70-78.

[61] Revilla, E., Carrasco, D., Benito, A., & Arroyo-Garcia, R. (2010). Anthocyanin composition of several wild grape accessions. *American Journal of Enology and Viticulture*, 61, 636-642.

[62] Revilla, E., Carrasco, D., Carrasco, V., Benito, A., & Arroyo-García, R. (2011). Composición antociánica de la vid silvestre (Vitis vinífera spp. sylvestris). Proceedings of the 34th World Congress of Vine and Wine, Porto, Portugal.

[63] Revilla, E., Carrasco, D., Carrasco, V., Benito, A., & Arroyo-García, R. (2012). On the absence of acylated anthocyanins in some wild grapevine accessions. Vitis (in press).

[64] Ranc, N., Munos, S., Santoni, S., & Causse, M. (2008). A clarified position for Solanum lycopersicum var. cerasiforme in the evolutionary history of tomatoes (Solanaceae). BMC Plant Biol 8:130

[65] Regner, F., Stadlbauer, A., Eisenheld, C., & Kaserer, H. (2000). Genetic relationships among Pinots and related cultivars. *American Journal of Enology and Viticulture*, 51, 7-14.

[66] Riahi, L., Soghlami, N., El -Heir, K., Laucou, V., Cunff, L. L., Boursiquot, J. M., Lacombe, T., Mliki, A., Ghorbel, A., & This, P. (2010). Genetic structure and differentiation among grapevines (Vitis vinifera) accessions from Maghred region. Genetic Resources and Crop Evolution, , 57, 255-272.

[67] Ryan, J. M., & Revilla, E. (2003). Anthocyanin composition of Cabernet Sauvignon and Tempranillo grapes at different stages of ripening. *Journal of Agricultural and Food Chemistry*, 51, 3372-3378.

[68] Sefc, K. M., Steinkellner, H., Lefort, F., Botta, R., Machado, A. D., Borrego, J., Maletic, E., & Glossl, J. (2003). Evaluation of the genetic contribution of local wild vines to European grapevine cultivars. American Journal of Enology and Viticulture , 54, 15-21.

[69] Snoussi, H., Slimane, M. H., Ruiz-Garcia, L., Martinez-Zapater, J. M., & Arroyo-Garcia, R. (2004). Genetic relationship among cultivated and wild grapevine accessions from Tunisia. Genome, , 47(6), 1211-19.

[70] Smulders, M. J. M., Cottrell, J. E., Le fever, F., van der Shoot, J., Arens, P., Vosman, B., et al. (2008). Structure of the genetic diversity in black poplar (Populus nigra L.) populations across European river systems: consequences for conservation and restoration. For Ecol Manag , 255, 1388-1399.

[71] This, P., Lacombe, T., & Thomas, M. R. (2006). Historical origins and genetic diversity of wine grapes. *Trends in Genetics*, 22, 511-519.

[72] Weising, K., & Gardner, R. C. (1999). A set of conserved PCR primers for the analysis of simple sequence repeat polymorphisms in chloroplast genomes of dicotyledoneous angiosperms. Genome,, 42, 9-19.

[73] Wenzel, K., Dittrich, H. H., & Heimfarth, M. (1987). Anthocyanin composition in berries of different grape varieties. Vitis , 26, 65-78.

[74] Zecca, G., De Mattia, F., Lovicu, Gm., Labra, M., Sala, F., & Grassi, F. (2010). Wild grapevine:silvestis, hybrids or cultivars that escaped from vineyards? Molecular evidence in Sardinia. Plant Biology, , 12, 558-562.

[75] Zinelabine, L. H., Haddioui, A., Bravo, G., Arroyo-Garcia, R., & Martinez-Zapater, J. M. (2010). Genetic origins of cultivated and wild grapevines from Morocco. American Journal of Enology and Viticulture, 61:1.

[76] Zohary, D. (1995). Domestication of the Grapevine Vitis vinifera L. in the. Near East. In: PE Mc Govern, SJ Fleming, SH Katz (eds) The Origins and Ancient History of Wine. Gordon and Breach Sciences Publisher, New York, USA, , 23-30.

Chapter 4

Inter- and Intra-Varietal Genetic Variability in *Vitis vinifera* L.

Stefano Meneghetti, Luigi Bavaresco,
Antonio Calò and Angelo Costacurta

Additional information is available at the end of the chapter

http://dx.doi.org/10.5772/52847

1. Introduction

Grapevine (*Vitis vinifera* L.) is the most economically important and widely cultivated fruit crop in the world and it is one of the oldest crops and the only Mediterranean representative of the *Vitis* genus (Clarke, 2001; Mullins, 1992; Galet, 2000). Its domestication produced cultivars suited to a wide diversity of climates and tastes (Levadoux, 1956; Royo, 1997). In effect this genus shows a wide morphological and genetic variability that is causing confusions and ambiguity for biotypes and clones identification, in particular considering varieties that are widely distributed and cultivated for centuries (Tessier, 1999). Ampelography, ampelometry, and biochemical traits analysis have been traditionally employed to identify the different biotypes in viticulture (Galet, 1979; Calò & Costacurta, 2004). However, these analyses are based on phenotypic characteristics which can be affected by environmental conditions (Meneghetti, 2011).

The DNA molecular analyses are essential for internationally accepted grapevine identification and the investigation of genetic differences among the *Vitis vinifera* L. clones (Meneghetti, 2009). Methods based on DNA analysis have been used with varying degrees of success. This might be affected by the variability level of examined grape varieties and by the types of markers systems employed to investigate genetic relationships.

Simple sequence repeat (SSR) markers are universally used for the identification of the grape varieties (This, 2004). Di-nucleotide repeats pose some problems for stuttering, adjacent alleles, and binning and a possible SSR development was proposed by using microsatellites with a longer core repeat (Cipriani, 2010).

The molecular approaches are also essential for internationally accepted grapevine identification and to investigate the genetic inter- and intra-varietal variability. Molecular markers have been used on *Vitis vinifera* in several studies to distinguish among clones of the same cultivars as RAPD random amplified polymorphic DNA, PCR specific analysis ISSR (inter-microsatellites), SNP (single nucleotide polymorphism), S-SAP (specific sequence amplified polymorphism), IRAP (inter-retrotransposon amplified polymorphism), REMAP (retro-transposon microsatellite amplified polymorphism), M-SAP (methylsensitive amplified length polymorphism), chloroplast DNA polymorphisms, SSCP (single-strand conformation polymorphism) (Moreno, 1995; Bavaresco, 2000; Imazio, 2002; Owens, 2003; Labra, 2004; D'onofrio, 2009).

A molecular strategy to obtain DNA polymorphisms of *Vitis vinifera* genotypes from the same cultivar to study the intra- and inter- varietal genetic variability, to discriminate accessions, clones, and biotypes of a same grape variety, and to analyze the relationships between molecular profiles and some environmental parameters (i.e., geographic site) or morphological traits was reported by Meneghetti et al. (2012a; 2012b; 2012c). This approach uses four different molecular marker systems (i.e., AFLP amplified fragment length polymorphism, SAMPL selective amplification of microsatellite polymorphic loci, M-AFLP microsatellites amplified fragment length polymorphism, and ISSR inter simple sequence repeat).

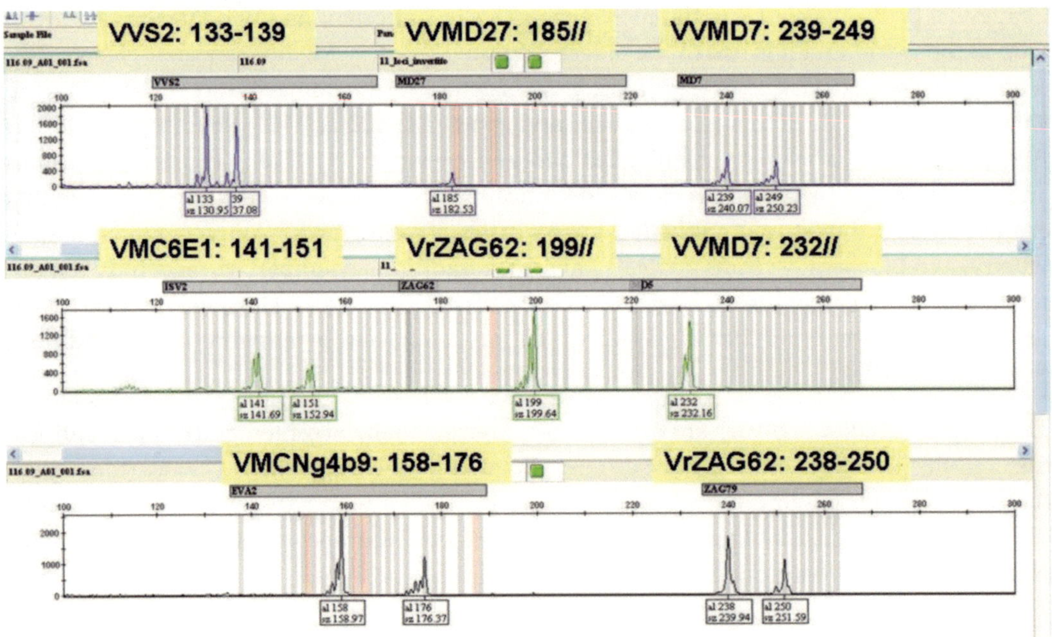

Figure 1. Example of Di-nucleotide SSR profile of Sagrantino cultivar generated by the 3130XL capillary sequencer at eight loci using different fluorochromes. The peaks indicate the alleles and its size on *Vitis vinifera* SSR BinSet (Meneghetti, 2012c).

2. Grapevine cultivar identification

Simple sequence repeat (SSR) markers are universally used for the identification of the grape varieties (Figure 1). Microsatellites consist of tandemly repeated simple sequence motifs with a high variation in repeat number among individuals. Applications of microsatellite markers include not only cultivar identification but also parentage testing, pedigree reconstruction and studies of population structure.

A strategy of grapevine cultivar identification is to analyze eleven di-nucleotide microsatellite loci as VVS2, VVMD5, VVMD7, VVMD27, VVMD28, VrZAG62, VrZAG79, VMC6E1, VMC6F1, VMC6G1 and VMCNG4b9 (Meneghetti, 2012c). PCR reaction mixture at 11 loci was performed by the workstations using SSR forward labeled primers with 6FAM, VIC, NED and PET dyes, and SSR reverse primers unlabeled each at 5 pmol/μl (Applied Biosystems). The PCR was performed in a GeneAmp PCR System 9700 (Applied Biosystems) and SSR polymorphisms were resolved on an ABI-3130XL capillary sequencer (Applied Biosystems) using GeneMapper version 4.1 (Applied Biosystems) with a *Vitis vinifera* microsatellite BinSet of 11 SSR standard loci. The important molecular polymorphisms were checked by the Sequi-Gen GT Sequencing Cell electrophoresis (Biorad) (Meneghetti, 2012a).

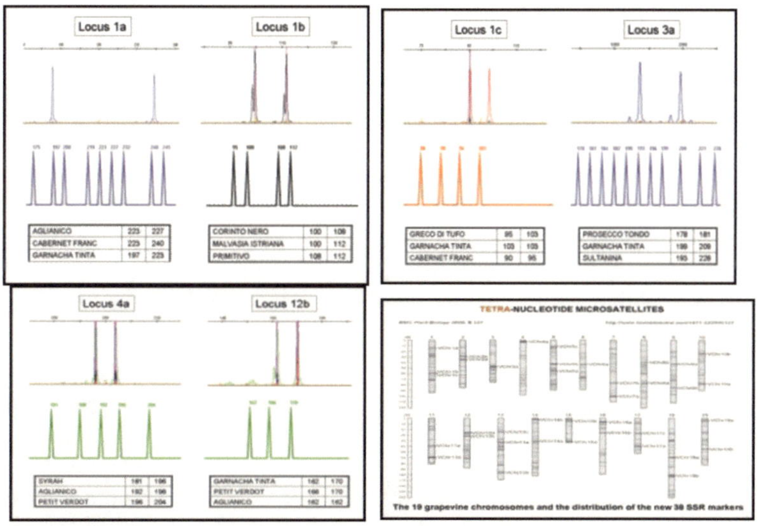

Figure 2. Examples of microsatellites with a longer core repeat in grapevine with the locus name and their position on *Vitis vinifera* genome (Meneghetti, 2012c).

3. Inter-varietal genetic variability

The genus *Vitis* is characterized by great morphological and genetic variability. It is necessary to increase the number of SSR loci used for cultivars identification to analyze the genetic inter-varietal variability by SSR polymorphisms in *Vitis*. For example: analyzing other microsatellite markers as VVS1, VVS29, VVMD8, VVMD17, VVMD21, VVMD24, VVMD25, VVMD26, VVMD32, VVMD36, VrZAG47, VrZAG64, VrZAG83, VMC6E4, VMC6H6, VMC1E12, VMC4G6, VMC2H9, VMC2A5, VMC3D7, VMC2G2, VMC6E10 (Bowers, 1999). It is also possible to use the microsatellite primers with a longer core repeat (Cipriani, 2010) (Figure 2) or different molecular markers as M-AFLP (Figure 3).

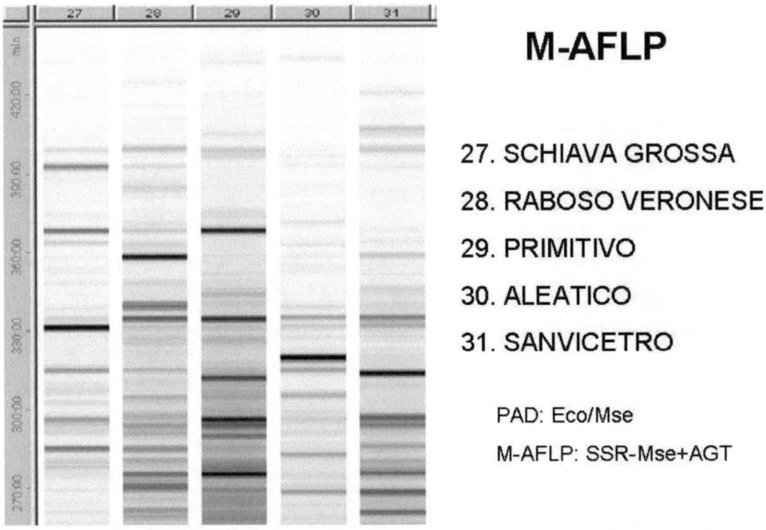

Figure 3. Examples of molecular polymorphisms of Schiava grossa, Raboso veronese, Primitivo, Aleatico and Sanvicetro grape cultivars by M-AFLP technique.

Genetic dissimilarity of SSR (GD) estimates between grapevine cultivars (inter-varietal genetic variability) were calculated by using the following formula:

$$GDij = -\ln(PS)$$

PS is the percentage of common microsatellite alleles within the i and j genotypes, according to Dangl (2001).

Dendrogram was produced by the Unweighted Pair-Group Arithmetic Average Method (UPGMA) clustering algorithm and the Numerical Taxonomy and Multivariate Analysis System (NTSYS-pc) Version 2.10 (Exeter Software Co., NY, USA).

In particular for grapevine SSR variability, an additional study was performed by the BAND Genetic Similarity (GS) coefficient of Lynch (1990) used for SSR data in diploid genomes according to the following formula:

$$GSij = Nij/(Ni - Nj)$$

Nij is the number of bands in common, Ni and Nj are the numbers of bands in the two individuals (i and j) being compared. Thus, GSij = 1 indicates the identity between i and j, whereas GSij = 0 indicates complete diversity. A pair of diploid individuals can have 0, 1, or 2 bands in common at each SSR locus. Dendrogram of the analyses were constructed from the symmetrical GS BAND matrix (NTSYS-pc).

An example of these molecular analyses can be explained by the grape Malvasia family. The name Malvasia has ancient origins and refers to a numerous and heterogeneous group of varieties growing in many European countries. Malvasias is spread in Italy from north to south and seventeen Malvasia cultivars are registered in the Italian National Catalogue (Calò & Costacurta, 2004). There are few Malvasia varieties with black berries, mostly grown in the North-Western Italian region of Piedmont (i.e., Malvasia di Casorzo, Malvasia nera lunga and Malvasia di Schierano). Malvasia nera di Brindisi/Lecce contributes to the Salento oenological production in the southern Italian region of Apulia (Lacombe, 2007; Crespan, 2006).

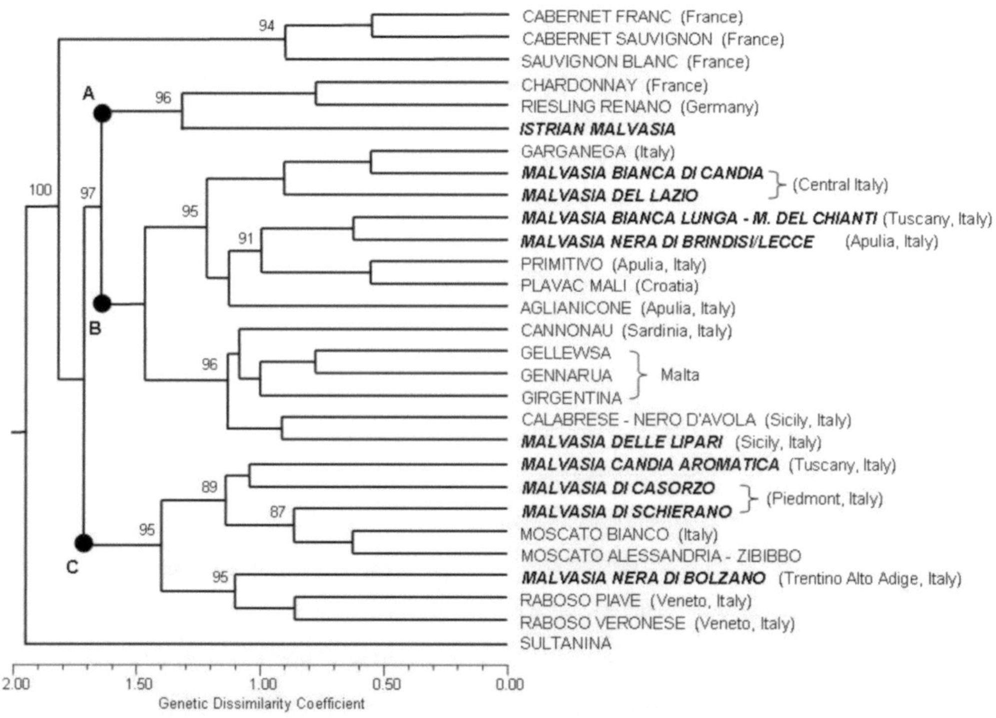

Figure 4. Dendrogram of ten Malvasia cultivars (i.e., Istrian Malvasia, M. delle Lipari, M. bianca di Candia, M. di Candia Aromatica, M. del Lazio, M. Bianca lunga / M. del Chianti, M. nera di Brindisi/Lecce, M. di Casorzo, M. di Schierano, M. nera di Bolzano) and 19 grapevine varieties (i.e., Cabernet Franc, Cabernet Sauvignon, Sauvignon Blanc, Chardonnay, Riesling Renano, Garganega, Primitivo, Plavac Mali, Aglianicone, Cannonau, Gellewsa, Gennarua, Girgentina, Calabrese / Nero d'Avola), Moscato Bianco, Moscato di Alessandria / Zibibbo, Raboso Piave, Raboso Veronese, Sultanina) based on Dangl's Genetic Dissimilarity (Meneghetti, 2012b).

Figure 4 reports a dendrogram of Malvasia cultivars by SSR molecular polymorphisms. The Malvasia cultivars were divided into three distinct groups: Istrian Malvasia was grouped with Riesling Renano and Chardonnay without the other Malvasias (cluster A). Sultanina was an out-group. The dendrogram showed clearly the genetic divergence of Malvasias family detected using only the SSR approach, in agreement with Calò and Costacurta 2005. The cluster analysis allowed to distinguish some variety groups cultivated in neighboring geographical regions: Cabernet Franc, Cabernet Sauvignon, Sauvignon Blanc, and Chardonnay from France; Malvasia bianca di Candia and M. del Lazio from Central Italy; Primitivo and Aglianicone from Apulia region (Southern Italy); Nero d'Avola and Malvasia delle Lipari from Sicily region (Southern Italy); Malvasia di Casorzo and M. di Schierano from Piedmont region (Northern Italy); Raboso Piave and R. Veronese from Veneto region (Northern Italy); Gellewsa, Gennarua, and Girgentina from Malta. The Istrian Malvasia was positioned in the A group, while Malvasia bianca di Candia, M. del Lazio, M. Bianca lunga (also known as M. del Chianti), M. nera di Brindisi/Lecce and M. delle Lipari accessions were clustered in the B group and M. di Casorzo, M. di Schierano, and M. nera di Bolzano in the C group.

The ten Malvasias shown in Figure 5 were further analyzed by Genetic Similarity BAND coefficient using the microsatellite polymorphisms. The dendrogram of Malvasias in Fig. 5 showed the grouping of the Malvasia as in dendrogram in Fig. 4. In fact, Malvasia bianca di Candia, M. del Lazio, M. Bianca lunga, M. nera di Brindisi/Lecce, and M. delle Lipari were clustered into B group while M. di Casorzo, M. di Schierano, and M. nera di Bolzano were grouped into the C group, while the Istrian Malvasia is positioned between the two main groups (Meneghetti, 2012b).

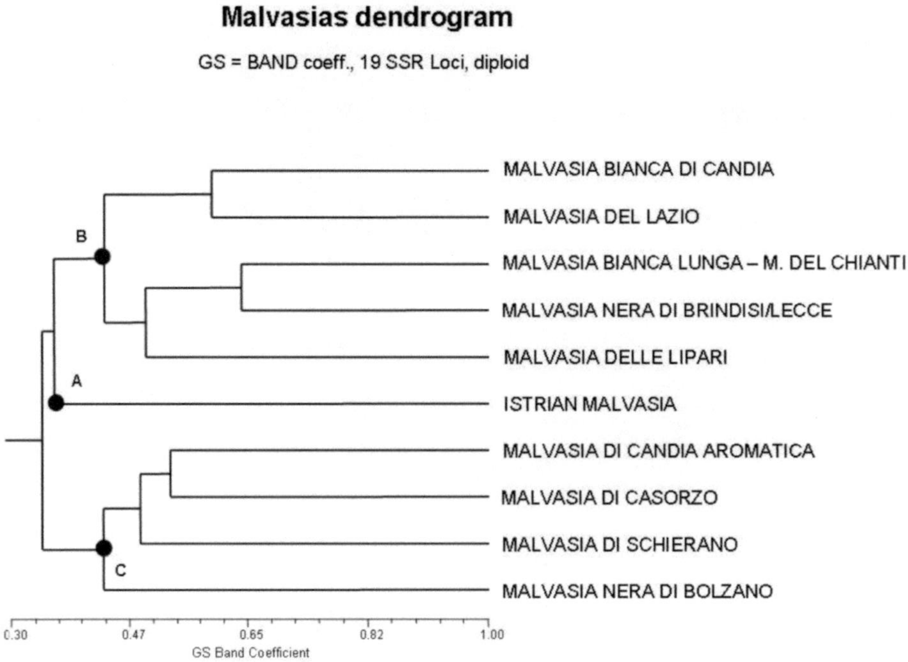

Figure 5. Dendrogram of ten Malvasias cultivars obtained using the Genetic Similarity BAND coefficient. The Genetic Dissimilarity analyses (Fig. 4) were confirmed by this approach and the three subgroups (A, B and C) were showed also in this dendrogram with Istrian Malvasia positioned between the two main groups (Meneghetti, 2012b).

http://dx.doi.org/10.5772/52847

4. Intra-varietal genetic variability

The genetic variability of accessions from the same grape cultivar can be investigated by means of AFLP, SAMPL, M-AFLP and ISSR molecular markers according to Meneghetti et al. (2012c).

The AFLP, SAMPL and M-AFLP analyses were performed using a Cy5-labeled EcoRI+3 (or PstI+2) primer and an unlabeled MseI+3 primer (three selective nucleotides). The amplification products were resolved on ReproGel High-Resolution pre-made acrylamide–bisacrylamide solutions (8% w/v) (GE Healthcare) in modified TBE buffer and detected on a semi-automated DNA sequencer, the ALFexpress-II DNA Analysis System (Amersham Pharmacia Biotech). Markers were visualized automatically by the ALF-win Fragment Analyses 1.09 software and checked by Quantity One 4.6.7 and PD Quest Basic 8.0.1 software (Biorad) (Meneghetti, 2011).

The Inter-microsatellite analysis was performed using the PCR protocol reported by Meneghetti et al. 2012a, with minor changes. ISSR experiment were carried out using the same procedure of AFLP.

A binary presence or absence (1 *vs.* 0) matrix was created for AFLP, SAMPL, M-AFLP and ISSR markers and for each grapevine accessions. Molecular markers were defined by a standard ladder using the ALF-win Fragment Analyses 1.09 software (Amersham Pharmacia Biotech) and two reference DNA genotypes and visualized automatically by the ALF-win software. The scoring was checked by using Quantity One 4.6.7 and PD Quest Basic 8.0.1 software (Biorad) (Meneghetti, 2012c).

Genetic similarity (GS) estimates among individuals were calculated in all possible pair-wise comparisons using the Dice's coefficient which was based on the probability that a marker from one accession will also be present in another and calculated using the following formula:

$$GSij = 2X/(2X + Y + Z)$$

X represents the number of shared amplification products scored between the pair of samples/fingerprints (i and j) considered, Y is the number of products present in i but absent in j, Z is the number of products present in j but absent in i (Dice, 1945).

Thus, GSij = 1 indicates identity between i and j, whereas GSij = 0 indicates complete diversity.

GS was calculated within (GS_W) and between (GS_B) cultivars and marker systems (AFLP, SAMPL, M-AFLP).

The cluster analysis of GS was performed according to the UPGMA algorithm using the NTSYS software.

Centroids of the grapevine accessions were plotted on a 2-dimensional graph according to the principal coordinates extracted from the GS matrices estimated by the three molecular marker systems. All calculations and analyses were conducted using the appropriate routines of the NTSYS Version 2.10 software.

The information content of each marker system in discriminating the accessions of the same variety was calculated using the marker index (Powell, 1996). The efficiency of dendrograms was tested by cophenetic correlation. The reliability of clusters was evaluated by the bootstrapping procedure using 100 random samples of molecular markers. The software used was PHYLIP 6.6 (http://evolution.genetics.washington.edu/phylip.html).

Hence it was reported and discussed using the molecular results of six grape cultivars (i.e., Garnacha tinta, Primitivo, Malvasia nera di Brindisi/Lecce, Negroamaro, Malvasia di Candia and Istrian Malvasia) on a few different aspects: genetic similarity, genotypes discrimination, biotypes discriminations and clones identification. There were correlations between geographic origins of materials and DNA fingerprinting plus relationships between morphological traits and molecular polymorphisms.

4.1. Garnacha tinta

Garnacha is one of the most widely planted grape varieties in the world (240,000 ha). It is known by local names in Mediterranean regions: Garnacha tinta and Grenache noir are the Spanish and French name, while in Italy this variety is known as Cannonao, Alicante, and Tocai rosso (three Italian synonymous) but also as Cannonau (Sardinia) and Gamay perugino (Tuscany) (Galet, 2000; Calò, 1990).

Fifty-three Garnacha accessions were investigated: 28 Italian accessions, 19 Spanish accessions, and 6 French accessions. The Italian accessions were 6 Tocai rosso from the Vicenza area, 8 Alicante from Sicily and Elba island, 4 Gamay perugino from Perugia province and 10 Cannonau from Sardinia. In order to verify the varietal identity, the analyses based on 11 SSR loci confirmed that only one SSR profile was obtained for the 53 accessions (Figure 6) (Meneghetti, 2011).

The study of intra-varietal genetic variability was performed using AFLP, SAMPL and M-AFLP molecular markers. The bi-dimensional plotting of centroids reported in Figure 6 showed six different groups: 1) Italian Alicante accessions from Sicily; 2) Italian Tocai rosso accessions from Vicenza area (Colli Berici); 3) Italian Gamay perugino accessions from Tuscany and Umbria; 4) Spanish Garnacha accessions from Andalucia, Aragón, Cataluña, Castilla y León, Madrid; 5) French Grenache noir accessions; 6) Italian Cannonau accessions from Sardinia. The first coordinate allowed to distinguish clearly Spanish, French and Italian accessions while the second one separated the 4 Italian geographic origins (Figure 6).

Genetic similarity of 53 Garnacha samples was calculated within groups (GS_W) and also between groups (GS_B) (Meneghetti et al. 2012c). The PCA analysis confirmed the high genetic variability within Italian genotypes on the base of their provenance, on the contrary the 19 Spanish accessions were clustered in a more homogeneous group that showed a high genetic similarity (GS_W= 0.9872).

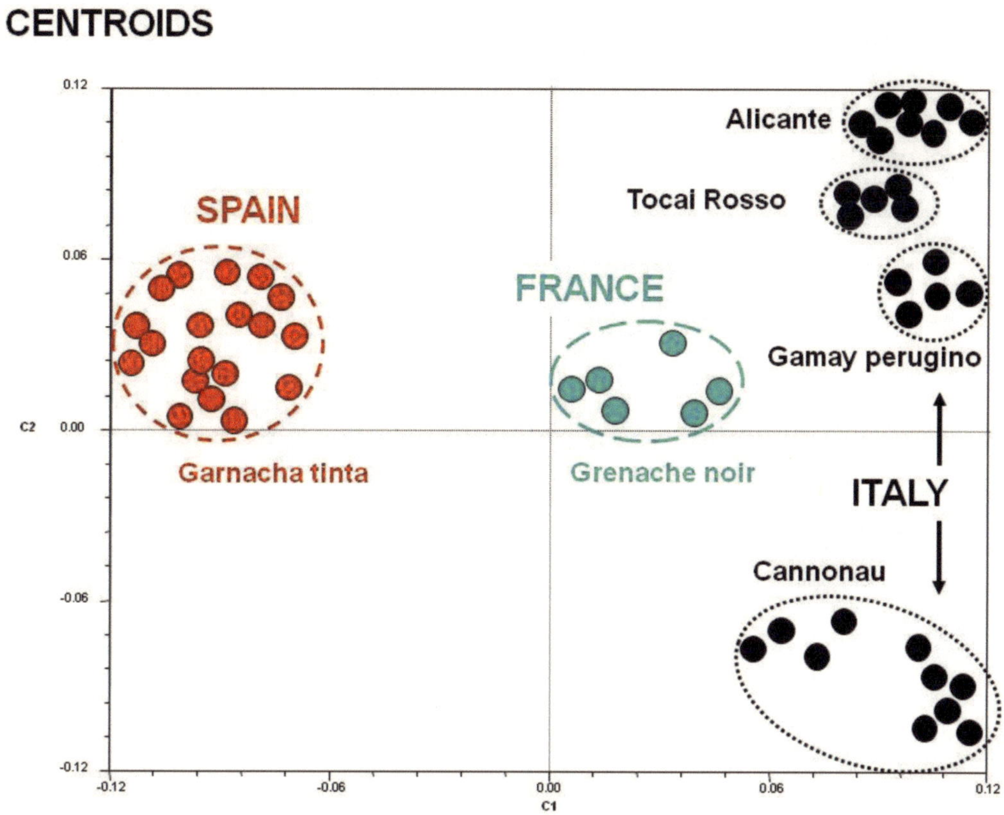

Figure 6. Centroids of Garnacha tinta from Spain, Grenache noir from France, Alicante from Sicily (Italy), Tocai rosso from Vicenza area (Italy), Gamay perugino from Tuscany (Italy), Cannonao from Sardinia (Italy). The Genetic Similarity analyses were confirmed by this approach and the materials with same SSR profile were distinct according to the six geographic origins and the three Countries (Meneghetti, 2011).

The molecular approach discriminates all genotypes of this cultivar. Italian samples showed a high genetic variability within genotypes (GS_W = 0.9481), while Spanish samples showed a high GS (GS_W = 0.9872). GS_W of Italian accessions (0.9481) was very similar to GS_B (0.9480), but the four Italian origins are clearly separated by these molecular markers (Meneghetti, 2011).

AFLP, SAMPL, and M-AFLP were able to clearly distinguish the 53 Garnacha accessions from Italy, Spain, and France. The large number of molecular markers and their high degree of polymorphism make them important tools for many genetic studies.

Provenance-specific molecular polymorphisms were reported in Figure 7 and AFLP analyses shown in Figure 8.

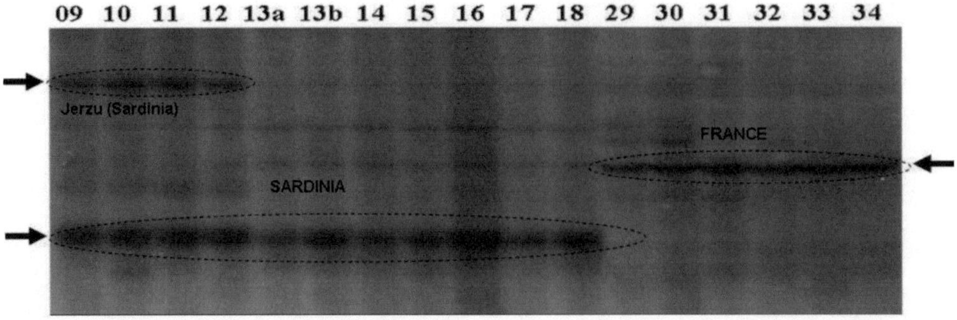

Figure 7. Provenance-specific polymorphisms by SAMPL molecular markers obtained for the 10 Sardinian Cannonau (CAN-09/18) and six French Grenache (GRE-30/34) using Silver Staining technique. The first four line (09-12) corresponding to Cannonau from Jerzu (CAN-09/12) with a specific amplification product (Meneghetti, 2011).

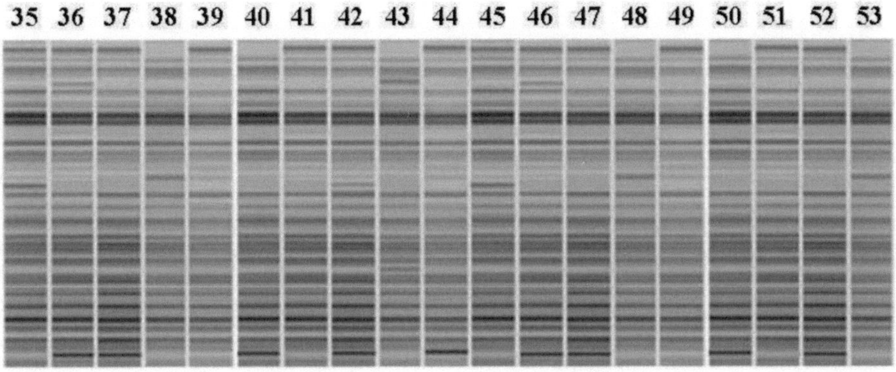

Figure 8. An example of a digitalized electropherogram of the AFLP profiles obtained for the 19 Spain Garnacha accessions (GAR-35/53) using an ALFexpress-II DNA Automated Sequencer. The majority of AFLP markers were monomorphic but there were some clearly differences: the line 36 and 46 are very similar (two G. blanca genotypes) and line 37 and 49 showed only a different marker (two G. peluda genotypes). Genotypes from left to right: 35= Garnacha tinta, 36= G. blanca, 37= G. peluda, 38= G. roja; from 39 to 45= Garnacha tinta; 46= G. blanca, 47= G. erguida, 48= G. roya, 49= G. peluda; from 50 to 53= G. tinta (Meneghetti, 2011).

4.2. Primitivo

Primitivo is a grapevine variety very important for Apulian viticulture and according to tradition it was first planted by Benedictine monks in Gioia del Colle (Bari, Apulia, Italy). Primitivo di Gioia is the best known variety used in Gioia del Colle DOC wine and is genetically equivalent to the Croatian Crljenak Kaštelanski and the American Zinfandel (Calò & Costacurta, 2004; Calò, 2008).

Fifty-nine different vines have been selected based on discriminating traits (i.e., shape, size, density, color of the skin of the bunch and of the berry). Five typologies called A, B, C, D, and E have been identified by means of ampelographic and phyllometric analyses. The morphological traits of the five biotypes (i.e., leaves, bunch, and berry) were maintained after repeated propagation of these biotypes in experimental vineyards. Thus, the morphological traits could have been fixed and stabilized during several centuries of cultivation at Gioia del Colle (Meneghetti, 2012a).

The identical SSR profiles of Primitivo biotypes are shown by a Reference Primitivo clone from Taranto (Apulia, Italy) and two Zinfandel accessions from USA.

Dice's GS matrix was used to perform the Principal Coordinate Analysis of all Primitivo accessions.

Molecular markers discriminated the five biotypes from Gioia del Colle (Bari, Italy) to those From Pulsano (Taranto, Italy) and Zinfandel accessions from USA (Figure 9).

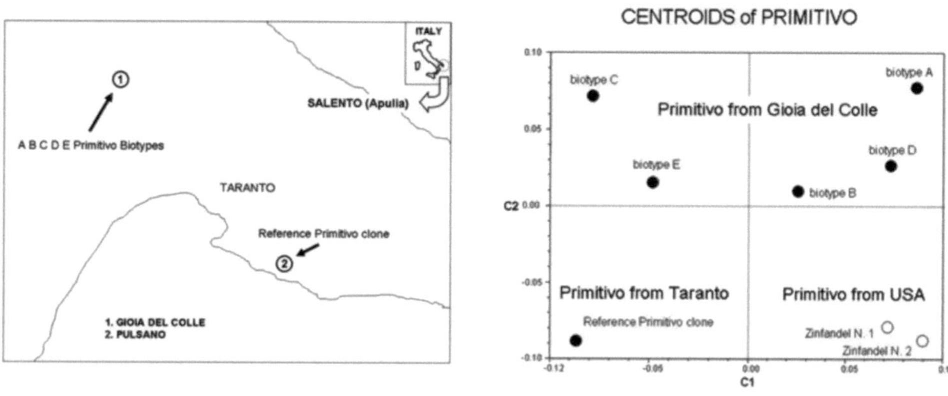

Figure 9. Geographic map of the five Primitivo biotypes from Gioia del Colle (Apulia, Italy) and the reference clone from Pulsano (TA). Centroids discriminated the genotypes according to the different geographic origins (i.e., Gioia, Pulsano, USA) (Meneghetti 2012a).

A total of 2,223 reproducible amplification products were obtained using four molecular marker systems (i.e., 837 AFLPs, 713 SAMPLs, 616 M-AFLPs and 57 ISSRs) and 1,156 products (52.0%) were polymorphics.

The molecular analysis displays a high genetic variability within Primitivo genotypes which is in agreement with the non-homogenous geographical areas of cultivation. The GS was 0.8129 among Primitivo biotypes from Gioia del Colle; the GSw was 0.9477 within American accessions; the GS_B was 0.7489 between Gioia del Colle biotypes and the Reference clone from Pulsano and the GS_B was 0.7013 between the five Apulian biotypes and the two Zinfandel accessions from USA.

Dice's GS matrix was used to perform the Principal Coordinate Analysis using all Primitivo accessions (Figure 9).

The molecular markers discriminated the Gioia del Colle biotypes from the Pulsano Reference biotype and to the two American Zinfandel accessions (Figure 9).

The first coordinate of the centroids allowed to distinguish the five different biotypes of Gioia del Colle. The second coordinate allowed to separate the biotypes of Gioia del Colle from the two American Zinfandel accessions and the Primitivo reference clone of Pulsano (Figure 9).

Thus, we could discriminate both the Primitivo accessions (i.e., the 5 biotypes from Gioia del Colle, the clone from Pulsano, the two American Zinfandel accessions) and the different geographical origins of the plants.

4.3. Malvasia nera di Brindisi/Lecce

Malvasias belong to a numerous and heterogeneous population of varieties growing in many European countries and their history is an intriguing enigma. Several types of grape varieties have been traditionally considered under the generic term of Malvasia, often with a complementing name related to geographic origin (Crespan, 2008; Calò & Costacurta, 2004).

In Italy, at the present time, Malvasias are spread from North to South and 17 Malvasia cultivars are registered in the Italian National Catalogue. Apulian Malvasia nera is a cultivar with black berries and belongs to the Apulian ampelographic assortment: this grape is very widespread in the Salento peninsula, from the Taranto area right across to the provinces of Brindisi and Lecce.

The Malvasia nera of Lecce and Brindisi, originated from the cross between Malvasia bianca lunga or Malvasia del Chianti and Negroamaro. It represents an important variety in the Apulia region. In the past, Malvasia nera of Brindisi and Malvasia nera of Lecce were considered two different cultivars, but this presumed synonymy has been ascertained with SSR markers and therefore these two Malvasia nera would be considered to be the same variety (Meneghetti, 2012a). Morphological analysis allows to differentiate accessions of this cultivar when we compare biotypes cultivated from the Lecce region with others from the Brindisi region. For this reason deeper molecular analyses have been conducted to investigate differential molecular traits between these two Malvasia cultivars with different geographical origin.

Thirteen accessions of Italian Malvasia nera from Brindisi (Salento, Apulia) and thirteen accessions of Italian Malvasia nera from Lecce (Salento, Apulia) were analyzed. All the accessions show the same SSR profile and were identified as Malvasia nera of Lecce and Brindisi. AFLP, SAMPL, M-AFLP and ISSR analyses were performed in order to study the intra-varietal variability.

A total of 2,049 reproducible amplification products were obtained with the four molecular marker systems, 756 AFLPs, 615 SAMPLs, 626 M-AFLPs and 52 ISSRs.

The discrimination among the 26 genotypes of Malvasia nera of Lecce and Brindisi from the two different geographic origins of Salento (Lecce and Brindisi) was possible using the four marker types as reported in Figure 10 where as MLB is Malvasia nera of Lecce and Brindisi.

The MLB genotypes with the numbers 1 to 13 were from Brindisi while samples with the numbers 14 to 26 were from Lecce.

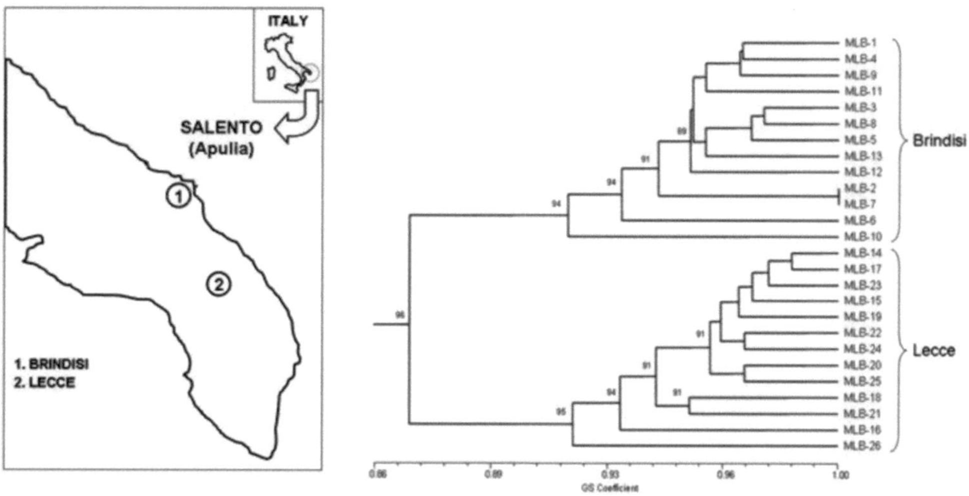

Figure 10. Geographic map of Lecce and Brindisi (Apulia, Italy) and dendrogram of genotypes of Malvasia nera di Brindisi/Lecce from Lecce and from Brindisi regions (Meneghetti, 2012a).

The cluster analysis clearly grouped the 26 accessions according to the two geographical origins, Lecce and Brindisi. Two accessions from Brindisi (number 2 and 7) showed the same molecular profile (i.e., identical genotype).

Genetic similarity (Dice, 1945) estimated within and between the two origins, Brindisi and Lecce, was confirmed that these two groups were not genetically identical. The GS_{TOT} was 0.8269, the GS_W within the 13 accessions from Brindisi was 0.9544 and the GS_W relating of the 13 accessions from Lecce was 0.9589; GS_B between the two origins was 0.7572.

The molecular approach was efficient to discriminate the Apulian Malvasia nera accessions from these two different provinces of the Salento area.

4.4. Negroamaro

Negroamaro is a grape variety native to Southern Italy and is grown almost exclusively in Apulia (Calò & Costacurta, 2004).

This grapevine cultivar is considered to have an even older origins in Apulia (i.e., possibly brought by ancient Greek settlers that colonized Southern Italy) and it is one of the most important popular wine varieties of the Salento area.

This variety produces the famous regional red and rosé wines 'Negroamaro Cannellino' that comes from a distinct biotype which is listed separately in the Italian Register of Grapevine Cultivar (Calò, 2000). Although the SSR markers don't distinguish it from the Negroamaro variety, the somatic mutation that allows a characteristic precocity of maturation (15 days) of 'Negroamaro Cannellino' affects a fundamental physiological distinctive trait. Therefore, it is not possible to consider these two Negroamaro biotypes from the same cultivar (Meneghetti, 2012a).

Forty-four accessions of Negroamaro from Apulia (Italy) analyzed at 11 microsatellite loci showed a microsatellite profile in agreement with the Negroamaro grapevine variety.

In order to define the intravarietal variability AFLP-based molecular markers and inter-microsatellites were used.

The Negroamaro accessions were from eight different geographic origins of Salento (Apulia, Italy): Alezio, Tuglie, Copertino, Veglie, Leverano, San Pancrazio, Cellino San Marco and Ceglie Messapica.

A total of 2,282 reproducible amplification products were obtained with the four molecular marker systems, 856 AFLPs, 756 SAMPLs, 620 M-AFLPs and 50 ISSR and 1,022 (44.8%) of these were polymorphics.

The Negroamaro accessions were separated according to their specific origins and according to a gradient "lowland-hill" or "North-South Apulia" as shown in Figure 11. The Negramaro accession from the Northern hilly origin, Ceglie Messapica, is shown as an outgroup.

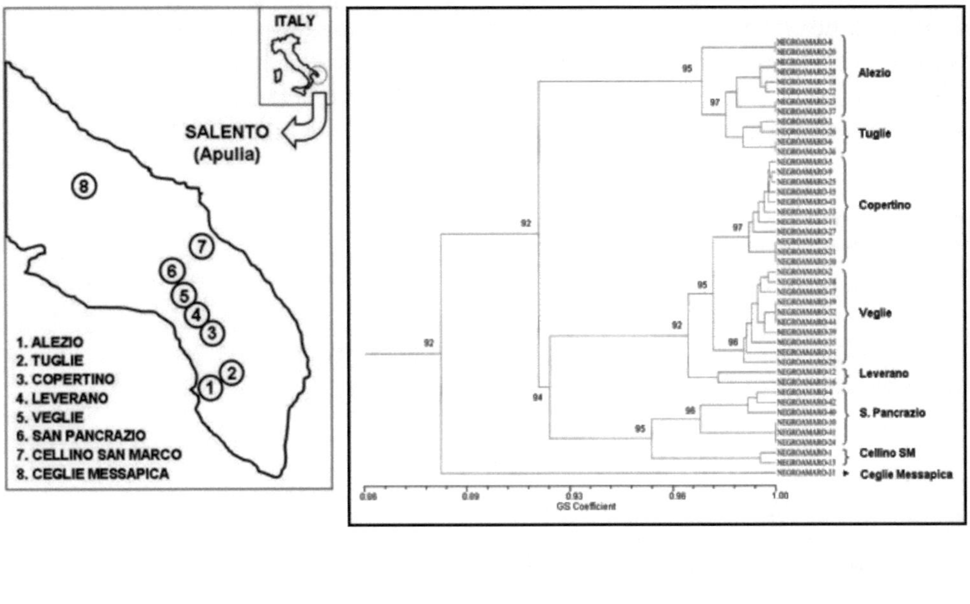

Figure 11. Geographic map of the 8 Negroamaro origins (Apulia, Italy) and Dendrogram of genotypes of Negroamaro from the different Salento areas (Meneghetti, 2012a).

The genetic variability among the Negroamaro materials showed an high correlation between the geographic origins (environmental variability) and the molecular profiles; this is important for the choice of the Negroamaro clones to be propagated in the Salento area.

4.5. Malvasia di Candia

The white Malvasia di Candia SS (i.e., Simple Savor, not aromatic) is a cultivar of the great and heterogeneous Malvasia family and represents one of the principal varieties of the Frascati DOC area. It is also known as 'Red Malvasia' due to the red shoots color (Calò & Costacurta, 2004).

Many biotypes of Malvasia di Candia with large sized berry bunches are present in the Frascati area after 1950. Thirty accessions of this cultivar were selected from 150 old vineyards from this area in an earlier study. Morphological and molecular analyses were performed to indentify the most interesting biotypes which revealed a large variability at morphological and molecular levels. The 30 accessions were identified as white Malvasia of Candia (SS) by SSR markers.

Ampelography and ampelometry analyses clustered four biotypes called AA, A, B, and AB (Figure 12).

Figure 12. Bunches of the four biotypes of Malvasia di Candia (Meneghetti, 2012c).

Biotype AA shows medium sized, long bunches with evident wings; medium irregular berry size (Figure 12). Biotype A was similar to biotype AA with smaller sized bunches and wings. Biotypes B has smaller, shorter, less compact bunches than biotypes AA and A. Biotype AB showed bunch with intermediary characteristics between biotypes A and B. AFLP, M-AFLP, and SAMPL molecular markers were used to analyze the intra-varietal genetic variability (Meneghetti, 2012a).

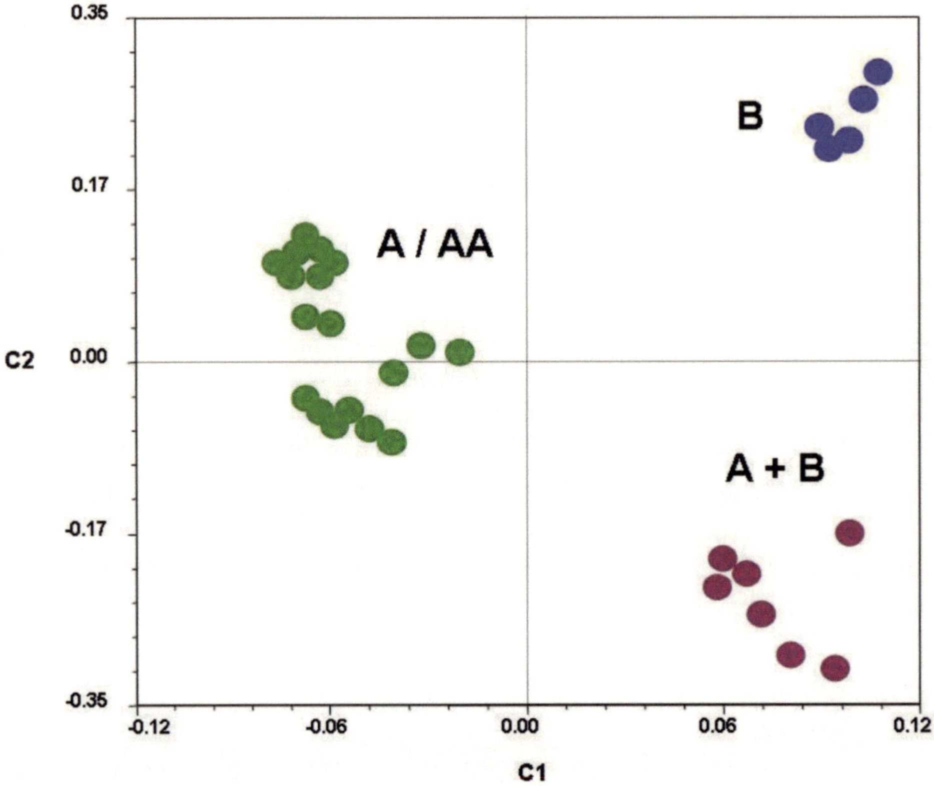

Figure 13. Centroids of Malvasia di Candia accessions that discriminated the 30 genotypes according to the different morphological traits of bunches (i.e., A, AA, B, AB).

Cluster analyses showed a correlation between molecular profile and morphological traits of bunches relating to Malvasia di Candia biotypes (Figure 13).

Biotypes B (smaller fruit size) were clearly discriminated from the remaining typologies (larger fruit size) even if the accessions with A and AA bunch types were grouped in the same cluster (A/AA) (Figure 13).

4.6. Istrian Malvasia

Istrian Malvasia is a cultivar from Northern Italy and the Istrian Peninsula (Calò & Costacurta, 2004; Crespan, 2006). It is known in Croatia as Malvazija istarska (Crespan, 2008). This cultivar is the most commercially important and widely cultivated grapevine variety in Istria (Croatia).

Several biotypes of this grapevine cultivar were selected in Italy during clonal selections by research institutes.

A study was carried out on 30 Istrian Malvasia genotypes consisting of eight Italian clones (i.e., ISV 1, ISV F6, VCR 4, VCR 113, VCR 114, VCR 115, ERSA 120, ERSA 121) and 22 autochthonous grapevine accessions grown in Istrian Peninsula (Croatia); the morphological and genetic intra-varietal variability of this cultivar was evaluated.

Ampelographic characterizations of accessions were performed using 20 OIV descriptors relative to young shoot, shoot, young leaf, mature leaf, inflorescence, bunch and berry (2nd edition of the OIV descriptor list for grape varieties and *Vitis* species). Dendrogram based on morphological data was performed using the absolute mean distances (Manhattan - City Block) and the Complete Linkage (Fabbris, 1997).

The microsatellite analyses confirmed the varietal identity of the 30 genotypes analyzed. SSR profile of Istrian Malvasia was reported in Figure 14.

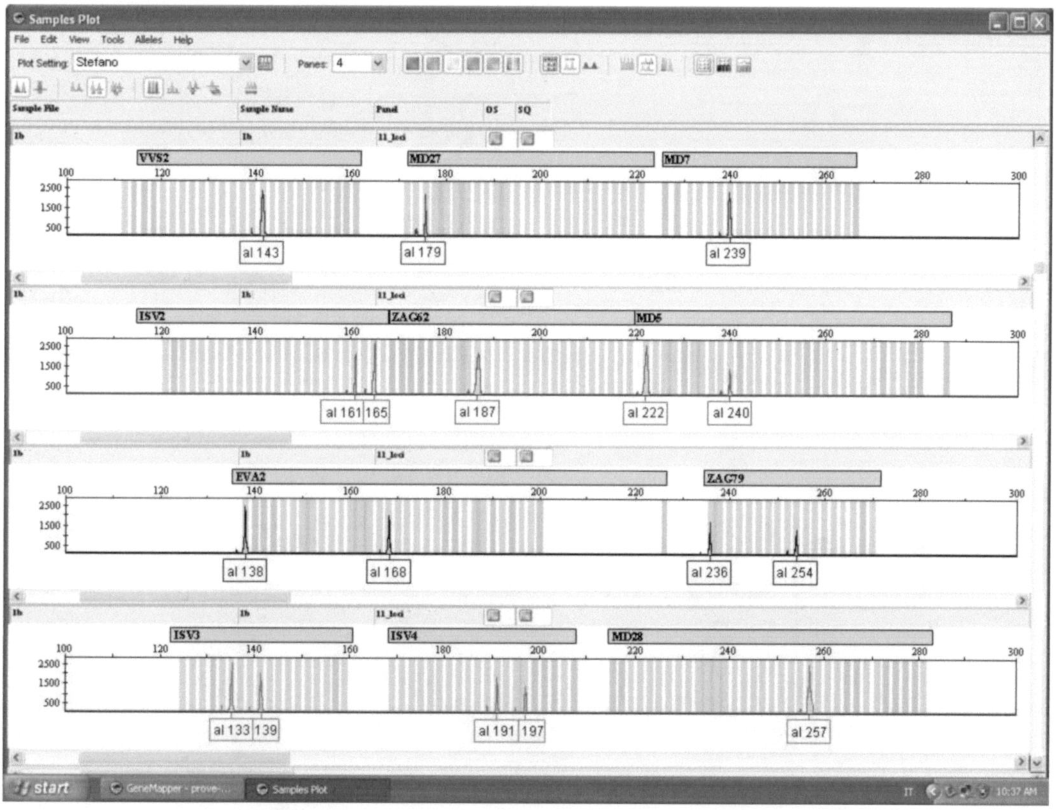

Figure 14. SSR profile at 11 loci of Istrian Malvasia cultivar by 3130XL Genetic Analyzer.

Malvasia dendrogram of morphological data in Figure 15 showed two distinct main groups: first consisted of the 22 autochthonous accessions from Croatia and second comprised the eight Italian clones.

Figure 16 reports the 16 geographic origins of the analyzed Istrian Malvasia accessions or clones.

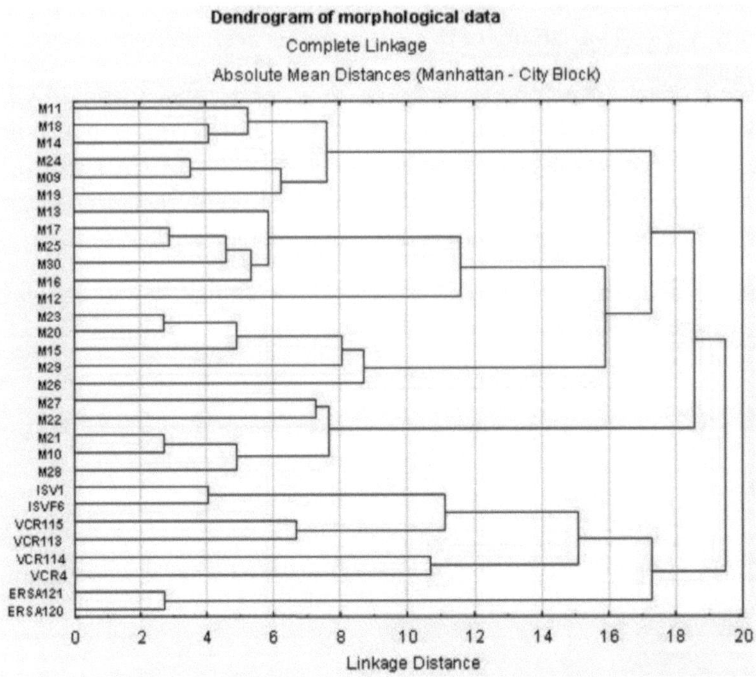

Figure 15. Dendrogram of morphological data of the 30 Istrian Malvasia accessions, of which 8 Italian clones of Istrian Malvasia (i.e., ISV 1 and ISV F6 clones, ERSA 120 and ERSA 121 clones, VCR113, VCR114, VCR115 and VCR 4 clones) and the 22 autochthonous accessions from Croatia.

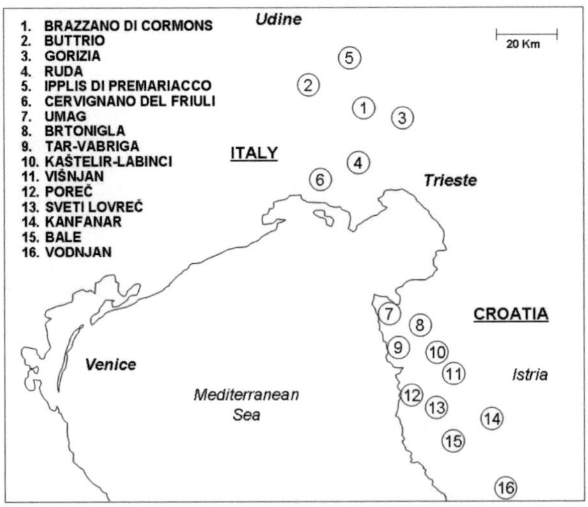

Figure 16. Geographic map of the geographic origins of the 30 Istrian Malvasia accessions from Italy (1-6) and Croatia (7-16) (Meneghetti, 2012b).

The morphological analyses performed using the OIV ampelographic descriptors (Figure 15) discriminated the Italian clones in accordance with the three different selectors: the two clones of the ISV (i.e., ISV 1 and ISV F6), the two clones of ERSA (i.e., ERSA 120 and ERSA 121) and the four clones of VCR (i.e., VCR 113, VCR 114, VCR 115 and VCR 4). Italian clones and Croatian accessions were separated by morphological traits.

In order to study the intra-varietal genetic variability of 30 mentioned accessions AFLP, SAMPL and M-AFLP molecular analyses were performed.

A total of 1,754 reproducible amplification products were obtained (i.e., 682 DNA fragments from AFLPs, 597 DNA fragments from SAMPLs and 475 DNA fragments from M-AFLPs). Results revealed 931 (70.1%) polymorphic molecular markers: 308 AFLPs, 302 SAMPLs and 321 M-AFLPs.

The GS_{TOT} values of the three marker types showed that all molecular systems applied were efficient to show molecular polymorphisms between the Istrian Malvasia genotypes.

The observed GS_{TOT} was 0.8974, the GS_W within the eight Italian clones was 0.8376 and the GS_W within the 22 Istrian samples was 0.9552. This result showed that the Istrian accessions were genetically more similar to each other than the Italian clones. GS_B was 0.8667 between Italian and Croatian accessions.

The GS_W values were 0.9302, 0.9478 and 0.9278 within ERSA, ISV and VCR clones respectively. The GS_B values were 0.8066, 0.8162 and 0.7806 between ERSA and ISV, ERSA and VCR, and ISV and VCR clones respectively.

Dice's GS matrix was used to perform the cluster analysis (Figure 17).

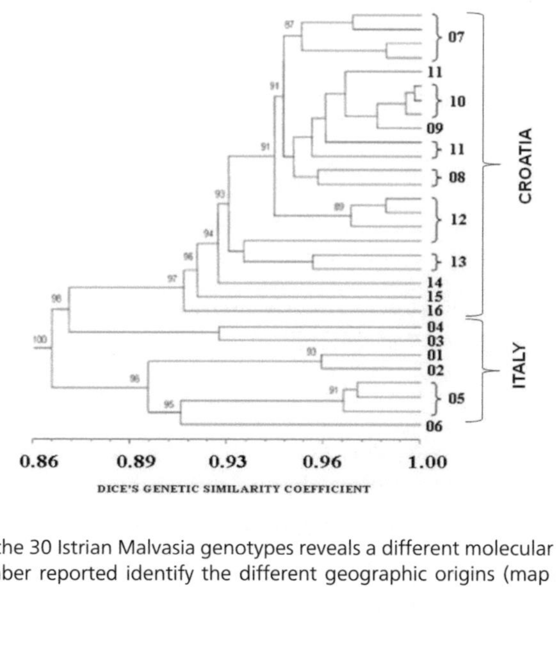

Figure 17. Dendrogram of the 30 Istrian Malvasia genotypes reveals a different molecular profile between Italian and Croatian samples. The number reported identify the different geographic origins (map of Figure 16) (Meneghetti, 2012b).

Figure 17 reports two distinctive groups: Croatian accessions and Italian clones. Results of the AFLP, SAMPL and M-AFLP analysis did not show a complete correlation with morphological observations. In fact the dendrogram obtained by molecular data (Figure 17) was not exactly equivalent with that of morphological observations (Figure 15). However, both cluster analyses showed a clear correlation between accessions and their selectors or country.

Furthermore, the Croatian accessions were distinct in ten sub-groups in agreement with their geographic origins (i.e., 7= Umag, 8= Brtonigla, 9= Tar-Vabriga, 10= Kaštelir-Labinci, 11= Višnjan; 12= Poreč; 13= Sveti-Lovreč, 14= Kanfanar; 15= Bale, 16= Vodjan).

A similar level of distinction could be observed for the three Italian sub-groups.

We could argue that the genetic similarities are in agreement with the distance of the geographic origins.

These results suggest the need to emphasize the environmental role on the selection of genotypes during the centuries. The emphasis on preserving the autochthonous grapevine biotypes is crucial to preserve the richness of the Istrian Malvasia germplasm.

The study confirmed the importance of choosing appropriate propagation material for future cultivation in order to save the genetic variability of local biotypes. The propagation of the same clone in different territories should be also avoided in order to preserve the good interaction among genotypes and their specific environments.

Further intra-varietal studies (i.e., DNA analysis, together with ampelographic investigations), allowed the identification of Italian clones and Croatian autochthonous accessions of Istrian Malvasia.

The results have highlighted the existence of genetic variability among the Istrian Malvasia accessions from different geographical cultivation areas. These molecular approaches allowed the identification of different clones within the Istrian Malvasia cultivar and the characterization of accessions according to their geographic origins.

5. Conclusions

In summary, the molecular and morphological analyses showed that *Vitis vinifera* is a species characterized by vast genetic variability.

Molecular analyses of DNA are essential for the grapevine identification using SSR markers.

These results showed also the wide genetic variability for the grape cultivars (intra-varietal level) suggesting the need for the preservation of autochthonous grapevine biotypes found in different areas by a proper selection of the grape multiplication materials.

In fact, this genetic variability accumulated during centuries of cultivations and selections, should be both recognized and preserved, being corroborated by scientific experimental results.

The importance of saving the genetic variability of the varieties is crucial in order to avoid to propagate the same clone in different cultivation areas.

It is highly recommended to promote the propagation of the typical autochthon biotypes, which are already wisely selected by grape vine growers.

Acknowledgment

This study is supported by both ASER and IDENTIVIT research grant from Ministero delle Politiche Agricole, Alimentari e Forestali MiPAAF, Rome, Italy.

Author details

Stefano Meneghetti[1], Luigi Bavaresco[1], Antonio Calò[2] and Angelo Costacurta[2]

1 CRA-VIT Centro di Ricerca per la Viticoltura, Italy

2 AIVV Accademia italiana della Vite e del Vino, Italy

References

[1] Bavaresco, L., Giachino, E., Pezzutto, S., Fregoni, C., & Fogher, C. (2000). PCR specific analysis of barbera clones. Bullettin de l'O.I. , 73-831, 296-311.

[2] Bowers, J. E., Dangl, G. S., & Meredith, C. P. (1999). Development and characterization of additional microsatellite DNA markers for grape. *American Journal of Enology and Viticulture*, 125-130.

[3] Calò, A., Costacurta, A., Cancellier, S., & Forti, R. (1990). Garnacha, Grenache, Cannonao, Tocai rosso, un unico vitigno. Vignevini, 9, , 45-48.

[4] Calò, A., Costacurta, A., Catalano, V., & Di Stefano, R. (2000). Negro Amaro precoce. Rivista Viticoltura Enologia, 3, , 27-44.

[5] Calò, A., & Costacurta, A. (2004). Dei vitigni italici. Treviso: Mattei Editor.

[6] Calò, A., Costacurta, A., Maraš, V., Meneghetti, S., & Crespan, M. (2008). Molecular Correlation of Zinfandel (Primitivo) with Austrian, Croatian, and Hungarian Cultivars and Kratošija, an Additional Synonym. *American Journal of Enology and Viticulture*, 205-209.

[7] Clarke, O. (2001). Encyclopedia of grape (Orlando: Harcourt Books., 91-100.

[8] Crespan, M., Cabello, F., Giannetto, S., Ibáñez, J., Kontić, J. K., Maletić, E., et al. (2006). Malvasia delle Lipari, Malvasia di Sardegna, Greco di Gerace, Malvasia de

Sitges and Malvasia dubrovačka-synonyms of an old and famous grape cultivar. Vitis, 45, , 69-73.

[9] Crespan, M., Coletta, A., Crupi, P., Giannetto, S., & Antonacci, D. (2008). Malvasia nera di Brindisi/Lecce' grapevine cultivar (Vitis vinifera L.) originated from 'Negroamaro' and 'Malvasia bianca lunga'. Vitis, 47-4, , 205 EOF-212 EOF.

[10] Cipriani, G., Spadotto, A., Jurman, I., Di Gaspero, G., Crespan, M., Meneghetti, S., Frare, E., Vignai, R., Cresti, M., Morgante, M., Pezzotti, M., Pe, E., & Testolin, R. ((2010).). The SSR-based profile of 1005 grapevine accessions uncovers new synonymy and parentages and reveals a large admixture among varieties of different geographic origin. TAG, 121-8: ., 1569-1585.

[11] Dangl, G. S., Mendum, M. L., Prins, B. H., Walzer, M. A., Meredith, C. P., & Simon, C. J. (2001). Simple sequence repeat analysis of a clonally propagated species: A tool for managing a grape germplasm collection. *Genome*, 432-438.

[12] Dice, L. R. (1945). Measures of the amount of ecologic association between species. *Ecology*, 297-302.

[13] D'Onofrio, C., De Lorenzis, G., Giordani, T., Natali, L., Scalabrelli, G., & Cavallini, A. (2009). Retrotransposon-based molecular markers in grapevine species and cultivars identification and phylogenetic analysis. *Acta Horticulturae*, 45-52.

[14] Fabbris, L. (1997). Statistica multivariata, analisi esplorativa dei dati. Milano: McGraw-Hill.

[15] Galet, P. (2000). Dictionnaire encyclopédique des cépages. Hachette. Paris.

[16] Galet, P. (1979). A practical ampelography: Grapevine identification. Ithaca, New York: University Press.

[17] Imazio, S., Labra, M., Grassi, F., Winfield, M., Bardini, M., & Scienza, A. (2002). Molecular tools (SSR, AFLP, MSAP) for clone identif. ication: The case of the grapevine cultivar Traminer. Plant Breeding, 121-6, , 531-535.

[18] Labra, M., Imazio, S., Grassi, F., Rossoni, M., & Sala, F. (2004). Vine-1 retrotransposon-based sequence-specific amplified polymorphism for Vitis vinifera L. genotyping. *Plant Breeding*, 180-185.

[19] Lacombe, T., Boursiquot, J. M., Laucou, V., Dechesne, F., Varès, D., & This, P. (2007). Relationships and genetic diversity within the accessions related to Malvasia held in the Domaine de Vassal Grape germplasm repository. American Journal of Enology and Viticulture , 124-131.

[20] Levadoux, L. (1956). Les populations sauvages et cultivees de Vitis vinifera L. A. nn. Amelior Plantes, 6, , 59-118.

[21] Lynch, M. (1990). The similarity index and DNA fingerprinting. *Molecular Biology and Evolution*, 478-484.

[22] Meneghetti, S., Costacurta, A., Crespan, M., Maul, E., Hack, R., & Regner, F. (2009). Deepening inside the homonyms of Wildbacher by means of SSR markers. Vitis: 48-3, , 123-129.

[23] Meneghetti, S., Costacurta, A., Frare, E., da, Rold. G., Migliaro, D., Morreale, G., Crespan, M., Sotés, V., & Calò, A. (2011). Clones identification and genetic characterization of Garnacha grapevine by means of different PCR-derived marker systems. Molecular Biothecnology, 48-3, , 244-254.

[24] Meneghetti, S., Costacurta, A., Morreale, G., & Calo`, A. (2012a). Study of intravarietal genetic variability in grapevine cultivars by PCR-derived molecular markers and correlations with the geographic origin. Molecular Biothecnology, 50-1: , 72-85.

[25] Meneghetti, S., Poljuha, D., Frare, E., Costacurta, A., Morreale, G., Bavaresco, L., & Calò, A. (2012b). Inter- and intra-varietal genetic variability in Malvasia cultivars. Molecular Biothecnology, 50: , 189-199.

[26] Meneghetti, S., Bavaresco, L., & Calò, A. (2012c). A Strategy to Investigate the Intravarietal Genetic Variability in Vitis vinifera L. for Clones and Biotypes Identification and to Correlate Molecular Profiles with Morphological Traits or Geographic Origins. Molecular Biothecnology, 52-1: , 68-81.

[27] Moreno, S., Gogorcena, Y., & Ortiz, J. M. (1995). The use of RAPD markers for identification of cultivated grapevine (Vitis vinifera L.). *Scientia Horticulturae*, 237 EOF.

[28] Mullins, L. G., Bouquet, A., & Williams, L. E. (1992). The Biology of the grapevine. Cambridge University Press. Cambridge, UK.

[29] Owens, C. L. (2003). SNP detection and genotyping in Vitis. *Acta Horticulturae*, 603, 139-140.

[30] Powell, W., Machray, G. C., & Provan, J. (1996). Polymorphism revealed by simple sequence repeats. *Trends in Plant Science*, 215-222.

[31] Royo, J. B., Cabello, F., Miranda, S., Gogorcena, Y., González, J., Moreno, S., Itoiz, R., & Ortiz, J. (1997). The use of isoenzymes in characterization of grapevines (Vitis vinifera L.). Influence of the environment and time of sampling. *Scientia Horticulturae*, 145-155.

[32] Tessier, C., David, J., This, P., Boursiquot, J. M., & Charrier, A. (1999). Optimization of the choice of molecular markers for varietal identification in Vitis vinifera L. Theoretical and Applied Genetics, 89, , 171-177.

[33] This, P., Jung, A., Boccacci, P., Borrego, J., Botta, R., Costantini, L., et al. (2004). Development of a standard set of microsatellite reference alleles for identification of grape cultivars. Theoretical and Applied Genetics, 109, , 1448-1458.

Section 2

Genetics in Service of National Germplasms Preservation

Centuries-Old Results of Cultivation and Diversity of Genetic Resources of Grapes in Azerbaijan

Mirza Musayev and Zeynal Akparov

Additional information is available at the end of the chapter

http://dx.doi.org/10.5772/ 52387

1. Introduction

The Azerbaijan Republic is an ancient country located on the South-East of the Caucasus Mountains and on the North-West of the Iranian Plateau, at the crossroads of Eastern Europe and Southwest Asia. Extreme diversity of the soil and climatic conditions of Azerbaijan support a very rich diversity of plant genetic resources. More than 4700 higher plants have been registered here, 237 of which are endemic. Historically wild fruits are used by people for food, as medicinal crops and for other purposes. Azerbaijan is considered one of the evolution centers of cultivated plants. Practically all present-day major cultivated plants appeared for the first time in Azerbaijan several millennia B.C. As an example, evidence of ancient horticulture was discovered in a settlement west of Goy-Gol in the early second millennium B.C. Fruit crops (apple, pear, apricot, pomegranate, quince, fig, almond, walnut, hazelnut etc.) and grape have been cultivated to meet the demands of the population for foodstuff and other products. Most of these crops are still considered major agricultural crops in the country. In the book of the ancient Greek scientist Strabon - "Geography" was indicated a high prevalence of fruits in Azerbaijan: "The whole country is rich in wild and cultural fruits, evergreens, even olive grows here".

On the territory of Azerbaijan are distributed 149 species of fruit crops belonging to 39 genera and 15 families. The big number of genera and species of wild fruit and fruit-berry plants sprouting in forests and rural regions of Azerbaijan provides the greatest diversity of fruit crops: *Amygdalus communis* L., *Armeniaca vulgaris* Lam., *Berberis vulgaris* L., *Castanea sativa* Mill., *Cerasus avium* (L.) Moench, *C.vulgaris* Mill., *Cornus mas* L., *Corylus avellana* L., *Crataegus orientalis* Pall. ex M. Bieb., *Cydonia oblonga* Mill., *Ficus carica* L., *Fragaria vesca* L., *Hippophae rhamnoides* L., *Juglans regia* L., *Malus domestica* Borkh., *Mespilus germanica* L., *Morus* L., *Persica vulgaris* Mill., *Pistacia mutica* Fisch. & C. A. Mey., *Pistacia vera* L., *Prunus cerasifera*

Ehrh., *Punica granatum* L., *P.domestica* L., *P.spinosa* L., *Elaeagnus angustifolia* L., *Pyrus communis* L., *Rubus* L., *Olea europaea* L., *Vitis vinifera* L. subsp. *sativa* D.C., *V.vinifera* L. subsp. *sylvestris* (C. C. Gmel.) Hegi. and etc. [1,2,3,4,5,6]

Vine-growing was one of the most ancient and widespread professions in the economic life of our people. As evidenced from archaeological excavations, paleobotanical studies, ampelographic information, folklore and written history sources, Azerbaijan has proven to be one of the cultural centers of vine-growing.

The territory of Azerbaijan has very favourable conditions for improvement of *Vitis* and development of vine-growing. Primitive men who primarily utilized hunting and fishing to provide sustenance also collected wild fruits and berries, including wild grape.

In 1963 in the western part of Bozdagh (Goy Gol region) while conducting geological investigations, Azerbaijan scientists discovered an abundance of plant remains in Absheron sediments which formed 1-2 million years ago. Most of the residues were impressions of wild grape leaves on stone. Formation of wild grape in this area (approximately 500. 000 years ago) was demonstrated by grape leaf impressions found in Nakhchivan [7]. This discovery proves that the region is one the ancient vine-growing centers (Figure 1.). These discoveries are very valuable, not only for historians, but also for specialists of other sciences – paleobotany, ampelographics, fruit-growers, geologists, soil scientists.

Figure 1. *V.vinifera* L. subsp. *sylvestris* (C. C. Gmel.) Hegi. Leaf impressions in stone. Discovered in the Eastern part of Araz River in Nakhchivan AR (according to I.M. Palibin, 1964)

Researchers indicated that the origin of cultivated grape was within geographic areas where wild grape was endemic. According to N.I.Vavilov, like animals, plant domestication is also possible in areas that are enriched with available wild species. Through his long term investigations, he determined that Azerbaijan and the Southern Caucasus Mountains are the main centers of crop origin, including grape [8,9,10].

Archaeological materials are considered important sources in highlighting the socio-economic life of people, and also for studying the historical development of vine-growing. There are a number of findings on the vine-growing culture of ancient Azeri people in the archaeological record. In some cases, parts of grape bunches dried in soil or rotted, but the majority of grape bunches only became charred and kept its form and size when discovered in ancient pots or storage vessels.

The oldest examples of findings on vine-growing in Azerbaijan territory date to the V-IV millennium BC. In these noted millenniums an important event occurred in the lives of ancient families, our ancestors passed from a hunter/gatherer society to a sedentary life and husbandry culture.

During archaeological excavations near Aghstafa region in 1962, various plant remains, including grape seed, were found in "Shomutepe" monument and were dated to V-IV millennium BC. Investigations had shown the culture and year parameters of grape seeds. Mainly on the basis of this finding it was shown that the history of cultured vine-growing in Azerbaijan has at least 7 millenniums.

Grape seeds, stone tools for vine production and different cultural material samples which were found at Uzerliktepe monument, Goy Gol region, and ancient monuments in Nakhchivan during archaeological excavations near Aghdam region showed that vine-growing had played an important role in farmers' lives. Grape seeds found in Uzerliktepe date to 3500 years ago. Scientists determined that they were table grape varieties. Grape seeds found in Uzerliktepe were of different sizes. The biggest seed was 6.5 mm, separate seeds of grape were nearly 18-20 mm. Findings in Ganjachay territory show that the people were occupied with horticulture, as well as vine-growing during the Bronze Age. Chemical analysis of remain which were found in earthenware crockery belonging to last Bronze Age proved that it consisted of wine sediments and grape seeds. Wine pitchers were also found in tombs from the end of the Bronze Age in Mingechevir region. Besides jewelleries, weapons, kitchen utensils and food, as well as wine bottles had been found in Bronze Age grave monuments in Goy Gol region and Haloylutepe. In the bottom of crockery, grape seed sediments and wine remains were observed. It was determined that all these were put near the deceased person in connection with confidence to 'The Hereafter.'

During archaeological excavations in old Saritepe settlement in the west part of Gazakh region, big pitchers were discovered that contained grape seed and seed fragments. Professor A.M. Negrul who investigated the grape seeds determined that they belonged to a cultivated grape variety. Those investigations concerned artifacts dated to II millennium BC. The big pitchers found in Saritepe have a great importance as proof of grape juice extraction and wine production. One of the grape juice pitchers with a height of 2m and 1000 litre capacity is being kept in the historical museum of Azerbaijan.

Cultivated grape seeds were also found at approximately 5 m depth in Meydantepe from II millennium BC in Kultepe settlement in Nakhchivan.

During a 1970 excavation, other cultural samples, as well as lots of burnt grape seed remains belonging to the first half of I millennium AD (app. IV-V centuries) were discovered in Galagah. In Galagah, a piece of stone crockery used in grape juice extraction was found. Separate seeds of grape and stone crockery proved that this place was a production center for wine-making.

In III-VIII centuries, wine-making played an important role in the economic life of the Mingechevir population. In Mingechevir the main archeological discoveries had been grape seeds. A number of wineskin remains were discovered here. In these wineskins there were grape seeds and ancient wine remains.

While conducting archaeological excavations it was determined that urban population of Beylagan were also occupied with vine-growing. First period stratums of Beylagan (VI-VIII centuries), were vine-growers as evidenced by numerous discoveries of grape seeds. At the result of 1962-68 excavation,s burnt grape bunches were discovered in a building complex characterizing XI-XIII century settlements.

Different fruit remains, as well as grape seeds, were found in ruins of Gabala region of Azerbaijan. Grape seeds were found in economic wells and basin-shaped earthenware crockery.

All of these archaeological excavations provide good evidence on grape remains, economic pitchers and tools (hoe, spade, hook, gardening shears, trough), as well as dishes (strainer, jug, basin, glass, bowl and etc.) which are endemic to any given region.

Populations used the grape for juice extract, prepared non-alcoholic drinks and made various food stuffs of these juices and syrups.

Grape squashing stones were found in cultured stratum of settlements and places belonging to III-VIII centuries of Ganja, Pirhesenli village of Aghsu region and Mingechevir.

Vine-growing provided the Azerbaijan population with grape juice and syrups, currants, bakmaz, vinegar, abgora and other products. The production of these required great experience, a large labour force and a variety of differently formed dishes. During excavations lots of potteries were unearthed. Clay strainers had been used in juice filtering. Clay strainers belonging to II millennium BC and VII century were found in II Kultepe and Julfa region, Bronze Age cultured stratum of Mingechevir and Gabala region. Earthenware crockery used as a strainer in wine-making which belonged to III-II centuries was discovered in a graveyard near Gubakhalili settlement of Ismayilli region. Such strainers were also obtained in Seyfeli village and places throughout the Shamkir region.

Paleobotanical findings, artificial irrigation ornamental remains, differently sized and formed potteries, also glass wares and tools which were found through archaeological excavations show the high degree of development of vine-growing in Azerbaijan.

Written sources of information (Latin, Greek, Syrian, Arabic, Persian, Turkish) are very important in determining the developmental history of vine-growing, as well as the husbandry culture of Azerbaijan population. There is plenty of information in ancient scholars' works on these issues. The Greek scholar Herodotus, who lived in V century BC, gave information http://dx.doi.org/10.5772/ 52387

about events relative to VI century on vine-growing in Azerbaijan. Abundance and quality of Azerbaijan grape is even more significant than in Iran, Babylon or Greece. The Roman scholar Great Plini (23-79 years BC) admired inexhaustible resources and advanced farming culture of Azerbaijan territory: "I have never seen such sweet grape anywhere. This nation can cultivate the land better than Egyptians".

The famous Greek geographer Strabon (I century BC- I century AD) noted the great role of vine-growing in the economic life of Azerbaijan: "There (in Azerbaijan) grapevines were cut off once in five years, new grapevines began to yield fruit from the second year, yield was higher and even some parts of grape stayed on grapevines".

In some zones of Azerbaijan planting of grapevines in winter is connected with climate. In some regions of Azerbaijan, especially in the Nakhchivan Autonomous Republic covering viticulture. In recent years, cultivation of grapevines is more widely spread in "Khiyaban" in Azerbaijan. Grapevines that are cultivated in "Khiyaban" not have to do pruning every year. A higher yield is obtained though cutting off the dried parts and useless shoots once every 5-6 years. Strabon's information on high yield of grape is also connected with grapevine productivity. It is known tha, grape mainly is harvested in the months of September-October. If taking into account comparative climate changes, in ancient Azerbaijan the grape was collected in September-October months too. In certain years, the productivity of grapevines was higher, and it had been impossible to harvest everything until December. When Strabon noted not collecting part of the grape harvest, he namely intended this issue.

In modern Azerbaijan vine-growing and wine-making are considered among the most profitable fields of agriculture. Although local grape varieties are cultivated on big farms, they are an insignificant part of the national grape collection. And this cannot provide sustainable and safe preservation of local valuable grape varieties. Therefore, local grape varieties and wild grape forms spread in our republic (in old vineyards, little peasant-farmer households, courtyards, etc.) should be collected, included in the collection, and evaluated for their possible utilization.

Recently, new ampelographic collections have been established and enriched with local and introduced grape varieties and wild grape species in Genetic Resources Institute of ANAS.

At the conclusion of our latest investigations, it was known that of the more than 600 local and introduced grape varieties spread throughout Azerbaijan, 75 of them had already been lost, with more than 100 varieties currently being threatened.

Most of threatened local grape varieties were collected as a result of expeditions (in old vineyards, peasant-farmer households and courtyards) organized in different regions of our Republic.

2. Material and Methods

Materials of research work consisted of grapevines and yields of local grape varieties and wild grape forms.

Ampelographic description of grape varieties and wild grapevines had been implemented on the basis of common methods [11,12,13,14,15].

Phytopathological and immunological descriptions and assessments of grapevines on natural background were carried out by appropriate methods [16].

Finally, varieties and forms were evaluated by modern methods [17, 18] for their reaction to various stresses.

Total genomic DNA was isolated from young grape leaves. The leaves were ground in liquid nitrogen. For DNA isolation the CTAB based extraction procedure was used [19]. When necessary, extracted DNAs were purified with GenElute columns (Sigma-Aldrich, St. Louis, MO). In the case of silica dried leaves DNA was isolated by Plant genomic DNA extraction miniprep system (VIOGENE, USA). Sequence diversity polymorphisms of wild grape samples were investigated at two non-coding plastid DNA regions (the *trnH-psbA* intergenic spacer and the *rpl*16 intron).

The *trnH-psbA* intergenic spacer was amplified with the primers "trnH" and "psbA". The *rpl*16 intron was amplified with the primers "*rpl*16-5'" and "*rpl*16-3'" [20,21]. The primers were synthesized by Integrated DNA Technologies, Inc. (Coralville, IA), and Sequencing Service of Institute of Biochemistry and Biophysics, Polish Academy of Science (Poland).

PCR conditions included 1 minute denaturing at 94°C, 30 cycles of 94°C denaturing (1 minute), 55°C annealing (1 minute), and 72°C extension (2 minutes), followed by a final extension step at 72°C (5 minutes). PCR products were purified with GenElute PCR Clean-Up Kits (Sigma-Aldrich, St. Louis, MO), dye-labeled using a Big Dye Terminator Kit (Applied Biosystems, Foster City, CA) and analyzed on either Applied Biosystems 3100 or 3700 genetic analyzers (Biology Department of Washington University, St.Louis, MO and Laboratory Services Division of the University of Guelph, ON, Canada). MEGA and SeqMan softwares were used for sequence analysis.

While on expeditions, the coordinates of wild grapevines areas were defined by GPS. Relevant photos were taken with digital cameras, some ampelographic indicators and phytocenotic traits were described.

3. Results and Discussion

Recently, world interest in wild grape has increased and this resulted in a widening of investigations in this field. By studying wild grape, we can infer some questions on grape phytogenesis and use the varieties which possessed positive bio-agricultural traits and different biotic-abiotic factors resistant genes as donor in grape selection. For this purpose, research on collection, improvement, investigation and sustainable utilization of genetic resources of wild grape are being implemented in Azerbaijan Republic and organized joint expeditions in different regions.

In our Republic wild grape samples are spread widely in large areas and along the banks and shores of river, lake and sea, and on mountain slopes of Absheron, Nakhchivan AR,

Ganja-Gazakh, Garabagh, Mil-Mughan, Shirvan and Talysh regions. A number of studies were implemented in Khachmaz, Guba, Khudat, Nabran, Gusar, Shamakhi, Ismayilli, Aghsu, Oghuz, Gabala, Shaky, Zagatala, Lankaranand Fuzuli regions for studying the genetic resources of grape.

At the same time it may be concluded that wild grape spread throughout the whole territory of Azerbaijan in a very ancient form. Wild grape - *V.vinifera* L. subsp. *sylvestris* (C. C. Gmel.) Hegi. of Azerbaijan is distinguished with specific characters. It grows on the territory of Azerbaijan from 12 m below sea-level (Kyur riverside, Salyan region) to 2000 m above sea-level (Gusar region). There are two kinds of wild grape in Azerbaijan: *typica* Negr. (with hairs) and *aberrans* Negr. (hairless).

Figure 2. Formation of wild grape in Nabran. Coastal Area of the Caspian Sea.

The geographic origins of grapevine domestication are not currently known. According to many researchers, the Caucasus region (north-western Turkey, northern Iraq, southern Russia, Azerbaijan, Georgia) and adjacent areas (Anatolia, modern day Syria, Lebanon, Israel), are the geographic areas where grapes were most likely were first domesticated [10,22,23,24]. Special climate conditions in this area occurred which were favorable for the diversification of wild varieties from which cultivated grapes were domesticated. At the same time, having extraordinary abundance of the wild species of grape (*V.vinifera* L. subsp. *sylvestris* (C. C. Gmel.) Hegi.), Azerbaijan and the whole South Caucasus region are regarded as the potential places for domestication of cultivated grapes. It is here that the natural distribution of *V.vinifera* most closely approaches the probable origin of Western agriculture [24]. This assumption is proven by the recent chemical analysis of archeological pottery from Georgia and Eastern Anatolia which showed that winemaking dates back to early VI millennium BC in these regions (McGovern, in preparation). Distribution area of *V.vinifera* L. subsp. *sylvestris* (C. C. Gmel.) Hegi. is very wide: It's a Europe, northern Africa and the Middle East, including Mediterranean, Black and Caspian Sea Basins from Spain to Turkmenistan [25].

The main goal of the proposed study was the investigation of plastid DNA sequence diversity in a geographically diverse set of South Caucasian *V.vinifera* L. subsp. *sylvestris* (C. C. Gmel.) Hegi. To date no study has broadly assessed DNA sequence variation of wild grapevines in this way. The greater Caucasus region is widely believed to be the area in which grape domestication began [22,26], and the study of genetic diversity in this region is viewed as key to understanding grape domestication in general. This information is of great interest from an ethno-botanical standpoint, but also relates to crop improvement. It's well known that cultivated varieties of grapevine differ greatly in their resistance to pests and diseases, and ancestral wild populations are obvious first targets for use in breeding and genetic engineering.

Figure 3. The distribution of haplotypes in the SouthCaucasus.

For clarification of some questions and characteristics of domestication of wild grapes of the South Caucasus, we studied samples of the wild grape of the South Caucasus region. Forty-five wild grape (*V.vinifera* L. subsp. *sylvestris* (C. C. Gmel.) Hegi.) samples from the South Caucasus were analyzed. This group included 19 samples from the Republic of Georgia, 10 samples from Azerbaijan, 2 samples from Armenia and 14 samples from Turkey. A plastid DNA sequence variation study revealed the presence of three polymorphic sites in DNA: one in *trnH-psbA* intergenic region and two in the *rpl*16 intron area. According to this observation investigated samples of Caucasian *V.vinifera* L. subsp. *sylvestris* (C. C. Gmel.) Hegi. were divided into four different haplotypes: AAA, ATT, GTA and ATA [27]. For each haplotype the first nucleotide represents single polymorphism at the *trnH-psbA* intergenic region and another two nucleotides at two targeted sites from *rpl16* intron area. The AAA haplotype is restricted to East Georgia and Azerbaijan, the ATA haplotype is distributed randomly across the entire study area, the ATT haplotype is distributed in the southern part of the study area from the Black Sea to the Caspian Sea. The single GTA haplotype was only found in the South-West part of Georgia (Figure 3).

The AAA haplotype is observed in both wild and cultivated (*V. vinifera* subsp. *vinifera*) grape samples from the Caucasus. This observation and the presence of all other plastid haplotypes observed in a previous study of worldwide set of grape cultivars highlight both unique and high levels of genetic variation in wild grape (*V.vinifera* L. subsp. *sylvestris* (C. C. Gmel.) Hegi.) from the greater Caucasus region.

Sequence Group AAA			
Population	**Geographic Region**	**River Basin**	**Coordinates**
Quba distr.,village Alpan	North Azerbaijan	Quruçay	N 41^{0} 21′ 17,2″ EO 48^{0} 22′ 01,2″
Quba distr.,village Ağbil	North Azerbaijan	Quruçay	N 41°26′ 03,7′ EO 48°33′ 49,1′
Quba distr.,village Ağbil	North Azerbaijan	Quruçay	N 41^{0} 26′ 03,7′ E0 48^{0} 33 ′ 49,1′
Quba distr.,village Susay-Qışlaq	North Azerbaijan	Quruçay	N 41^{0} 28′ 02,3′ E0 48^{0} 34′ 43,3′
Sequence Group ATT			
Population	**Geographic Region**	**River Basin**	**Coordinates**
Ağsu distr.	Central Azerbaijan	Girdmancay	N 40^{0} 55′ EO 48^{0} 15′
Qobustan distr., in gorge	East Azerbaijan	-	N 40^{0} 10′ E 49^{0} 20′
Sequence Group ATA			
Population	**Geographic Region**	**River Basin**	**Coordinates**
Quba distr., village Susay-Qışlaq	North-East Azerbaijan	Qusarçay	N 41°28.25,5′ EO 48°36.14,0′
Balakan distr.	North-West	Balakancay	N 41^{0} 43′ EO 46^{0} 25′
Qabala distr.	North-West	Turyancay	N 40^{0} 47′ 814″ EO 47^{0} 38′ 334″
Zaqatala distr.	North-West	Alazan	N 41^{0} 25′ EO 46^{0} 45′

Table 1. Sample information of sequenced samples from Azerbaijan

In Nabran forests of Guba-Khachmaz region black and dark purple coloured grape forms were found.

While exploring in Guba-Khachmaz region it was discovered that Guba region is enriched with wild grape. In forests of this region (Uzunmeshe, Alpan, Khujbala, Digah, Aghbil, Susay Gishlag, Dallakand villages) along Guruchay, Gusarchay, Gudyalchay rivers lots of wild grape forms were found.

In forests of Khachmaz (Pir forest), Shaky (Oraban), Lankaran (Seligavul) and Gabala (Shongar) regions small seedy black-skinned wild grape varieties were also observed.

Figure 4. Samples of wild grapes in forest number 1 (Khachmaz d.)

On the banks of Kondalanchay River in Fuzuli region black, dark red and dark purple coloured grape (seeded forms) were observed.

In general, more than 3000 samples of wild grapes were found in explored regions and phytocenotic features of their geographic areas were described.

In Azerbaijan while investigation of areal of wild grape – *V.vinifera* L. subsp. *sylvestris* (C. C. Gmel.) Hegi. it was determined that various forms of wild grape spread widely together with the following fruit-berry and forest plant varieties and species: medlar – *Mespilus germanica* L., cornel – *Cornus mas* L., walnut – *Juglans reqia* L., hazelnut – *Corylus avellana* L., pomegranate – *Punica granatum* L., chestnut – *Castanea sativa Mill.*, quince – *Cydonia obonga* Mill., apricot – *Armeniaca vulgaris* Lam., caucasian hawthorn – *Crataegus caucasica* C. Koch., eastern hawthorn - *Crataegus orientalis* Pall., red hawthorn – *Crategus kyrtostyla* Fingerh., blackberry – *Rubus caucasicus* Focke., sea buckthorn – *Hippophae rhamniodes* L., willow – *Salix caucasica* Andress., poplar – *Populus gracilis* A. Grassh., hornbeam – *Carpinus caucasica* A. Grossh., elm – *Ulmus foliaceae* Gilib., oak – *Quercus iverica* Stev., birch– *Acer campestre* L., *Paliurus spina Christi* Mill., tamarisk – *Tamarix ramosissima* Led., horse tail – *Equisetum* L., elder berry – *Sambucus nigra* L.

Figure 5. Sample of wild grape in Gabala (Shongar).

It was determined that different populations of wild grape in our republic spread mainly in two formation - tugay (streamside forest) and typical broad-leaved forests. On the banks of Kungut River (Oraban village) of Sheki, Guruchay, Gusarchay, Gudyalchay rivers (Uzunmeshe, Alpan, Khujbala, Digah, Akbil, Susay Gishlag, Dallakand villages) of Guba region wild grapevines spread mainly in tugay forests densely and widely. But typical forest formation of wild grape was found in Agharehimoba, Godekli, Gimilgishlag, Gadashoba and Nerecan villages and forests (forest number 1, Pir forest) of Khachmaz region, Seligavul forest of Lankaran region and Shongar spring of Gabala region.

Wild grape samples distinguish each other for their biomorphological traits. As a rule, male grapevines are strong, functional female grapevines are weak. All samples of wild grape can be divided into 4 groups for leaves size: very small (length up to 4,0-8,0 cm), small (length up to 8,0-12,0), medium (length 12,0-15,0 cm) and large leaved (length more than 15 cm). Most of studied varieties involved small and medium leaved group. Wild grape samples can be divided into 3 groups for leaves sub-sections: whole, medium and cross-section leaves. Some samples are covered with white net-shaped blooms, but in some cases lower leaf surfaces are bare. Samples are distinguished by leave margins. Sides are mainly sharp, triangular and round shaped. Stalk hollows are namely lira-shaped, but rarely sides are parallel and bottoms are flat. Wild grape samples are two of two sexes, that is they have male or female flower groups [28].

Self-pollinated perfect flowered groups of wild grape samples were not observed. According to some researchers' opinions, types of flower groups of wild grape are very important morphological trait for defining grape origin, because wild grape is divided into two subspecies. Bunch flowers of wild grape can be distinguished from each other through their forms, they are small or medium sized. As a rule, the bunch flowers of male grapevines are big and cone-shaped. But bunch flowers of female grapevines are small, cone-shaped-cylindrical or cylinder-shaped.

Bunches of wild grape are small, the length being 7,0-13 cm and the width from 6-8 cm. There are 1-2 bunches on productive shoots. Bunches are mainly set on 3rd-5th churn-stuffs of new shoots. Skin of grape is black or reddish black. Seeds are oval-shaped. The surface is covered with a thick wax layer. Most wild grape varieties are resistant to mildew and oidium disease.

More famous local varieties of grapevine are cultivated in Absheron, Garabagh, Ganja-Gazakh, Shirvan, Guba-Khachmaz regions and Nakhchevan AR of Azerbaijan. Hundreds (according to some sources, more than 600) of landraces of grapevine are grown in the Republic. At the present time the total area of vineyards in Azerbaijan more than 16,000 hectares.

White, red, black and pink colored table, technical and seedless grapevine varieties:Agh shani, Absheron's gyzyl uzumu, Alvan, Amiri, Askari, Agh Sahibi, Agh Aldara, At uzum, Aghri, Arnaqrna, Bandi, Rishbaba, Chilal, Kishmishi, Tulkuguyrugu, Huseyni, Madrasa, Marmari, Qara Aldara, Qoc uzumu, Tabrizi, Molla Ahmadi, Novrast, Karimgandi, Durna gozu, Davagozu,Kechiamcayi, Khazri, Khalili, Gara shani, Gizil uzum, Chil uzum, Beylagani, Kharci, Khan uzum, Pishras, Malayi, Mahmudabi, Misgali, Khindogny, Hafizeli, Hachabash, Haji Abbas, Hamashara, Sarigila, Shiray, Shirvanshahi, Shireyi, Shirshira, Shafeyi, Shakarbura, Shahangir, Shakari, Sisag and others are cultivated here. Most of them are only grown in definite areas and private courtyards by amateur gardeners [29,30].

A number of grape varieties in current use were the material resources of our ancestors. Each biomorphological trait of these varieties was selected corresponding with land and climatic condition of our republic. These varieties are named for their size, colour, view of bunches, form, taste and quality, as well as names of areas, villages and persons.

http://dx.doi.org/10.5772/ 52387

Ancient experience in grapevine cultivation allowed the Azeri to improve their secrets about vine care. Every viticultural technique is a product of local experience in different regions and historical periods. Each one has been adapted to the local conditions and this is why we have such a variety of training systems, like "Khiyaban", "Molla cheperi", "Keleser", "Serilen forma", "Yarimgovs", "Chardak" and others.

High qualitative products-jams, "doshab", vinegar, "abgora", "sucuq", "kishmish", "movuc", "lavashana", juice, syrup, vines, alcohol which made from grapevine in different regions of Azerbaijan show that grapevine-growing has developed expediently.

From its colour – Agh shani, Agh sahibi, Ala shani, Benovsheyi, Gara shani, Garagile, Girmizi chileyi, Goy gezendayi, Gara serme, Gizili, Mermeri uzum, and etc.

From its quality, taste, aroma – Gulabi, Kishmishi, Shekeri, Tembeyi, Shireyi, Kerimgendi and others.

From its view, trunk size – Gushureyi, Misgali, Tulkuguyrughu, Devegozu, Tulagozu, Kechimemesi, Inekemceyi, Pishik uzum, Goyungozu, Ayiboghan, Gelinbarmaghi and others.

From its skin thickness – Dash uzum, Galingabig, Nazikgabig.

From its seed size and bunch form – Sapdadurmaz, Hachabash, Bendi.

From names of villages and regions – Beylegani, Tebrizi, Shabrani, Derbendi, Ordubadi, Shirvanshahi, Tatli, Merendi, Medrese, Shakhtakhti, Nakhchivan gara uzumu, Beneniyari, Agh aldere, Zeyneddin uzumu.

From producer's name – Khelili, Huseyni, Asgari, Sekine xanum, Mukhtari, Jelali, Khatini, Khanimi, Meshedi Ali.

From name of old tribes – Khalaj variety in Mil-Mughan region.

Locally selected varieties of grape can be found outside their historical formation areas and today they are grown by amateur gardeners and in peasant-farmer households. These varieties are met in the following areas:

Absheron region – here approximately more than 50 valuable local grape varieties are grown. Agh shani, Gara shani, Ala shani, Sarigile, Haji Abbas, Khatuni, Pishraz, Gavangir, Goybendam, Rishbaba, Khalbasar, Absheron Gelinbarmaghi, Absheron Gizil uzumu, Gara gushureyi, Absheron kechiemceyi, Nadirgulu, Gara Derbendi, Salyani, Zabrat uzumu, Sirkeyi, Movuju, Gala kishmish, Shireyi, Turabi, Shabrani, Gaz khani, Merendil, Garachi, Seyid Amiri, Sebze, Mashtagha khatunisi, Yalanchi shani, Alimemmed, Gargha dili, Sikhsalkhim, Beledi, Gilami and other varieties are the most qualitative and valuable grape varieties.

Ganja-Gazakh region– Tebrizi, Bayanshire, Tatli, Khircha kishmish, Shal uzum.

Shirvan region– Medrese, Shirvanshahi, Devegozu, Shamakhi merendisi, Sisag, Khezri, Chil uzum, Kechiemceyi, Khan uzumu, At uzumu, Beylegani, Shekeri, Khungi, Elvan.

Guba-Khachmaz – (there are approximately 50 grape varieties) - Devechi Agh chileyisi, Devechi giziluzumu, Shabrani, Chileyi, Girmizi chileyi, Derbendi, Khetmi, Khaldar.

Garabagh region – Amiri, Ari merendi, Gara merendi, Gushureyi, Aghdam giziluzumu, Zeynebi, Gul merendi, Aghdam kechiemceyi, Kal uzum, Aghdam khazarisi, Khindogni, At-merendi.

Nakhchivan AR – (more than 100 valuable grape varieties are cultivated) - Ayiboghan, Agh khalili, Agh uzum, Agh kurdeshi, Bendi, Girmizi Inekemceyi, Gara kurdeshi, Gara khalili, Gara shafeyi, Gizili sebze, Girmizi tayfi, Girmizi shafeyi, Girmizi huseyni, Inekemceyi, Kechiemceyi, Kehraba, Miskali, Nakhchivan girmizi shanisi, Nakhchivan agh tayfisi, Nakhchivan gara shanisi, Nakhchivan gizil uzumu, Nakhchivan huseynisi, Nebi, Nekhshebi, Sari shafeyi, Khatinbarmaghi, Khatini, Hachabash, Gulabi, Abbasi, Agh aldere, Badamli, Batikh, Beneniyar, Talibi, Goyungozu, Durzali, Zeyneddin uzumu, Meshedi Ali, Narinjigile Pishik uzumu, Sari aldere, Sahibi, Teberze, Khanimi, Gara khazani, Shangirey, Hafizeli and others.

Up to the period of adoption of Islam the vine-growing was mainly developed in direction of wine-making, therefore technical varieties dominated in vineyards. At that time Medrese, Meleyi, Agh aldere, Gara uzum, Khetmi, Henegirna and other varieties were cultivated widely in these areas. After adoption of Islam wine-making was prohibited and cultivation of table grapes was stopped. In historical sources it was noted that in these areas lots of kishmishi and table varieties had been cultivated. Some of them (Agh Shani, Gara Shani, Sarigile, Tebrizi, Kishmishi, Khelili, Kurdeshi, Bendi, Nahkchivan huseynisi, Misgali, Nakhchivan gizil uzumu, Shefeyi, Gulabi, Inekemceyi and other varieties) are national selection samples of our ancestors. Different products such as dried raisins and movuc (dried grapes with seeds) were produced and even these products were exported to Near Eastern countries. Formerly, a number of grape varieties had been observed by travellers, merchants and these varieties had spread widely to other regions, several countries of the world and had been named with appropriate synonyms.

Physiological complete maturity period is a characteristic inherited for each variety. Varieties, clones and new forms studied in genefund are distinguished from each other by their maturing periods. It was determined by investigations that maturing periods of fruits of local grape varieties in Azerbaijan Republic can be divided into the following groups:

The earliest maturing (approximately 120 days) varieties: Girmizi huseyni, Agh khelili, Gara khelili, Agh kurdashi, Gara kurdashi;

Early maturing (120-130 days) varieties: Gara pishras, Salyani, Agh chileyi, Gara kishmishi, Gehveyi kishmishi, Yumrugile sari kishmishi;

Middle fast growing (131-140 days) varieties: Agh kishmishi, Nakhchivan huseynisi;

Middle growing (141-150 days) varieties: Agh Shani, Asgari, Beylagani, Gavangir, Gulabi, Gara shani, Sarigile, Fatmayi, Absheron gelinbarmaghi, Tebriz, Shireyi, Girmizi kishmishi, Xirdagile kishmishi, Sari shafeyi, Shekerbura;

Middle late maturing (161-170) varieties: Gavangir, Ala shani, Haji Abbas, Bendi, Goybendem, Julu merendi, Inekemceyi, Shamakhi merendisi, Girmizi shafeyi, Nakhchivan agh tayfisi, Kechiemceyi, Gulabi, Bendi, Negshebi, Miskali, Nakhchivan gara shanisi, Hachabash, Khatinbarmaghi, Khezani, Meshedi Ali, Zeyneddin uzumu, Shangirey, Narinjigile, Henegirna, Shakhtakhti, Jelali, Gara serme, Meleyi, Bilev uzumu, Ayiboghan, Khanimi, Hafizeli, Talibi, Mukhtari, Sari shireyi;

The latest maturing (171 days and more) varieties: Agh derbendi, Devechi gizil uzumu, Khezeri, Kechiemceyi, Nakhchivan girmizi shanisi, Nakhchivan gizil uzumu, Khatini, Kehreba uzum, Agh uzum, Nebi, Gizili sebze, Beneniyar, Durzali, Sari aldere, Agh aldere, Goyungozu, Sahibi, Abbasi, Gara aldere, Batikh, Gara henegirna, Khetmi, Khanlari, Zalkha, Dashgara, Rizagha, Agh uzum, Chol uzumu, Zereni gorasi, Nakhchivan gara uzumu, Girmizi henegirna, Goy uzum, Innabi, Khan uzumu, Agh kelenpur, Girmizi gemeri, Khalli uzum, Pishik uzumu, Badamli.

Existing local grape varieties are distinguished from each other by their usage in our Republic. Here table, technical and universal varieties are known. Between them the table grape is more dominant.

Table grape varieties: These grape varieties are used fresh. Absheron gelinbarmaghi, Absheron khatini, Absheron kechiemceyi, Absheron gizil uzumu, Agh gavra, Absheron merendisi, Salyan uzumu, Shireyi, Agh kishmishi, Agh Beylagani, Khalaj, Khalbasar, Khan uzum, Aghdam khezerisi, Fatmayi, Gavangir, Haji Abbas, Gul merendi, Aghdam kechiemceyi, Agh goybendem, Gara kishmishi, Girmizi kishmishi, Gehveyi kishmishi, Yumrugile sari kishmishi, Mermeri, Sari aldere, Sarigile, Seyid Amiri, Siyezen agh uzumu, Shabrani, Sari kishmishi, Khirdagile sari kishmishi, Asgari, Ayiboghan, Agh khalili, Agh uzum, Agh kurdashi, Bendi, Girmizi Inekemceyi, Gara kurdashi, Gara khelili, Gara shafeyi, Gizili sebze, Girmizi shafeyi, Girmizi chileyi, Girmizi kherji, Girmizi merendi, Girmizi huseyni, Inekemceyi, Kechiemceyi, Kehraba, Miskali, Nakhchivan girmizi shanisi, Nakhchivan agh tayfasi, Nakhchivan gara shanisi, Nakhchivan gara shanisi, Nakhchivan gizil uzumu, Nakhchivan huseynisi, Nebi, Nekhshebi, Sari shafeyi, Shamakhi merendisi, Khatinbarmaghi, Khatini, Gara salyan uzumu, Hachabash, Gulabi are table grapes.

Technical grape varieties: They are used in making different alcoholic and non alcoholic drinks, total juice extract exceeds 75,0%. Arayatli gara uzum, Ari merendi, Bayanchire, Shirvanshai, Medrese, Tatli, Aghdam gizil uzumu, Sherabi, Arazvari, Agh kelenpur, Hamashara, Khindogni, Gara khatuni, Agh Almerdan, Bilev uzumu, Gara serme, Gara henegirna, Goch uzumu, Girmizi gemeri, Girmizi henegirna, Dashgara, Dagh uzumu, Dabbi gulabi, Jelali, Zalkha, Zereni gorasi, Innabi, Mukhtari, Meleyi, Nakhchivan gara uzumu, Rizagha, Sari shireyi, Sari uzum, Gara aldere, Tulagozu, Talibi, Khalli uzum, Khan uzumu, Khanlari, Kherji, Khetmi, Chol uzumu, Shahtakhti, Shahangul, Shekerbura, Haji Ahmadi, Henegirna, Goy uzum are technical varieties.

Figure 6. Shamakhi merendisi

Figure 7. Gara henegirna

Figure 8. Agh Aldere

Universal varieties: These are both table and technical varieties distinguished by their biomorphological and agrobiological traits. They ripen in different times and possess separate agrobiological parameters. These varieties are used fresh and for technical purposes. Abbasi, Agh aldere, Agh gulabi, Khungi, Gara merendi, Gara okuz gozu, Gara sebze, Boz merendi, Badamli, Batikh, Beneniyar, Talibi, Goyungozu, Durzali, Zeyneddin uzumu, Mehsedi Ali, Mahmudu, Narinjigile, Pishik uzumu, Sari aldere, Sahibi, Tabarza, Khanimi, Gara khazani, Shangirey, Shekerbura, Hafizeli are universal varieties.

Expeditions and investigations were implemented for the purpose of identification, collection and inventory of local grape varieties in Azerbaijan. Areas of local grape varieties and wild grapevine expansion were found through expeditions and investigations, etiquette of grapevines were noted and their morphological-biological and immunological characteristics were determined and mechanical and chemical investigation (in lab condition) of yield were carried out.

During expeditions and studies arranged in Absheron region, Gavangir, Fatmayi, Haji Abbas, Sarigile, Absheron gelinbarmaghi and Ala shani table grape varieties were sampled fresh. Gavangir and Sarigile varieties exhibited higher juice extraction yield than others. Therefore doshab and grape juice are produced of them. It was known that bunches and seeds of these varieties are medium and large-sized and this is characteristic for table varieties. The biggest individual seeds belong to Absheron gelinbarmaghi (berries size – 18-23x16-22 mm), Haji Abbas (berries size - 20-26x19-24 mm), Ala shani (berries size – 16-24x15-23 mm) varieties. This preference was reflected on the weight of 100 individual grape seeds. Sweetness of individual seeds of grape was 17,2 (Gavangir) -27,9 gr/100 cm^3 (Sarigile). Average weight of bunches was lower in Sarigile (170 gram) and Fatmayi (180 gram) varieties, in Absheron gelinbarmaghi (250 gram), Ala shani (240-278 gram), Haji Abbas (286 gram) was medium, but in Gavangir variety, average bunch weight was higher (386,4 gram) [28].

№	Region and names of varieties	Bunches size, cm	Seeds size, mm	Number of seeds	Weight of 1000 seeds, gr.	Average weight of bunches, gr.	Sweetness of seeds, gr/ 100cm³	Seed acidity gr/dm³	Vegetation period, day
	Garabagh-Mil region (Fuzuli, Beylagan region)								
1	Agh Beylagani	18-22x11-15	18-23x17-22	104	266	276,0	19,6	4,62	166
2	Gelinbarmaghi	18-26x12-16	28-36x20-22	84	542	386,5	18,6	5,76	177
3	Nubari	8-17x5-8	10-15x10-15	52	216	126,8	15,9	6,05	120
4	Arı uzumu	11-21x7-10	15-17x15-16	102	224	200,0	19,2	5,70	146
5	Arayatlı gara uzum	13-16x7-9	15-18x14,5-17,5	96	307	180,0	18,2	6,00	139
6	Agh Gavra	20-28x16-20	26-32x19-22	88	396	335	18,6	5,27	177
7	Surmeyi	16-26x11-15	22-26x15-18	72	423,7	210,6	16,2	5,89	147
8	Fuzuli kechimemesi (Kehrabayi)	13-27x8-14	27-35x19-20	96	527,8	441	15,2	6,41	152
9	Gizil uzum	18-21x7-8	15-18x14,5-17,5	108	298,5	234,5	17,5	5,18	176
10	Gozel uzum	15-27x7-12	20-27x14-19	88	424	322	17,0	6,04	171
11	Alikhanli kechimemesi	13-15x8-12	20-27x13-17	82	336	253	16,0	5,97	155
12	Bey uzumu	17-28x12-15	23-28x18-21	130	421	564,8	17,5	5,15	155
	Absheron region								
1	Gavangir	15-20x10-14	14-20x14-19	152	230,8-286,4	386,4	17,2	6,60	162
2	Fatmayi	18-24x10-14	15-21x14-20	125	210	180,0	18,5	5,25	150
3	Haji Abbas	18-25x12-19	20-26x19-24	92	336	286,0	18,2-24,6	5,62-3,46	162-168
4	Sarigile	15-22x10-15	15-21x12-20	84	240	170,0	21,8-27,9	3,9-7,3	146
5	Absheron gelinbarmaghi	17-22x14-18	18-23x16-22	80	406	250	20,3	5,7	152
6	Ala shani	14-22x12-16	16-24x15-23	112	325	240-278	18,5	6,5	156

Table 2. Some morphological and technological traits of local grape varieties collected through expedition

It was known at the result of phenological observations that studied varieties ripen averagely (Sarigile, Fatmayi, Absheron gelinbarmaghi) and lately (Gavangir, Haji Abbas, Ala shani) (table 2).

Agh shani is one of the oldest and most widely spread valuable table grape varieties of Azerbaijan. While investigating a population of Agh shani variety 4 variations in seed size/ shape were observed: – oblong; big seedy – grew in middle period; lately maturing, more and medium-sized seeds of grape; pea-shaped seeds. Through study of several morphological, biological and technological traits of noted variation, it was determined that they are sufficiently distinguished from each other for most of their parameters (table 3).

№	Variations of Agh shani variety	Bunches size, cm	Seeds size, mm	Correlation of seeds length to the width L/W	Number of seeds	Average weight of bunches, gram	Weight of 100 seeds, gram	Amount of pea-shaped seeds, %	Part of seeds on bunch %	Sweetness of seeds, gr/ 100cm³	Seeds acidity, gr/dm³	Vegetation period, day	Number of bunches on grapevines, number	Productivity, kq
1	Longish Agh shani	12-18 x8-11	21-24x 15-18	1,31-1,55	70	234	340	2,5	95,7	22,6	4,62	136	22	4,8
2	Big seedy Agh shani	15-22 x10-14	23-28x 21-23	1,1-1,3	87	382	446	3,2	94,5	20,4	5,86	148	16	6,0
3	Agh shani which has more seeds	14-20 x10-12	16-20x 14-16	1,16-1,25	140	296	236	4,6	96,2	18,8	6,02	160	20	6,3
4	Pea-shaped Agh shani	8-18x 5-12	18-22x 16-20	1,05	97	126	346	68,0	86,5	20,6	5,72	152	26	2,4

Table 3. Some morphological and agro-technological traits of variations of Agh shani variety

It was determined through immunological assessments of local grape varieties from Absheron that they were resistant to oidium disease (2-2,5 points) and tolerant (3-3,5 points). The climate of Absheron is dry-subtropical and therefore in most cases development of mildew disease is not a major problem there. Thus mildew disease was not observed in evaluated varieties. At the result of observations it was known that Gavangir and Fatmayi (3-3,5 points), Haji Abbas, Sarigile, Absheron gelinbarmaghi, Ala shani varieties (2,5 point) were tolerant to grey rot disease (table 4).

№	Regions and varieties	mildew		oidium		grey rot
		leave	fruit	leave	fruit	fruit
	Garabagh-Mil region					
1	Agh Beylagani	4	4	3	3	3
2	Gelinbarmaghi	4	4	3	3	3
3	Nubari	3	3	3	3	2,5
4	Ari uzumu	3	3	3	3	2,5
5	Arayatlı gara uzum	3,5	3,5	3	3	2,5
6	Agh Gavra	3	3	3	3	2,5
7	Surmeyi	2,5	2,5	3,5	3,5	5
8	Fuzuli kechimemesi	3	3	3	3	3
9	Gizil uzum	3,5	3,5	3,5	3,5	3,5
10	Gozel uzum	3,5	3,5	3	3	2,5
11	Alikhanli kechimemesi	3	3	3	3	2,5
12	Bey uzumu	3	3	3	3	2,5
	Absheron region					
1	Gavangir			3	3	3
2	Fatmayi			2,5	2,5	3
3	Haji Abbas			3	3	2,5
4	Sarigile			3	3	2,5
5	Absheron gelinbarmaghi			2,5	2,5	2,5
6	Ala shani			2,5	2,5	2,5
7	Note: 0-point-immune 1 point – more resistant 2-2,5 points - resistant 3-3,5 points - tolerant 4-4,5 points – not resistant 5 points –not more resistant					

Table 4. Resistance of local grape varieties to main fungus diseases on the natural background found through expedition, point

Of the 25 local and 2 introduced grape varieties that were found while exploring in Garabagh-Mil region, 12 of them were low spread local varieties. Agh Beylagani, Gelinbarmaghi, Nubari, Ari uzumu, Arayatli gara uzumu, Agh Gavra, Surmeyi, Fuzuli kechimemesi (Kehrabayi), Gizil uzum, Alikhanli kechimemesi, Bey uzumu are the low spread local grape varieties. It was known during morphometric measurements that their bunches were medium (Nubari, Arayatli gara uzum, Alikhanli kechimemesi) and large-sized (Agh Beylagani, Gelinbarmaghi, Agh Gavra, Surmeyi, Fuzuli kechimemesi, Gizil uzum, Gozel uzum, Bey uzumu). Separate seeds of studied varieties were different-coloured, formed, mainly small (Nubari), medium (Ari uzumu, Arayatli gara uzumu, Gizil uzum), large (Agh beylagani) and largest (Gelinbarmaghi, Agh gavra, Surmeyi, Fuzuli kechimemesi, Gozel uzum, Alikhanli kechimemesi, Bey uzumu) sized (table 2).

It was determined through phytopathological evaluation of above-mentioned varieties against mildew, oidium and grey rot diseases in natural situation that Agh Beylagani and Gelinbarmaghi varieties were not resistant to mildew disease (4 points), but showed average resistance (3 points) to oidium and grey rot diseases. Surmeyi variety was tolerant (3,5 points) to mildew and oidium diseases, but bunches were intolerant (5 points) to grey rot disease. Other varieties showed resistance (3-3,5 points) to mildew and oidium diseases. It was also defined that Nubari, Ari uzumu, Arayatli gara uzumu, Agh Gavra, Gozel uzum, Alikhanli kechimemesi, Bey uzumu varieties were resistant (2,5 points) to grey rot disease (table 4). Above-mentioned varieties are local and they are mainly used fresh. Agh Beylagani, Gelinbarmaghi, Agh Gavra, Fuzuli kechimemesi, Gozel uzum, Alikhanli kechimemesi, Bey uzumu can be stored for a long time and sometimes clusters are kept on grapevines till winter. Black-seeded Arayatli gara uzumu and Ari uzumu varieties possess high juice extraction and sweetness; therefore, red table wines are made of these varieties by local people [28].

Research studies on evaluation of biological-agricultural traits of grape varieties and forms (local, introduced) cultivated in ampelographic collection gardens and experimental fields were implemented. While evaluating disease and pests resistance of 74 studied varieties and forms, it was determined that a number of varieties were infected by oidium disease. Among them 17 varieties – Agh uzum, Fuzuli kechimemesi, Gara Asma, Parkent, Sari Karan, Oktyabrski, Vishnyoviy, Tozlayici, and others showed tolerance (3-3,5 points). Only Bayanshire variety was tolerant to mildew disease. 4 varieties and forms –Nakhchivan gulabisi, Gara Nakhchivan Khatini, Kishmish Khishrau and form number 2 were resistant to pests and were less infected (1 point).

Salt and drought resistance of 25 table and seedless (kishmish) grape varieties were studied for their main physiological traits (stress depression of pigment complex in osmotic solution (sucrose 2% NaCL) in complete formation stage of leaves). It was known that studied varieties demonstrated different reaction to stress factors and plants showed unlike attitude to salt and drought. And it was possible to select resistant varieties on these bases. Experimentation yielded those varieties with sufficient salt and drought resistance were Gırmızı kishmish, Kishmish Yangiyer, Belqradskiy bessemyannıy, Qara Qushureyi, Ruşaki, Kishmish Batır, Zerefshan kishmishi, Kishmish Batır, Sarı kishmish, Gırmızı turkmen kishmishi varieties and these varieties were distinguished for non-stress depression in chlorophyll [31,32,33].

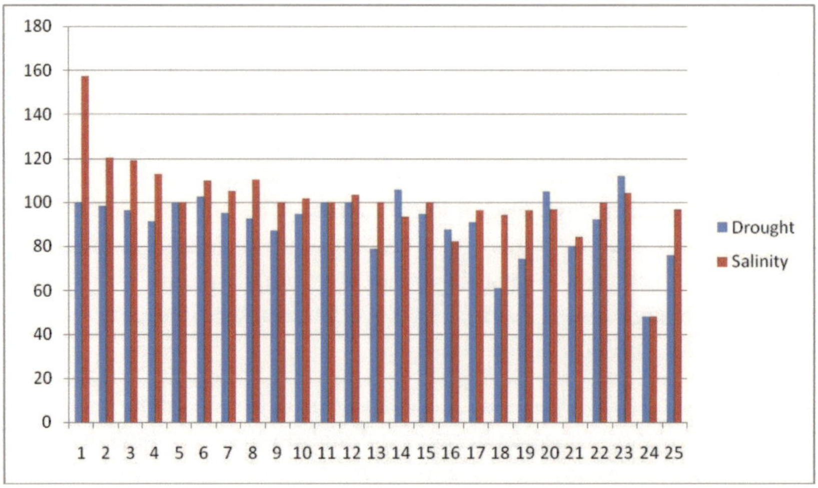

Figure 9. Changing the amount of chlorophyll in some grape varieties and forms of stress due to salinity and drought. 1.Gırmızı kishmish., 2.Zerefshan kishmishi., 3.Kishmish Yangiyer., 4.Ruşaki., 5.Polubessemyannıy., 6.İrtişar., 7.Bidane., 8.Vatkana., 9.Belqradskiy bessemyannıy., 10.Kishmish Xişrau., 11.Tezyetishen., 12.Kishmish Terakli., 13. Soqdiana., 14.Sarı kishmish., 15.Oktiyabiskiy., 16.Agh kishmish,. 17.Vishnyovıy., 18.Girde kishmish., 19.Form 21-18-36., 20.Gırmızı turkmen kishmishi., 22.Qara Qushureyi., 23.Kishmish Batır., 24. Samarkand kishmishi., 24.Vatkana-2., 25.Form 1.

It was defined that, since ancient times people had engaged with cultivation of new varieties and forms possessed different biological-agricultural traits. This tradition is also being continued today. Though the abundance of grape varieties of traditional breeding in Azerbaijan Republic, selection of new, of highly productive varieties, with big berries, with a high biologically active substances in berries, with a valuable economic characteristics, resistant to pests and diseases, as well as to stressful environmental factors were continued and some results were obtained in this field.

4. Conclusions

For the purpose of collecting ancient naturally selected varieties and their wild relatives of grape, a number of expeditions were organized in different regions of our Republic, their areal was determined, biological-agricultural traits of collected varieties and forms were evaluated, and for the first time ampelographic descriptions of newly threatened varieties were given and collected varieties were certificated and included in database. Some phytopathological, immunological and physiological parameters of grape varieties conserved in collection were evaluated and new resistant varieties were selected. Collected materials were included in gene pool, enriching the collections. Taking into account economic efficiency the growing of above mentioned grape varieties can have great perspectives, not only in Azerbaijan, but also in countries with similar climatic conditions. Therefore conservation of plant genetic diversity of grapes existed in Azerbaijan Republic, selection of productive samples, evaluation and protection are one of the most important problems in modern time.

Author details

Mirza Musayev* and Zeynal Akparov

*Address all correspondence to: m_musayev4@yahoo.com

Laboratory of Subtropical Plants and Grapevine, Genetic Resources Institute of the Azerbaijan National Academy of Sciences, Baku, Azerbaijan

References

[1] Mammadov, M, Asadov, K, & Mammadov, F. M. (2000). Dendrology. *Baku.*, 388, (in Azeri).

[2] Asadov, K, & Asadov, A. (2001). Wild fruits of Azerbaijan. *Baku*, 252, (in Russ).

[3] Hasanov, Z, & Aliyev, C. (2011). Horticultire. *Baku*, 520, (in Azeri).

[4] Akparov, Z, Imamaliev, Q, & Musayev, M. (2003). Diversity of the genetic fund of fruit plants in Azerbaijan. *Journal Azerbaijan & Azerbaijanis*, [3-4], *Baku* [3-4], 98-100.

[5] Maghradze, D, Akparov, Z, Bobokashvili, Z, Musayev, M, & Mammadov, A. (2012). The importance, usage, and prospective of crop wild relatives of fruits, grapevine, and nuts in Georgia and Azerbaijan. Proceedings of the 1st International Symposium on Wild Relatives of Subtropical and Temperate Fruit and Nut Crops., Davis, California, USA, March 19-23, 2011., *Acta Horticulturae*, 33-40.

[6] Akparov, Z, & Musayev, M. (2012). Diversity of the fruit plant genetic resources in the Azerbaijan. Proceedings of the 1st International Symposium on Wild Relatives of Subtropical and Temperate Fruit and Nut Crops. Davis, California, USA, March 19-23, 2011., *Acta Horticulturae*, 948, ISHS,May, 33-40.

[7] Babaev, T. (1988). Azerbaijan is an ancient wine-growing land. *Baku* , 86 .

[8] Vavilov, N. I. (1926). Centres of origin for cultivated plants. *Proceedings of Applied Botany, Genetics and Breeding.*, 16(2), in Russ.

[9] Vavilov, N. I. (1931). Wild relatives of fruit trees of Asia part of the USSR and the Caucasus and problems of fruit trees origin. Proc. of Appl. Bot. Genet. and Plant Breeding, 26(3), in Russ.

[10] Zhukovskii, P. (1964). Cultivated plants and their wild relatives. 2nd edition. Leningrad. Publishing house "Kolos". (in Russ), 790.

[11] Lazarevsky, M. A. (1963). The study of grape varieties. Rostov University Publishing, (in Russ), 152.

[12] Makarov, S. N. (1964). Scientific basis of method of experimental work in viticulture. *Proceedings, Volume IX. Chisinau: Map Moldovenyaske*, 280, in Russ.

[13] Morozova, G. (1987). Viticulture with the basics ampelography. M. Kolos, (in Russ), 251.

[14] Prostoserdov, N. (1963). The study of grapes for determining its use. M., Food promizdat, 80, in Russ.

[15] Smirnov, K, Kalmykova, T, Morozova, G, & 1987. Viticulture. M Agropromizdat in Russ, 367.

[16] Chisinau, S. (1985). New methods of phytopathological and immunological studies in viticulture.(in Russ). 138.

[17] Kushnirenko, M D, et al. (1976). Methods for assessing plant resistance to unfavorable environmental conditions. Leningrad: Kolos (in Russ), 87.

[18] Udovenko, G. (1988). Diagnosis of plant resistance to stress. Leningrad, 22-46, in Russ.

[19] Lodhi, M. A, Ye, G, Weeden, N. F, Reisch, B. I, & 1994. A Simple and efficient method for DNA ex-traction from grapevine cultivars and Vitis species. *Plant Molecular Biology Reporter*, 12, 6-13.

[20] Schaal, B, Beck, J, Hsu, S, Beridze, C, Gamkrelidze, T, Gogniashvili, M, Pipia, M, Tabidze, I, This, V, Bacilieri, P, Gotsiridze, R, & Glonti, V. M. (2010). Plastid DNA sequence diversity in a worldwide set of grapevine cultivars (Vitis vinifera L. subsp. vinifera). Abstracts of 10th International Conference on Grapevine Breeding and Genetics., 65, Geneva, New York.

[21] Beridze, T, Pipia, I, Beck, J, Hsu, S-C, Gamkrelidze, M, Gogniashvili, M, Tabidze, V, This, P, Bacilieri, R, Gotsiridze, V, Glonti, M, & Schaal, B. (2011). Plastid DNA sequence diversity in a worldwide set of grapevine cultivars (Vitis vinifera L. subsp. vinifera). *Bulletin of the Georgian National Academy of Sciences*, 5(1), 91-96.

[22] Negrul, A. (1946). Ampelography of USSR. v.I, Moscow. (in Russ).

[23] Sauer, J. D. (1993). Historical Geography of Crop Plants. CRC Press, Boca. Raton USA.

[24] Jackson, R. S. (1994). Wine Science Principles and Application. Academic Press New York.

[25] Arroyo-garcia, R, Ruiz-garcia, L, Bolling, L, Ocete, R, Lopez, M. A, Arnold, C, & Ergul, A. (2006). Genetic evidence for multiple centers of grapevine (Vitis vinifera L.) domestication. *Molecular Ecology*, 15, 3707-3714.

[26] Phillips, R. (2000). A Short History of Wine. Harper Collins, New York.

[27] Pipia, I, Gogniashvili, M, Tabidze, V, Beridze, T, Gamkrelidze, M, Gotsiridze, V, Meliyan, G, Musayev, M, Salimov, V, Beck, J, & Schaal, B. (2011). Plastid DNA Sequence

Diversity in Wild Grape Samples (Vitis vinifera L. subsp. sylvestris) from the Caucasus Region". Materials of XXXIV World Congress of Vine and Wine, Porto, Portugal,.

[28] Akparov, Z, Musayev, M, Mammadov, A, & Salimov, V. (2010). Study of the genetic resources of grapevine in Azerbaijan. *Journal "Agricultural science in Azerbaijan* [1-2], *Baku* [1-2], 40-44, in Azeri.

[29] Salimov, V, & Musayev, M. (2007). Genetic resources of grapevine in Azerbaijan. http://www.vitis/ru/pdf/rs.

[30] Musayev, M. (2003). Crapevine genetic resources in Azerbaijan. Report of a Working Group on Vitis. First Meeting. 12-14 June, Palic, Serbia and Montenegro. Rome. Italy., *Bioversity International.*, 2008., 57, www.ecpgr.cgiar.org/workgroups/vitis/Vitis1_WEB.pdf.

[31] Aliyev, R, Huseynova, T, & Musayev, M. (2007). Definition of drought resistance in some varieties of grapes. *Conference materials. Baku,*, 107-108, in Azeri.

[32] Musayev, M, & Huseynova, T. (2007). Ecological and physiological diagnosis of some varieties of grapes. *Conference materials Baku,*, 226, in Azeri.

[33] Aliyev, R, Huseynova, T, & Musayev, M. (2007). Biodiversity of grapes on the plant resistance to drought and salinity. VIII International Symposium "New and nonconventional plants and the prospect of their use." Proceedings of the Symposium. Volume II. Moscow,, 33-35, in Russ.

[34] Allahverdiyev, R. K., Suleymanov, C. S., et al. (1973). Ampelography Azerbaijan SSR. *Baku*, 490.

Chapter 6

Portuguese *Vitis vinifera* L. Germplasm: Accessing Its Diversity and Strategies for Conservation

Jorge Cunha, Margarida Teixeira-Santos,
João Brazão, Pedro Fevereiro and
José Eduardo Eiras-Dias

Additional information is available at the end of the chapter

http://dx.doi.org/10.5772/52639

1. Introduction

1.1. Economical, cultural and historical importance of grapevine in Portugal

Grapevine (*Vitis vinifera* L.) is the most widely cultivated and economically important fruit crop in the world. In the different Portuguese agro-ecosystems, grapevine plays an important role either as a border culture or as an extensive crop. The surface area used by vineyards amounts to 4.9 % of the arable land [1], representing 240,000 ha, being the 7th largest area in the world and the 4th in the European Union [2]. In 2011 Portugal produced 5.9 million hectoliters of which 2.9 million hectoliters were exported, making the country the 12th world wine producer [2]. There are fourteen wine regions with Protected Geographical Indication (Figure 1) and 31 wine areas with Designation of Origin status including Porto, established since 1756, the oldest legally established wine production region in the world. Each one of the wine regions has a particular set of grapevine cultivars adapted to its specific *terroirs*. Officially there are 343 cultivars allowed to be use in wine production in Portugal [3].

Grapes were eaten by Neolithic and Bronze Age populations of the Iberian Peninsula since the 3rd millennium BCE as proven by archaeological remains [4, 5, 6]. Consumption and production of wine is thought to have started by the Iberian populations in contact with the Phoenicians and Greeks trading ports. It further expanded during the Roman occupation and reach important religious prominence with the Christianization of population. It even continued during the Muslim caliphate since part of the population maintain the Christian faith. After the 10th century convents and monasteries spread again grapevine cultivation and implemented new tools for wine production. Since the 12th century, Portugal produces wine not only for local consumption but also for export, especially to northern Europe. This remote history of grapevine cultivation allowed the building up of great diversity. The number of cultivars increased until the tree waves of destruction from North American pest and diseases: powdery mildew (*Uncinula necator* Schweinf. Burrill) in 1851, phylloxera (*Dactylosphaera vitifoliae* Fitch) in 1863 and downy mildew [*Plasmopara viticola* (Berk. & M.A. Curtis) Berl & de Toni] in 1880. Until these severe pathological events grapevine was multiply simply by self-rooting of cutting our seed germination. Since the introduction of phylloxera the use of rootstocks from hybrids of other *Vitis* species is mandatory, except in areas were the phylloxera cannot survive. Such a case occurs in the Designation of Origin Colares wine region where the vineyards are settled in sandy soil and the roots are over tree meters deep. As early as the 19th century attempts to improve grape production result in a number of cultivars as Tinta do Aurélio (red cultivar selected by someone called "Aurélio"). However a truth breeding program to obtain new varieties was only started in the mid of the 20th century by José Leão Ferreira de Almeida and two of the obtain cultivars, Dona Maria (table grape) and Seara Nova (wine grape), occupy today a significant acreage [7]. The exact number of cultivars in use is unknown but from the 340 allowed for wine production, 240 are thought to be autochthonous [8, 9].

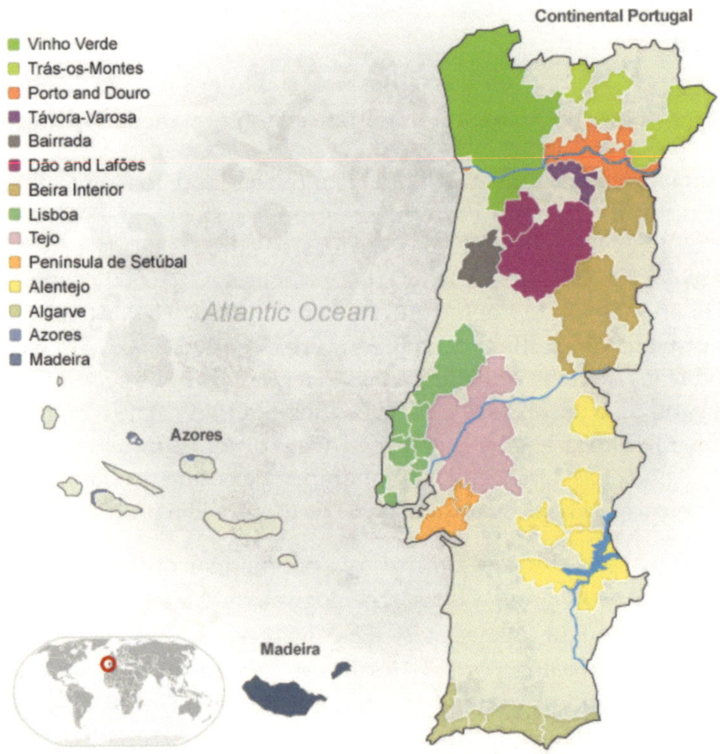

Figure 1. Location of the Portuguese wine regions. (Source: Wines of Portugal - http://www.winesofportugal.info/pagina.php?codNode=18012).

Traditionally morphological descriptors were used to characterize cultivars until the advent of molecular markers. Presently these have been successfully used in a wide range of applications such as assessing genetic diversity [10], linkage mapping [11], cultivar identification and pedigree studies [12], [13]. Microsatellites (SSR) are being used to characterize grapevine cultivars and wild vines [10, 14] and to carry out genetic diversity analyses [15]. Usually six *loci* are sufficient for differentiating between genotypes [16], but closely related cultivars require a larger number of *loci* [17]. Sequence variation at the chloroplastidial *loci* has been extensively used to assess phylogenetic relationships among plant *taxa,* based on their low rate of sequence evolution, the almost absent recombination and single parent inheritance [18]. All this range of tools is useful to make decisions on the strategies for conservation.

2. Diversity of the Portuguese grape germplasm

2.1. Wild vine populations: Geographical distribution, morphological and molecular characterization

Wild vine populations of *Vitis vinifera* L. subspecies *sylvestris* [(Gmelin) Hegi)] is closely related to the cultivated grapevine (*Vitis vinifera* subsp. *vinifera*), first domesticated 10,000 years BP around the Caspian Sea [19]. In Portugal these wild vine populations are distributed along riparian woods and flooded river banks in the southern part of the country in what is the most western habitats of this subspecies. From the Atlantic coasts of southwest Europe and northwest Africa this subspecies is distributed in patches adjacent to rivers along the Mediterranean basin, Central Europe and in Asia between the Black Sea and the Hindu Kush [20]. Once this subspecies occupied a larger area as a result of the its expansion after the last Quaternary glaciations [21, 22] but today´s remaining areas are refuges from human pressure and North-American pest and diseases introduced during the 19th century. Human populations since the early settlements in the Iberian Peninsula collected and consumed wild grapes [6] and this resource continued to be used until the late 20th century in folk medicine [20].

The wild vine populations found up to now in Portugal live in riparian woods along small streams (Figure 2) belonging to three large river basins – Tagus (Tejo in Portuguese), Guadiana and Sado (Table 1). The first two rivers are common to Portugal and Spain and the populations along these basins, even if found in patches, could be considered as a continuum [23, 24].

In these riparian woods the plants species most frequently found as tutors of V*itis vinifera* L. ssp. *sylvestris* are: *Adenocarpus complicatus, Alnus glutinosa, Fraxinus angustifolia, Nerium oleander, Olea europea, Quercus faginea* subsp. *Broteroi, Quercus suber, Rubus ulmifolius, Salix atrocinerea, Salix neotricha* and *Salix salvifolia* subsp. *salvifolia* [23, 25] . The thirteen populations found until now (Table 1) thrive in a typically Mediterranean environment. Fifty three plants belonging to four of these populations were characterized morphologically using the OIV [26] and GENRES-081 [27] descriptors [23, 28, 29].

Figure 2. *Vitis vinifera* subspecies *sylvestris* male plant from the São José/ Toutalga population in its natural habitat, a riparian forest along a small stream from the Guadiana river basin.

Population	River basin	Reference Code	Latitude	Longitude	Elevation (meters)	Estimated size of the population	PopRisk
Stª Sofia - Montemor-o-Novo	Tagus	01*[a]	38°36'41"N	08°05'24"W	306	[30-40]	3
Pônsul - Castelo Branco	Tagus	02*[a]	39°45'16"N	07°26'06"W	119	[30-40]	7
Guadiana - Mourão	Guadiana	03[a]	38°24'10"N	07°22'36"W	128	0	9
Vale do Guiso - Alcácer do Sal	Sado	04*[a]	38°14'46"N	08°22'30"W	49	[10-20]	3
Portel	Guadiana	05*	38°16'46"N	07°38'07"W	197	[20-30]	7
Ardila - Barrancos	Guadiana	06	38°07'56"N	06°57'41"W	208	[20-30]	5
Vendinha - Évora	Guadiana	07	38°27'18"N	07°41'02"W	163	[10-20]	5
Pintada - Montemor-o-Novo	Tagus	08	38°37'59"N	08°11'31"W	204	[10-20]	5
Fronteira	Tagus	09	39°02'38"N	07°42'14"W	93	[10-20]	5
Anta do Silval - Évora	Tagus	10	38°36'45"N	08°03'29"W	292	<10	5
Q. do Pinheiro - Montemor-o-Novo	Tagus	11	38°37'58"N	08°10'31"W	234	[20-30]	5
S.José/Toutalga - Moura	Guadiana	12	38°02'37"N	07°15'54"W	176	[20-30]	5
Enxota tordos - Grândola	Sado	13	38°13'27"N	08°30'22"W	34	>50	3

* Wild populations studied by [28, 29]/ [a] Wild populations studied by [33]

PopRisk (survival risk of the population): 1= No Risk; 3= Some Risk; 5= Medium Risk; 7= At Risk; 9= Extinct

Table 1. *Vitis vinifera* ssp. *sylvestris* Portuguese populations data: River basin; geographic coordinates, elevation (in meters) estimated size of the population, and risk of extinction.

The characterized wild vine plants featured the particularly morphological characteristics of the subspecies *sylvestris*: i) open young shoots, which is a characteristic allowing to differentiate between *Vitis vinifera* and the other *Vitis* species and hybrids; ii) the presence of male and female plants in each population (dioecious plants) (hermaphrodite plants are rare in wild vine populations and the rule in cultivated grapevines); iii) Stummer's Index (breadth/length ratio x 100) [30] of pips is equal or greater than 75 in wild vines. The morphological characteristics of the leaves, shoots and bunches were used to distinguish different phenotypes in the field. Until now only blue black berries were found and the ratio of male to female plants varies from population to population [28]. The 53 different wild vine accessions collected were genotyped using the six nuclear microsatellites suggested by the OIV [31, 32]. The diversity founded in wild vine genotypes (Table 2) reveals that the observed Heterozigocity (Ho) was less than the expected Heterozigocity (He) in all *loci*, confirming the result obtain in a different group of accessions from the same populations using a set of 11 SSRs [33] .

Locus	N	Na	Ne	Ho	He	F
VVMD5	53	10	2.428	0.585	0.588	0.005
VVMD7	53	9	2.869	0.547	0.651	0.160
VVMD27	53	8	3.417	0.509	0.707	0.280
VRZag 62	53	7	2.917	0.585	0.657	0.110
VRZag79	53	8	2.884	0.642	0.653	0.018
VVS2	53	11	5.021	0.736	0.801	0.081

Table 2. Diversity obtained in 53 Portuguese wild vines: *locus*, accessions number (*N*), number of alleles (*Na*), number of effective alleles (*Ne*), observed Heterozygosity (Ho), expected Heterozygosity (*He*) and Fixation Index (*F*).

The values of the Fixation Index (F) range from 0.005 to 0.28, showing the existence of inbreeding in some wild vine populations, since F is expected to be close to zero under random mating [34].

An Analysis of Molecular Variance (AMOVA) performed on the same molecular data showed that the genetic diversity was attributable to differences among individuals within populations (93.0%), but Fst values among populations are still significant (*Fst* = 0.071; P, 0.001), showing a low inter-population differentiation (Table 3). The morphological and molecular data confirmed that some of the collected plants were clones due to vegetative propagation (asexual propagation), but that the majority were different genotypes arising from seeds (sexual propagation).

Chloroplastidial microsatellites (cpSSRs) have been used to study the genetic relationships among grapevine cultivars [35], wild vines [36] and relations between both subspecies [37, 38]. Analysis of chloropastidial microsatellites (Figure 3) revealed the expected situation for the Iberian Peninsula [37] with the presence of chlorotypes A and B, being chlorotype A the most frequent within the wild vine populations (66%) of Portugal.

Variance component	Degrees of freedom	Sum of Squares	Variance components	Percentage of variation
Among Populations	3	17.0	0.15	7%
Within Populations	102	198.1	1.94	93%
Total	105	215.1	2.09	
Fixation index (Fst)			0.071 (P<0.001)	

Table 3. AMOVA analyses of six nuclear microsatellites data of 53 Portuguese wild vines on four distinct Southern Portuguese populations.

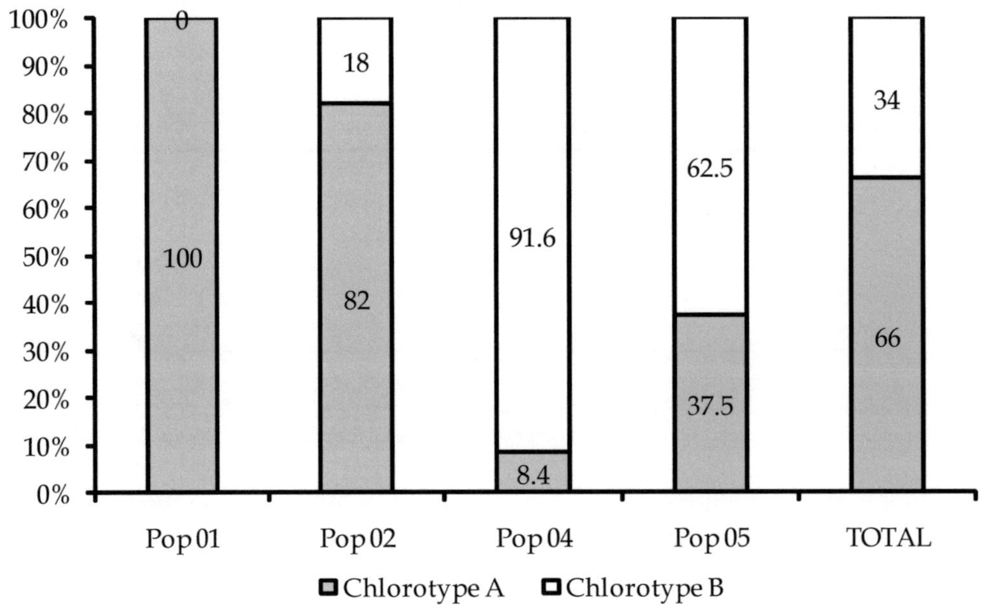

Figure 3. Chlorotypes identified in each Portuguese wild vine population. Chlorotype nomination according to [37].

Chlorotype A is the most frequent in Western Europe and absent in Near East where the domestication of *Vitis vinifera* occurred. The distribution of chlorotypes in four Southern Portuguese populations is heterogeneous. Only chlorotype A was found in plants of the population of Sta Sofia – Montemor-o-Novo. In the populations of Vale do Guiso - Alcácer do Sal, Pônsul – Castelo Branco and Portel both A and B chlorotypes were found but with distributions of 91.6%, 18% and 62.5% of chlorotype B respectively (Figure 3).

2.2. Cultivated grapevine: Morphological and molecular diversity

Portugal, a small country on the outer edge of Europe, has nonetheless a very rich diversity of grapevine cultivars build up over the centuries and back to the 19th century, 1482 different cultivar names were known. To organize the disarray that the different names caused to the wine sector the Ministry of Agriculture promoted a program to sort out the synonyms and homonyms using morphological descriptions [39, 40, 41, 42, 43, 44, 45, 46, 47, 48, 49]. Before Portugal joined the EEC (European Economic Community) in 1986, the Ministry of Agriculture finally drew up a list of "authorized" and "recommended" grapevine cultivars for each and every wine production areas (Figure 1). These efforts lead to the establishment of the Portuguese National Ampelografic Collection (in Portuguese "Coleção Ampelográfica Nacional" – CAN; international code PRT051) in 1988 after an extensive survey and collection of accessions all over the country. All CAN accessions were grafted into SO_4 rootstock and each access is represented by seven plants from the same original mother plant. This collection holds 691 accessions of *Vitis vinifera* ssp. *vinifera*; 30 accessions of *Vitis vinifera* ssp. *sylvestris*; 24 accessions of rootstocks and nine of other *Vitis* species. The sanitary status of the collection was also assessed for the principal viruses of grapevine (*Arabis* mosaic virus (ArMV), grapevine fanleaf virus (GFLV), grapevine fleck virus (GFKV), grapevine leafroll associated viruses 1, 2, 3 and 7 (GLRaV 1, 2, 3 and 7) grapevine virus A (GVA) and grapevine virus B (GVB) [50].

The molecular characterization of the Portuguese grapevine cultivars was initiated in 1999 by Lopes and collaborators and a number of known synonyms and homonyms as well as pedigrees were confirmed [51, 52, 53]. A systematic characterization of all the 340 varieties admitted for wine production in Portugal, including 243 autochthonous grape cultivars (Table 4) was done with the six nuclear SSRs recommended by OIV [8, 9]. These studies come to prove the synonyms and homonyms that previous morphologic description had established in the past and also allowed the finding out of new ones.

The diversity present in the 243 autochthonous grapevine cultivars analyzed based on the six nuclear SSRs genetic markers (Table 5) reveals that the observed Heterozigocity (*Ho*) was slightly higher than the expected Heterozigocity (*He*) in all *loci*. The Fixation Index (*F*) is negative for all *loci*, indicating an excess of Heterozigocity, probably due to the strong barrier caused by the vegetative propagation commonly used in grapevine.

Four chlorotypes (A, B, C and D) were found in the autochthonous grapevine cultivars so far genotyped (roughly one quarter of the 243) (Figure 4). Chlorotype A is the most frequent, and it is present in 75% of the cultivars, followed by chlorotype D with 19%. Chlorotypes B and C are each present in a very restricted number of cultivars [29, 32, 37]. These results support the presumption that most of the Portuguese cultivated grapevine germplasm may have derived from local domestication, but that some are the result of introgressed with foreign material as exemplified by important wine cultivars like Touriga Franca and Trincadeira that show the presence of the D chlorotype.

Access number	Grape cultivar	Origin	Access number	Grape cultivar	Origin
40403	Seara Nova	E.A.N.	41703	Malvasia Preta Roxa	Douro.
40404	Assaraky	E.A.N.	41705	Roxo de Vila Flor R	Douro.
40501	Promissão	Douro.	41702	Gouveio Roxo	Douro.
40502	Branco Valente B	Douro.	41707	Deliciosa	E.A.N.
40505	Sercial	Madeira.	41708	Bastardo Roxo	Douro.
40603	Malvasia Babosa B	Madeira.	41709	Donzelinho Roxo	Douro.
40604	Malvasia São Jorge	Madeira.	41806	Campanário	E.A.N.
40606	Granho	Alentejo.	50104	Ferral	unknown
40609	Tinta Aurélio	Douro.	50201	Complexa	E.A.N.
40701	Alvarinho Lilaz B	E.A.N.	50216	Terrantez do Pico	Pico - Açores.
40702	Castália	E.A.N.	50218	Arintaçor	Terceira - Açores.
40703	Naia	E.A.N.	50309	Castelo Branco	E.A.N.
40704	Malvasia de Oeiras B	E.A.N.	50314	Branca de Anadia	E.A.N.
40708	Cornichon	Alentejo.	50317	Verdelho	Açores.
40808	Generosa	E.A.N.	50602	Tinta Martins	Douro.
40809	Rio Grande	E.A.N.	50604	Tinta Mesquita	Douro.
41002	Pé Comprido	Douro.	50605	Português Azul	Douro.
41103	Esganinho	Vinhos Verdes.	50607	Tinta Gorda N	Douro.
41105	Branco Gouvães	Douro.	50608	Tinta Malandra N	Douro.
41107	Branco Desconhecido	Douro.	50611	Lameiro	Vinhos Verdes.
41202	Branjo	Vinhos Verdes.	50615	Àgua Santa	E.A.N.
41203	Galego	Vinhos Verdes.	50616	Gouveio Real	Douro.
41204	Labrusco	Vinhos Verdes.	50617	Gouveio Estimado	Douro.
41205	Melhorio	Vinhos Verdes.	50702	Mondet	Douro.
41206	Transâncora	Vinhos Verdes.	50703	Tinta Aguiar	Douro.
41208	Verdial Tinto	Douro.	50705	Touriga Fêmea	Douro.
41209	Alvarelhão Ceitão	Douro.	50706	Tinta Miúda de Fontes N	Douro.
41301	Moscatel Galego Tinto	Douro.	50707	Tinta Roseira N	Douro.
41302	Barreto de Semente T	Douro.	50708	Lourela	Douro.
41303	Casteloa	Douro.	50802	Gonçalo Pires	Douro.
41304	Farinheira	Douro.	50806	Padeiro de Basto N	Vinhos Verdes.
41305	Gouveio Preto	Douro.	50807	Tinta Pomar	Douro.
41306	Mourisco de Trevões	Douro.	50808	Tinta Varejoa N	Douro.
41309	Tinta Melra T	Douro.	50901	Casculho	Douro.
41502	Alentejana N	E.A.N.	50902	Concieira	Douro.
41503	Lusitano	E.A.N.	50904	Doçal	Vinhos Verdes.
41504	Tinta de Alcobaça N	E.A.N.	50905	Doçal de Refoios N	Douro.
41505	Agronómica	E.A.N.	50907	Tinta Pereira	Douro.
41508	Portalegre N	E.A.N.	50909	Malvasia Trigueira R	Douro.
41509	Triunfo	E.A.N.	50912	Malvasia Branca	Açores.
41601	Monvedro de Sines N	Sines	50914	Caracol	Madeira.
41603	Manteúdo Preto	Alentejo.	50915	Esganoso	Vinhos Verdes.
41605	Listrão	Madeira.	50916	Mourisco Branco	Douro.
41607	Mindelo	E.A.N.	50917	Rabigato Moreno	Douro.

Continued

50918	Roxo Rei	Douro.	51608	Tinta Valdosa N	Douro.
51002	Castelã	Douro.	51609	Dona Joaquina	Estremadura
51003	Amor-não-me-deixes	Alentejo.	51611	São Mamede	Vinhos Verdes.
51007	Pical-Polho N	Vinhos Verdes.	51613	Rabigato Franco	Douro.
51008	Tinta Engomada N	Douro.	51617	Perrum	Algarve.
51011	Sercialinho	E.A.N.	51701	Mourisco	Vinhos Verdes.
51012	Trincadeira Branca	Estremadura.	51708	Tinta do Rodo N	Douro.
51016	Caramela	Douro.	51715	Praça	Douro.
51017	Estreito Macio	Douro.	51803	Preto Martinho	Douro.
51018	Branco Guimarães	Douro.	51804	Monvedro	Dão.
51103	Tinta Ricoca N	Douro.	51806	Verdelho Tinto	Vinhos Verdes.
51108	Bastardo Espanhol N	Beira Interior.	51808	Beba	Algarve.
51113	Larião	Alentejo.	51816	Carrega Branco	Douro.
51115	Luzidio	Dão.	51901	Sousão	Vinhos Verdes.
51117	Bastardo Branco	Douro.	51902	Vinhão	Vinhos Verdes.
51202	Tinta Negra	Madeira.	51905	Tinta Caiada	Alentejo.
51205	Tintinha	Alentejo.	51910	Tamarez	Ribatejo.
51207	Corvo	Estremadura.	51914	Síria	Beira Interior.
51208	Tinta Roriz de Penajóia N	Douro.	52002	Marufo	Beira Interior.
51209	Dedo de Dama	Estremadura.	52003	Alfrocheiro	Dão.
51211	Uva Cavaco	Beira Interior.	52004	Cornifesto	Douro.
51212	Malvasia Cabral	Douro.	52005	Nevoeira	Douro.
51216	Branco Especial	Douro.	52006	Patorra	Douro.
51217	Pintosa	Vinhos Verdes.	52007	Alvarinho	Vinhos Verdes.
51304	Coração de Galo	Dão.	52011	Rabo de Ovelha	Alentejo
51307	Tinta Tabuaço	Douro.	52014	Rabigato	Douro.
51308	Tinta de Cidadelhe N	Douro.	52016	Bical	Estremadura
51314	Roupeiro B	Estremadura.	52017	Boal Espinho	Estremadura
51316	Sarigo	Douro.	52101	Tinta da Barca N	Douro.
51317	Côdega de Larinho	Douro.	52104	Arjunção	Algarve.
51402	Mourisco de Semente	Douro.	52105	Pedral	Vinhos Verdes.
51403	Sevilhão	Douro.	52106	Rufete	Dão.
51404	Cidreiro	Dão.	52111	Boal Vencedor B	Estremadura
51405	Corropio	Alentejo.	52112	Gouveio	Douro.
51410	Douradinha B	Dão.	52114	Alvadurão	Estremadura
51411	Dorinto	Douro.	52116	Boal Branco	Estremadura
51412	Arinto do Interior	Dão.	52117	Dona Branca B	Dão.
51413	Manteúdo	Alentejo.	52201	Tinta Carvalha	Douro.
51415	Uva Cão	Dão.	52202	Negra Mole	Algarve.
51417	Moscadet	Douro.	52203	Ramisco	Estremadura
51513	Verdelho Roxo	Açores.	52205	Touriga Franca	Douro.
51514	Folha de Figueira	Beira Interior.	52206	Touriga Nacional	Dão.
51516	Samarrinho	Douro.	52207	Encruzado	Dão.
51517	Cascal	Vinhos Verdes.	52210	Terrantez	Dão.
51602	Grangeal	Douro.	52213	Loureiro	Vinhos Verdes.
51604	Espadeiro Mole	Vinhos Verdes.	52216	Trincadeira das Pratas	Ribatejo.
51606	Pilongo	Pinhel.	52301	Moreto	Alentejo.

Continued

52304	Santareno	Douro.
52306	Donzelinho Tinto	Douro.
52307	Donzelinho Branco	Douro.
52309	Boal Ratinho B	Estremadura
52310	Avesso	Vinhos Verdes.
52311	Arinto	Bucelas.
52313	Almafra	Estremadura
52314	Fonte Cal	Beira Interior.
52316	Antão Vaz	Alentejo.
52402	Camarate	Estremadura
52407	Barcelo	Dão.
52410	Cerceal Branco	Douro.
52412	Cercial	Bairrada.
52502	Tinta Francisca	Douro.
52503	Jaen	Dão.
52505	Benfica N	E.A.N.
52506	Tinto Pegões	E.A.N.
52507	Batoca	Vinhos Verdes.
52512	Malvasia Fina	Douro.
52513	Diagalves	Estremadura
52515	Jampal	Estremadura
52605	Carrasquenho	Estremadura
52606	Baga	Bairrada.
52612	Malvasia Fina Roxa	Dão.
52614	Vital	Estremadura
52615	Castelão Branco	Estremadura
52702	Parreira Matias	Estremadura
52705	Preto Cardana	Ribatejo.
52706	Castelino	Estremadura
52708	Folgasão Roxo	Beira Interior.
52709	Folgasão	Douro.
52710	Trajadura	Vinhos Verdes.
52714	Malvasia	Estremadura
52715	Viosinho	Douro.
52803	Bastardo	Douro.
52807	Borraçal	Vinhos Verdes.
52809	Azal	Vinhos Verdes.
52810	Fernão Pires	Bairrada.
52815	Fernão Pires Rosado	Ribatejo.
52902	Carrega Burros	Ribatejo.
52903	Rabo de Anho	Vinhos Verdes.
52904	Espadeiro	Vinhos Verdes.
52905	Tinta Barroca	Douro.
52906	Tinta Grossa	Alentejo.
52908	Amaral	Vinhos Verdes.
52913	Galego Dourado	Estremadura
53006	Trincadeira	Douro.
53013	Malvasia Rei	Douro.
53015	Moscatel Nunes	Setúbal.
53102	Primavera	E.A.N.
53103	Cabinda	E.A.N.
53106	Castelão	Ribatejo.
53204	Amostrinha	Estremadura
53205	Malvasia Preta	Douro.
53206	Valbom	E.A.N.
53207	Alvarelhão	Dão.
53307	Tinto Cão	Douro.
53308	Malvarisco	Setúbal.
53312	Marquinhas	E.A.N.
53407	Mulata	E.A.N.
53806	Roal	Setúbal.
53807	Teinturier	Estremadura
54006	Almenhaca	*
54007	Alvar	*
54008	Alvar Roxo	*
54009	Arinto Roxo	*
54010	Boal Barreiro	*
54011	Branco João	*
54012	Cainho	*
54013	Calrão	*
54015	Corval	*
54016	Crato Espanhol	*
54017	Esgana Cão Tinto	*
54018	Galego Rosado	*
54019	Leira	*
54020	Malvasia Romana	*
54021	Malvia	*
54022	Perigó	*
54023	Pero Pinhão	*
54025	Pexem	*
54026	Rabo de Lobo	*
54027	Santoal	*
54028	Zé do Telheiro	*
54029	Tinta	*
54030	Tinto Sem Nome	*
54031	Valveirinho	*
54032	Verdial Branco	*
54033	Xara	*

* Recent Introduction in PRT051

Table 4. Autochthonous grapevine cultivars used in wine production in Portugal: Access number in the PRT051 collection, name of the grapevine cultivar, origin of grapevine accession.

Locus	N	Na	Ne	Ho	He	F
VVMD5	243	11	6.673	0.881	0.850	-0.036
VVMD7	243	12	3.965	0.765	0.748	-0.024
VVMD27	243	8	4.968	0.831	0.799	-0.041
VRZag 62	243	7	3.834	0.761	0.739	-0.030
VRZag79	243	12	4.051	0.765	0.753	-0.016
VVS2	243	15	5.822	0.881	0.828	-0.063

Table 5. Analyses of diversity in 243 Portuguese autochthonous cultivars: *locus*, accessions (*N*), number of alleles (*Na*), number of effective alleles (*Ne*), observed Heterozygosity (*Ho*), expected Heterozygosity (*He*) and Fixation Index (*F*).

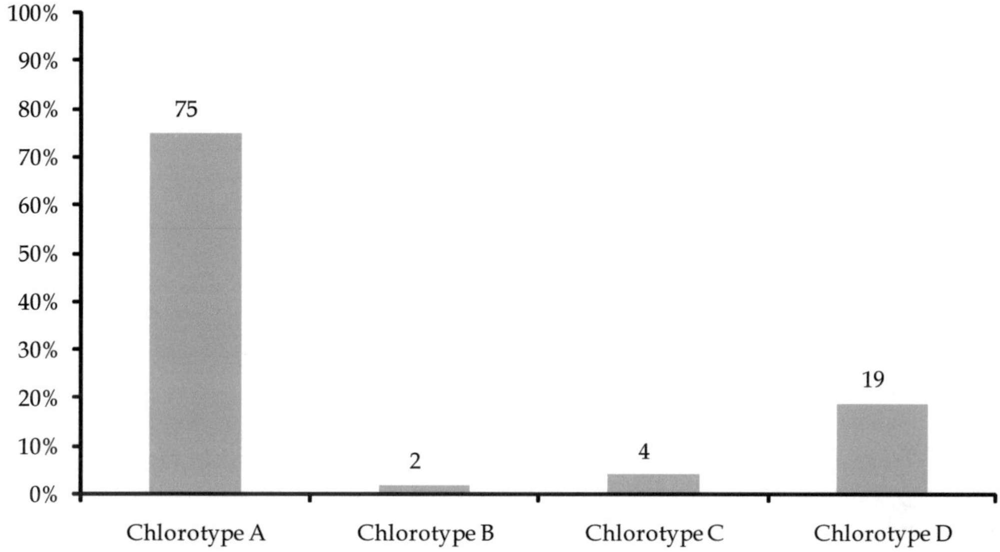

Figure 4. Chlorotypes of the Portuguese autochthonous grapevine cultivars. Chlorotype nomination according to [37].

The obtained results reinforce the suggestion that the Iberian Peninsula was a secondary center for grapevine domestication [37] despite the initial contribution of the Eastern gene pool some 3000 years ago and the more recent introgression from materials coming from central Europe.

Since 1978 a network of public and private associations lead by Antero Martins carried out an extensive work aiming at quantifying the intravarietal genetic variability within each of 45 Portuguese grapevine cultivars [54]. The static methods used were recently reviewed in [55]. These studies lead to the selection of a number of clones from Portuguese cultivars. In parallel and using the Geisenheim method of grapevine selection, a private nursery leaded by Jorge Böhm also selected a number of clones. Both groups registered a total of 122 clones from 27 different cultivars in the national grapevine catalogue (Table 6).

Variety	Obtainers Plansel clones JBP	 UTL clones ISA	 INIAV clones EAN
Alfrocheiro T	41		
Alvarinho B	42; 43	44; 45; 46; 47	
Antão Vaz B	50		
Aragonez T	106; 110; 111; 114; 117		54; 55; 56; 57; 58; 59; 60
Arinto B	34; 35; 107		36; 37; 38; 39; 40
Bastardo T	48		
Bical B	119		
Castelão T	5; 25; 26		29; 30; 31; 32; 33
Cerceal Branco B	120		
Fernão Pires B	1		68; 69; 70; 71; 72; 73; 74
Gouveio B	121; 122; 123		
Jaen T		91; 92; 93; 94; 95; 96; 97	
Loureiro B		81; 82; 83; 84; 85	
Malvasia Fina B	127	98; 99; 100; 101; 102; 103; 104	
Moreto T	51		
Perrum B	128		
Sercial B	49; 105		
Síria B			75; 76; 77; 78; 79; 80
Tinta Barroca T	9; 129		
Tinta Caiada T	115; 116; 118		
Touriga Franca T	24		
Touriga Nacional T	16; 108; 112	17; 18; 19; 20; 21; 22; 23	
Trajadura B		86; 87; 88; 89; 90	
Trincadeira das Pratas B	124; 125; 126		
Trincadeira T	6; 7; 8; 109		10; 11; 12; 13; 14; 15
Vinhão T		61; 62; 63; 64; 65; 66; 67	
Viosinho B	53		
PE 1103 P	4		
PE 110 R	2		
PE 140 Ru	113		
PE 99 R 3	3		

Plansel/ JBP - Plansel (Wine and Nursery Company) / Jorge Böhm Plansel
UTL/ISA – Universidade Técnica de Lisboa / Instituto Superior de Agronomia
INIAV/ EAN – Instituto Nacional de Investigação Agrária e Veterinária/ Estação Agronómica Nacional

Table 6. List of the certified Portuguese clones of grapevine cultivars.

2.3. Overall diversity of the Portuguese grapevine germplasm

Portuguese wild vine populations are in an apparent geographic fringe of the species distribution but the country richness in cultivar diversity [8, 9] and the importance in allele contribution to the overall diversity of grapevine [56] tell another story. Figure 5 represents a Principal Coordinate Analysis of the diversity computed with the six nuclear SSRs used to genotype the 243 autochthonous cultivars and 53 wild vines, calculated with the program GenAlex6 [57] The two first coordinates represent 44.12% (1st coordinate - 24.08% and 2nd coordinate - 20.04%) of the total variance. Both subspecies are spread between the four quadrants although most wild vines are in the right quadrants. Even the plausible occurrence of feral forms cannot explain the overall dotting of the four quadrants since the alleles found in the wild vines population include private and particular alleles (data from [32]). When a Multiple Discriminant Analysis was used to assign the accessions to the different wild vine populations or to the cultivated group, most plants were correctly assigned and only three wild vines were assigned to the *vinifera* subspecies. On the other hand eight cultivars were assigned to the *sylvestris* subspecies [58]. This seems to corroborate the assumption that the part of the Portuguese germplasm was locally domesticated and contributes to the hypothesis that the Iberian Peninsula has been a secondary center for grapevine domestication [37].

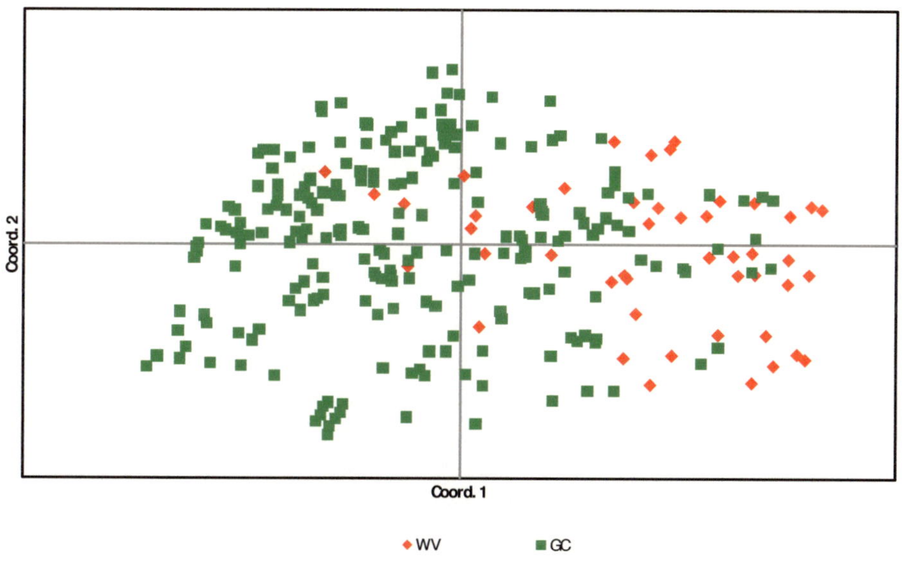

Figure 5. Scatter plot of a Principal Coordinate Analysis of six microsatellite loci from 243 Portuguese grapevine cultivars (GC, in green) and 53 wild vines (WV, in red) from four Portuguese populations.

3. The present situation of germplasm conservation in Portugal

Different strategies are needed to preserve the germplasm of the two grapevine subspecies. One obvious strategy is to maintain the natural habitats where the wild vines are present and keep them subjected to the selection pressures of the natural environment. For the cultivated subspecies the ideal situations should be maintaining the agro-systems where its diversity was buildup. However these *in situ* dynamic strategies must be accompanied by more static *ex situ* strategies, since natural habitats undergo a number of hazards and even the risk of disappearance, and today's commercial agro-systems tend to rely in a very small number of genotypes. Knowledge of the available diversity by multiple tools as reported above is the first step to decide on the strategies of conservation.

In situ conservation of wild vines populations is the leading choice to be considerate. There are a number of different problems that arise from this option: the land ownership where the plants subsist; the legal protection status of the subspecies; natural hazards, like fire; hazards caused by humans, like brutal cleaning of river banks; etc. Most of the populations are located in private owned land even when situated in areas where there is some kind of legal environment protection (populations 02 and 12). The first approach is to contact the land owner and explain the importance of wild vine populations and of the riparian habitats. In Portugal all contacted owners were willing to cooperate in the process of preserving the populations and some were even enthusiastic. Any major occurrence is usually reported like river bank cleaning or fire. Another important action is to contact the municipal authorities responsible for stream cleaning in order to adjust their actions to protect the riparian habitat. A good outcome of this policy was the case when the area where the population 04 inhabits was clean under the supervision of trained staff. Despite the positive results of these approaches some situations prove to be out of hand like the building of a dam, floods and fire. Population 03 was destroyed due to the construction of the Alqueva dam and part of population 12 was uprooted due to severe flooding. Populations 02 suffered a major fire in its habitat although with little loss in the total number of plants that recovered subsequently. To prevent the loss of the existing diversity an *ex situ* collection was started in 2005 at the CAN location (PRT051) with thirty wild vine accessions from three populations. Plants from other populations have been added to this collection.

Even though some European countries like France and Germany have a legal protection status for the subspecies *sylvestris,* in Portugal no such protection exists. An formal requirement was sent to the Portuguese agency for wildlife protection to establish a similar protected status for the Portuguese populations of *Vitis vinifera* subspecies *sylvestris* based on the information described in the previous sections.

Until the middle of the 20th century, most Portuguese farmers used to grow a mixture of vine cultivars as a way to overcome the effects of biotic and abiotic stresses but this situation was became increasingly rare and the vineyards are now mostly monovarietal. Nevertheless a recent report on *in farm* conservation, still found a considerable diversity in cultivated vineyards [59]. This is particularly observed when there is a weak relationship between the owner and the wine market, and a farm agro-ecological heterogeneity [59]. Today worldwide viticulture relies in a very restricted number of cultivars an even in a country like Portugal that has not abandoned its autochthonous cultivars, only 25 cultivars are planted in 80% of the new vineyards. The majority of the ancient cultivars is thus neglected and needs to be preserved *ex situ*.

Ex situ collection of grapevine cultivars were settled initially in the 19th century after the arrival in Europe of *Dactylosphaera vitifoli*ae in order to be post philoxera repositories of local cultivars. Today two types of collections exist in Portugal: typical ampelographic collections (Table 7) and collections with a large number of different accessions of the same cultivar. These later were established as a result of a grapevine selection group network leaded by Antero Martins and today managed by PORVID - a public/private consortium. The methodology used to establish these collections was recently reviewed in [55].

Management	**Owner**	**Coordinates Lat/Long**	**Number of accessions**	**Observations**	**International Code**
INIAV	Public	39° 04′ N 9° 18′ W	754	in renovation	PRT 051
INIAV	Public	38° 41′ N 9° 19′ W	180	duplicate in PRT051	PRT 010
DRAPAlg Tavira	Public	37° 07′ N 7° 39′ W	129 wine 76 table	in renovation	PRT 068
DRAPN Santa Bárbara	Public	41° 10′ N 7° 33′ W	170		PRT 078
DRAPC Nelas	Public	40° 31′ N 7° 51′ W	65		PRT 079
DRAPC Lamaçais	Public	40°18' N 7°23' W	local cultivars		-
DRAPC Anadia	Public	46°26' N 8°26' W	local cultivars		-
DRAPN Sergude	Public	41°22'N 8°10'W	local cultivars		-
JMF, Wine Company	Private	38° 32′ N 8° 58′ W	439		-
ESPORÃO, Wine Company	Private	38° 23′ N 7° 33′ W	180	being installed	-
PORVID	Consortium	38° 38′ N 8° 38′ W	12	each variety with 300 clones	-
UTAD	Public	41° 17′ N 7° 44′ W	local cultivars		-
CVRVV	Public	41°48' N 8°24' W	local cultivars		-

Table 7. National and regional public and private ampelographic collections existing today.

The existing collections continue to perform several functions. These functions were initially related to the characterization and identification of cultivars using classic ampelography including: i) standardization of the morphological descriptors of *Vitis*; ii) morphological description of the cultivars iii) production of illustrate catalogues of cultivars iv) and sorting out synonyms and homonyms. These roles have evolved with the availability of new tools particularly the use of molecular markers that allowed the confirmation of suspected pedigrees and finding unsuspected ones. It also allowed tracing the remote history of grapevine domestication including the existence of several secondary domestication centers. The availability in one location of large number of genotypes of a highly heterozygous species also allow the development of genetic association studies like the one developed by Cardoso [60] that establish a candidate gene association with berry colour and anthocyanin content in 149 red and rose grapevine cultivars. Field performance of large numbers of cultivars in one spot as is the case of Esporão collection (Table 7) will help in the decision of what cultivar to plant and how to develop new wine types on the climate change scenario. Finnaly, morphological, molecular and field performance data will be useful in establishing core collections aiming a better management of the germplasm available.

Acknowledgements

This work was funded by: "Fundação para a Ciência e Tecnologia" (SFRH/BPD/ 74895/2010) and "Ministério da Agricultura, do Mar, do Ambiente e do Ordenamento do Território" (PRODER - Ação 2.2.3.1. - PA 18621).

Author details

Jorge Cunha[1,2], Margarida Teixeira-Santos[3], João Brazão[1], Pedro Fevereiro[2,4] and José Eduardo Eiras-Dias[1]

1 INIAV, Quinta d'Almoinha, Dois Portos, Portugal

2 Universidade Nova de Lisboa, ITQB, Oeiras, Portugal

3 INIAV, Quinta do Marquês, Oeiras, Portugal

4 Universidade de Lisboa, Faculdade de Ciências, Lisboa, Portugal

References

[1] INE. Recenseamento Agrícola 2009. Análise dos principais resultados. Instituto Nacional de Estatística. Lisboa: I.P. Ed.; 2011.

[2] OIV. Statistical report on world vitiviniculture. Paris: International Organisation of Vine and Wine; 2012. http://www.oiv.int/oiv/info/enizmiroivreport (accessed 2 July 2012).

[3] Portaria nº, 428/2000 de 17 de Julho. Diário da Republica. 1ª Série, Nº 163.

[4] Rivera D, Walker MJ. A review of paleobotanical findings of early Vitis in the Mediterranean and on the origin of cultivated grape-vines, with special reference to new pointers to prehistoric explotation in the Western Mediterranean. Rewiev of Paleobotany 1989; 6: 205-237.

[5] Rego PR, Rodriguez AMJ. A palaeocarpological study of Neolithic and Bronze Age levels of the Buraco da Pala rock-shelter (Bragança, Portugal). Vegetation History and Archaeobotany. 1993; 2: 163-172.

[6] Buxó R. The agricultural consequences of colonial contacts on the Iberian Peninsula in the first millennium B.C.. Vegetation History and Archaeobotany 2008; 17: 145–154.

[7] Ghira JC, Carneiro LC, Carvalho HP, Garcia IS, Vinagre JS. Estudo Vitícola e Enológico de Castas Novas da EAN. Lisboa: Ministério da Agricultura, Comércio e Pescas; 1982.

[8] Almadanim MC, Baleiras-Couto MM, Pereira HS, Carneiro LC, Fevereiro P, Eiras-Dias JE, Morais-Cecilio L, Viegas W, Veloso MM. Genetic diversity of the grapevine (Vitis vinifera L.) cultivars most utilized for wine production in Portugal. Vitis 2007; 46: 116-119.

[9] Veloso MM, Almadanim MC, Baleiras-Couto MM, Pereira HS, Carneiro LC, Fevereiro P, Eiras-Dias JE. Microsatellite database of grapevine (Vitis vinifera L.) cultivars used for wine production in Portugal. Ciência e Técnica Vitivinícola / Journal of Viticulture and Enology 2010; 25: 53-61.

[10] Sefc KM, Lopes MS, Lefort F, Botta R, Roubelakis-Angelakis KA, Ibanez J, Pejic I, Wagner HW, Glössl J, Steinkellner H. Microsatellite variability in grapevine cultivars from different European regions and evaluation of assignment testing to assess the geographic origin of cultivars. Theoretical and Applied Genetics 2000; 100: 498-505.

[11] Doligez A, Bouquet A, Danglot Y, Lahogue F, Riaz S, Meredith CP, Edwards KJ, This P. Genetic mapping of grapevine (Vitis vinifera L.) applied to the detection of QTLs for seedlessness and berry weight. Theoretical and Applied Genetics 2002; 105: 780–795.

[12] Schneider A, Carra A, Akkak A, This P, Laucou V, Botta R. Verifying synonymies between grape cultivars from France and northwestern Italy using molecular markers. Vitis 2001; 40: 197-203.

[13] Crespan M. Evidence on the evolution of polymorphism of microsatellite markers in varieties of Vitis vinifera L. Theoretical and Applied Genetics 2004; 108: 231-237.

[14] Sefc KM, Regner F, Turetschek E, Glössl J, Steinkellner H. Identification of microsatellite sequences in Vitis riparia and their applicability for genotyping of different Vitis species. Genome 1999; 42: 1–7.

[15] Aradhya MK, Dangl GS, Prins BH, Boursiquot JM, Walker MA, Meredith CP, Simon CJ. Genetic structure and differentiation in cultivated grape, Vitis vinifera L. Genetical Research 2003; 81: 179–192.

[16] This P, Jung A, Boccacci P, Borrego J, Botta R, Costantini L, Crespan M, Dangl GS, Eisenheld C, Ferreira-Monteiro F, Grando S, Ibañez J, Lacombe T, Laucou V, Magalhães R, Meredith CP, Milani N, Peterlunger E, Regner F, Zulini L, Maul E. Development of a standard set of microsatellite reference alleles for identification of grape cultivars. Theoretical and Applied Genetics 2004; 109: 1448–1458.

[17] Meredith CP, Bowers JE, Riaz S, Handley V, Bandman EB, Dangl GS. The Identity and Parentage of the Variety Known in California as Petite Sirah. American Journal of Enology and Viticulture 1999; 50(3): 236-242.

[18] Vendramin GG, Lelli L, Rossi P, Morgante M. A set of primers for the amplification of 20 chloroplast microsatellites in Pinaceae. Molecular Ecology 1996; 5: 595–598.

[19] McGovern PE. Ancient Wine: The Search for the Origins of Viticulture. New Jersey: Princeton University Press; 2003.

[20] Ocete RR, López Martínez MÁ, Izquierdo MÁP, Moreno TR, Lara BM. Las poblaciones españolas de vid silvestre. Características de un recurso fitogenético a conservar. Madrid: INIA; 1999.

[21] Scossiroli RE. Origine ed evoluzione della vite. Atti dell'Istituto Botànico e del laboratorio Crittogámico dell'Universittà di Pavía. Pavia: 1988; 7: 35-55.

[22] Arnold C, Schnitzler A, Douard A, Peter R, Gillet F. Is there a future for wild grapevine (Vitis vinifera subsp silvestris) in the Rhine Valley?. Biodiversity and Conservation 2005; 14: 1507-1523.

[23] Cunha J, Cunha JP, Lousã M, Eiras-Dias JE. Os bosques ribeirinhos, fonte de diversidade genética de Vitis vinifera L. Ciência e Técnica Vitivinícola / Journal of Viticulture and Enology 2004; 19: 51–59.

[24] Ocete RR. Vitis sylvestris en Iberia. In: J. Böhm (Ed) Atlas das Castas da Península Ibérica: História, Terroir, Ampelografia. Lisboa: Dinalivro; 2011 p96-101.

[25] Moreira I, Saraiva MG, Aguiar F, Costa JC, Duarte, MC, Fabião A, Ferreira T, Loupa Ramos I, Lousã M, Pinto Monteiro F. As Galerias Ribeirinhas na Paisagem Mediterrânica. Reconhecimento na Bacia Hidrográfica do Rio Sado. Lisboa: ISA Press; 1999.

[26] Office International de la Vigne et du Vin. Code des caractères descriptifs des variétés et espèces de Vitis. Paris: O.I.V.; 1983.

[27] GENRES 081. Primary and Secundary descriptor list for grapvine cultivars and species (Vitis L.). Siebeldingen: Institut Fur Rebenzuchtung Geilweilerhof; 1999.

[28] Cunha J, Baleiras-Couto M, Cunha JP, Banza J, Soveral A, Carneiro LC, Eiras-Dias JE. Characterization of Portuguese populations of Vitis vinifera ssp. sylvestris (Gmelin) Hegi. Genetic Resources and Crop Evolution 2007; 54: 981–988.

[29] Cunha J, Santos MT, Carneiro LC, Fevereiro P, Eiras-Dias JE. Portuguese traditional grapevine cultivars and wild vines (Vitis vinifera L.) share morphological and genetic traits. Genetic Resources and Crop Evolution 2009; 56: 975-989.

[30] Stummer A Zur urgeschichte der Rede und des Weinbaues. Mitteilungen der Anthropologischen Gesellschaft in Wien 1911; 41:283-296.

[31] Office International de la Vigne et du Vin. 2ND Edition of the OIV descriptor list for grape varieties and Vitis species. Paris: O.I.V.; 2007.

[32] Cunha J, Teixeira Santos M, Veloso MM, Carneiro LC, Eiras-Dias JE, Fevereiro P. The Portuguese Vitis vinifera L. germplasm: genetic relations between wild and cultivated vines. Ciência e Técnica Vitivinícola / Journal of Viticulture and Enology 2010; 25 (1): 25-36.

[33] Lopes MS, Mendonça D, Rodrigues dos Santos M, Eiras-Dias JE, Câmara Machado Ad. New insights on the genetic basis of Portuguese grapevine and on grapevine domestication. Genome 2009; 52: 790-800.

[34] De Andrés MT, Benito A, Pérez-Rivera G, Ocete R, Lopez MA, Gaforio L, Muñoz G, Cabello F, Martínez Zapater JM, Arroyo-García R. Genetic diversity of wild grapevine populations in Spain and their genetic relationships with cultivated grapevines. Molecular Ecology 2011; 21: 800-816.

[35] Imazio S, Labra M, Grassi F, Scienza A, Failla O. Chloroplast microsatellites to investigate the origin of grapevine. Genetic Resources and Crop Evolution 2006; 10: 1–9.

[36] Grassi F, Labra M, Imazio S, Ocete Rubio R, Failla O, Scienza A, Sala F. Phylogeographical structure and conservation genetics of wild grapevine. Conservation Genetics 2006; 7: 837–845.

[37] Arroyo-García R, Ruiz-García L, Bolling L, Ocete R, López MA, Arnold C, Ergul A, Söylemezo"lu G, Uzun HI, Cabello F, Ibáñez J, Aradhya MK, Atanassov A, Atanassov I, Balint S, Cenis JL, Costantini L, Gorislavets S, Grando MS, Klein BY, Mcgovern PE, Merdinoglu D, Pejic I, Pelsy F, Primikirios N, Risovannaya V, Roubelakis-Angelakis KA, Snoussi H, Sotiri P, Tamhankar S, This P, Troshin L, Malpica JM, Lefort F, Martinez-Zapater JM. Multiple origins of cultivated grapevine (Vitis vinifera L. ssp. sativa) based on chloroplast DNA polymorphisms. Molecular Ecology 2006; 15: 3707–3714.

[38] De Mattia F, Imazio S, Grassi F, Baneh HD, Scienza A, Labra M. Study of Genetic Relationships Between Wild and Domesticated Grapevine Distributed from Middle East Regions to European Countries. Rendiconti Lincei 2008; 19(3): 223-240.

[39] Almeida CR. Catálogo das Castas. Região Demarcada da Bairrada. Lisboa: Instituto de Gestão e Estruturação Fundiária, Direcção Regional de Agricultura da Beira Litoral; 1986.

[40] Antunes AA, Costa JF. Catálogo das Castas. Região de Pinhel. Lisboa: Instituto de Gestão e Estruturação Fundiária, Direcção Regional de Agricultura da Beira Interior; 1986.

[41] Banza JP. Catálogo das Castas. Região do Alentejo. Lisboa: Instituto de Gestão e Estruturação Fundiária, Direcção Regional de Agricultura do Alentejo; 1986.

[42] Faustino RA. Catálogo das Castas. Região Demarcada do Algarve. Lisboa: Instituto de Gestão e Estruturação Fundiária, Direcção Regional de Agricultura do Algarve; 1986.

[43] Mota MT, Silva MF. Catálogo das Castas. Região Demarcada dos Vinhos Verdes. Lisboa: Instituto de Gestão e Estruturação Fundiária, Comissão de Viticultura da Região dos Vinhos Verdes; 1986.

[44] Pereira CD, Duarte AP. Catálogo das Castas. Região Demarcada do Dão. Lisboa: Instituto de Gestão e Estruturação Fundiária; 1986.

[45] Pereira CD, Sousa AC. Catálogo das Castas. Região Demarcada do Douro. Lisboa: Instituto da Vinha e do Vinho, Centro de Estudos Vitivinicolas do Douro; 1986.

[46] Duarte MTT, Eiras-Dias JE. Catálogo de Porta-enxertos mais utilizados em Portugal. Instituto da Vinha e do Vinho, Centro Nacional de Produção Agricola, Estação Vitivinicola Nacional; 1990.

[47] Vaz JT Catálogo de Castas. Uvas de Mesa cultivadas em Portugal. Instituto de Gestão e Estruturação Fundiária, Direcção-Geral do Planeamento e Agricultura; 1987.

[48] Eiras-Dias JE, Pereira CA, Cunha JP. Catálogo das Castas. Região do Ribatejo, Oeste e Península de Setúbal. Lisboa: Instituto da Vinha e do Vinho, Estacão Vitivinícola Nacional; 1988.

[49] Rocha ML, Barão AG, Martins JM. Catálogo de Novas Castas de Uva de Mesa obtidas na Estação Agronómica Nacional. Lisboa: Instituto da Vinha e do Vinho, Estação Agronómica Nacional; 1990.

[50] Santos MT, Brazão J, Cunha J, Eiras-Dias JE. Renewing and enlarging and the Portuguese Ampelographic Collection: screening for nine viruses by Elisa. Proceedings of the 17th Congress of the International Council for the Study of Virus and Virus-like Diseases of the Grapevine (ICVG), Davis, California, USA, October 7-14, 2012; 272-273.

[51] Lopes MS, Sefc KM, Eiras Dias E, Steinkellner H, Laimer da Câmara Machado M, Câmara Machado A. The use of microsatellites for germplasm management in a Portuguese grapevine collection. Theoretical and Applied Genetics 1999; 99: 733–739.

[52] Lopes MS, Santos MR, Eiras Dias JE, Mendonça D, Câmara Machado A. Discrimination of Portuguese grapevines based on microsatellite markers. Journal of Biotechnology 2006; 127: 34–44.

[53] Magalhães R, Faria MA, Maria dos Santos NM, Eiras-Dias JE, Magalhães N, Meredith CP, Ferreira Monteiro F. Verifying the Identity and Parentage of Cruzado de Rabo de Ovelha with Microsatellite Markers. American Journal of Enology and Viticulture 2003; 54: 56-58.

[54] Martins A. Variabilidade genética intravarietal das castas. In: J. Böhm (Ed.) Portugal vitícola, o grande livro das castas. Lisboa: Chaves Ferreira Publicações; 2007 p53-56.

[55] Gonçalves E, Martins A 2012. Genetic Variability Evaluation and Selection in Ancient Grapevine Varieties. In: Ibrokhim Y. Abdurakhmonov (Ed.) Plant Breeding. Rijeka: In Tech; 2012 p333-352.

[56] Le Cunff L, Fournier-Level A, Laucou V, Vezzulli S, Lacombe T, Adam-Blondon AF, Boursiquot JM, This P. Construction of nested genetic core collections to optimize the exploitation of natural diversity in Vitis vinifera L. subsp. sativa. BMC Plant Biology 2008; 8:31.

[57] Peakall R, Smouse PE. GENALEX 6: genetic analysis in Excel. Population genetic software for teaching and research. Molecular Ecology Notes 2006; 6: 288–295.

[58] Cunha J. Biological diversity of Vitis vinifera L. in Portugal: the genetic contribution of subsp. sylvestris to the origin of the Portuguese grapevine cultivars (subsp. vinifera). PhD thesis. Universidade Nova de Lisboa (Instituto de Tecnologia Química e Biológica) Oeiras; 2009.

[59] Cardoso S, Maxted Nigel. Regional and Crop-Specific Survey: Grapevine Landraces in Douro and Colares, Portugal. In: Veteläinen M, Negri V and Maxted N. 2009. European landraces onfarm conservation, management and use. Bioversity Technical Bulletin nº. 15. Rome: Bioversity International; 2009 p203-222.

[60] Cardoso SC. Genetics of berry colour and anthocyanin content variation in grapevine (Vitis vinifera L. subsp. vinifera). PhD thesis. Universidade Nova de Lisboa (Instituto de Tecnologia Química e Biológica) Oeiras; 2011.

Chapter 7

Genetic and Phenotypic Diversity and Relations Between Grapevine Varieties: Slovenian Germplasm

Denis Rusjan

Additional information is available at the end of the chapter

http://dx.doi.org/10.5772/51773

1. Introduction

Slovenia is a small central European country situated between the Alps and the Adriatic Sea, at the crossing between Italy (to the West), Austria (to the North), Hungary (to the East) and Croatia (to the South), a historical place of significant grape and wine trading. Grapevines have been cultivated in Slovenia since ancient times, although the first literal proofs on grape growing and winemaking date back merely to the Austro-Hungarian Empire [1,2]. However, the oldest *Vitis vinifera* ssp. L. (grape) pips (seeds) found during the archaeological excavation of the late Neolithic (Copper Age) pile-dwelling settlement of Hočevarica at the Ljubljansko barje moor and date back to the 37th/36th century B.C. [3].

The geographical position of the country as well as the regarding climate conditions, and the socio-political development throughout the nation's history, have contributed to a diverse assortment of grapevine [1,2]. At least around 100 old and less known varieties are enumerated especially in the western part of Slovenia (Sub Mediterranean) and accompanied by the widespread European, allochthon, autochthon/landrace and local varieties, a large number of grapevines is cultivated in Slovenia nowadays [4,5].

The intense trade in grapes and wine, especially due to the Venetians (Venetian Republic from the end of 7th to the end of 18th Century; Figure 1), and the varying climatic and geological conditions contributed to a great diversity of *Vitis vinifera* L. varieties. Additionally, the multiculturalism and multilingualism of the area as well as the turbulent historical events have contributed to identical genotypes having different names [1,2].

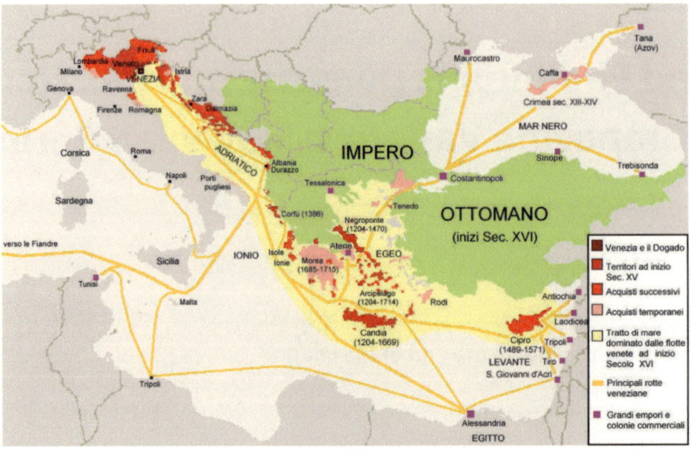

Figure 1. Territories of the Republic of Venice (697–1797). (http://en.wikipedia.org/wiki/File:Repubblica_di_Venezia.png)

Nowadays, Slovenian winegrowers (27.802) produce on 15.973 ha an annual of around 54.3 mio litres of wine; 62% whites and 38% reds. Fifty grapevine varieties (*Vitis vinifera* L.), international, allochthon, local and autochthon as well, are registered in the official Slovenian varietal list, but numerous accessions are still observed and non-descripted in many parts of Slovenia [1,6]. Although some of them form the basis of the renown regional varieties ('Rebula' in Goriška brda, 'Zelen' and 'Pinela' in Vipavska dolina, 'Šipon' in Ljutomer-Ormož etc.), many face extinction, since in some areas only a few plants are reported. However, Slovenia is facing the threat of the rapid erosion of the native/local germplasm on account of the introduction of foreign varieties, such as 'Merlot', 'Cabernet', 'Chardonnay' and 'Pinot'. Very often, however, the problems with the identification of these local, unknown varieties, occur, because of the presence of various synonyms of the same variety throughout the country [2,6,7].The gene pool of the varieties cultivated in Slovenia is composed of old allochthon, autochthonous, domestic/local varieties, which can all be additionally divided into two groups. The first group consists of commercially used varieties which are planted on more than 100 ha each, such as 'Rebula', 'Žametovka', 'Zelen', 'Pinela', 'Šipon', 'Radgonska Ranina', and 'Refošk'. The second group of varieties, such as 'Vitovska grganja' and 'Belina', are cultivated on less than 100 ha each and are known as local varieties. Moreover, around 50 well-known and rare varieties/accessions exist in Slovenia today; the majority of which have not yet been listed in the International List of Vine Varieties and Their Synonyms of the O.I.V. (International Organisation of Vine and Wine). Even though some morphological and agronomical descriptions related to these accessions exist, we are talking about mostly unexplored plant material and some of these varieties had survived only in less productive vineyards or owing to germplasm collections. The systematically collection of these endangered accessions started in 1980. Nowadays, there are five grapevine collections in the whole country: Slap at Vipava, Ampelografski vrt in Kromberk, the grapevine collection at Dobrovo in Goriška brda, Meranovo near Maribor and one near Ormož that include old varieties, clonal candidates and clones (Table1).

N	Variety	Synonym/Original name	Clone Code
1	'Barbera'		SI-36
2	'Beli pinot'	Pinot blanc	SI-19; SI-20
3	'Chardonnay'		SI-21; SI-39; SI-40
4	'Laški rizling'	Welschriesling, Graševina	SI-11; SI-12; SI-13; SI-41
5	'Istrska malvazija'	Malvasia d'Istria, Malvasia istriana	SI-37
6	'Pinela'	/	SI-28
7	'Ranina'	Bouvier Traube, Muscat de Saumur	SI-4; SI-5; SI-6; SI-7
8	'Ranfol'	Štajerska belina	SI-38
9	'Rebula'	Ribolla gialla	SI-30; SI-31; SI-32; SI-33; SI-34
10	'Refošk'	Refosco, Teran, Refosco d'Istria	SI-35
11	'Renski rizing'	Riesling, Rheinriesling	SI-22; SI-23; SI-24
12	'Sauvignon'	Sauvignon blanc	SI-1; SI-2; SI-3
13	'Šipon'	Furmint, Moslavac	SI-14; SI-15; SI-16; SI-17; SI-18
14	'Traminec'	Traminer	SI-8
15	'Zelen'	/	SI-26
16	'Žametovka'	Köllner blauer, Kavčina, Žametna črnina	SI-25

Table 1. List of varietal clones and clone candidates selected in Slovenia.

Nowadays, varieties can be characterized by several methods: (i) by means of a morphological description of plant parts at different phenological stages; (ii) morphometry based on the measurements of the parameters of the plant organs and (iii) quantitative or qualitative analysis of biochemical compounds. Furthermore, traditional methods of varietal description based on vegetative and reproductive (ampelography) parts of plants, contributed greatly to the full description of the identities and relationships among *V. vinifera* L. varieties; what also suggests that ampelographic characterization according to the characters put forward by the O.I.V. is the first step in the examination of grapevine varieties/accessions. Several authors described various analyses of primary and secondary metabolites and methods based on DNA polymorphism as outstandingly useful methods to complete the morphological identification of grapevine varieties [8,9].

2. Slovenian germplasm

One of the problems in the management of these germplasm collections is the use of synonymic and homonymic designations. The lack of order caused by synonyms and homonyms is caused by inadequate documentation and poor preservation of historical facts related to grape growing and trade. The identification and comparison of plant material by ampelographic methods often results in misinterpretations [10]. In contrast, DNA-based markers are independent of environmental factors and are therefore more appropriate for varietal identification [9,11]. In the last decade, more than 60 SSR primers from the genomic libraries of *Vitis vinifera* L. have been developed and used for identification purposes [12,13]. The genetic characterization of grapevine varieties (*Vitis vinifera* L.) using microsatellite markers and ampelometric methods has been done by many countries and regions already. The integration of the resulting molecular analyses and detection of synonymous grapevine varieties has already been performed among countries which share a common grapevine assortment, such as Croatia, North Italy, Austria, Germany, France, Spain, Portugal, Greece, etc. [5,9,14,15]. In the last decade many genetic studies of local varieties have been conducted in Slovenia [4,5,15-18] as well, and have consequently revealed the genetic biodiversity of grapevine varieties grown and cultivated in the country (Table 2).

Variety / Accession	Synonym	Homonym	Origin
Bela glera	/	Prosseco, Briška glera	G
Beli teran	/	Vitovska grganja, Vitouska	G
Belina	Heunish		L
Bianchera	Erbaluce, Albaluce, Albalucent, Bianco Rusti, Erbalus, Erbalucente, Uva Rustia		Al
Borgonja bela		Istrska malvazija	Al
Borgonja rdeča	Burgonja istarska, Gamay Beaujolais, Borgogna	Cipro	Al
Briška glera	Glera	Bela glera, Glera	G
Cipro	Muškat ruža Porečki, Muškat ruža omiški, Moscato rosa, Rosenmuskateller blauer, Likvor, Rdeča muškateljka	Borgonja rdeča	Al
Cividin	Cividino bianco		Al
Cohovka	/		A
Danijela	/		G
Dolga petlja	/		G
Drenik	/		G
Duranja	/		Al
Glera	Prosecco, Prosekar	/	Al
Grganc			G
Guštana	Auguštana		G
Istrijanka			
Istrska malvazija	Malvazija, Malvasia d'Istria, Istarska malvasia, Istrijanka, Malvasia Istra, Borgonja bela	/	L
Kanarjola		Canaiolo	Al
Klarnica	Klarnca, Klarna		A
Laščina			G
Maločrn	Piccola nera, Negra Tenera, Petite Raisin		Al

Variety / Accession	Synonym	Homonym	Origin
Medena glera	/	Glera	G
Pagadebiti	Curzola, Plavina		Al
Pergolin	/		Au
Pinela	/		Au
Planinka	/		G
Plavina	Brajdica, Curzola, Pagadebiti	Plavac mali	Al
Pokalca	Rdeča rebula, Pocalza, Ribolla Nera, Schioccoletto, Schiopetino, Schioppettino, Scoppiettino		Au
Pokov zelen	/		G
Poljšakica	/		Au
Pregarc	/		L
Prosecco	Ghera, Glera, Grappolo Spargolo, Prosecco Balbi, Prosecco Bianco, Prosecco Tondo, Proseko, Sciorina or Serprina	Briška Glera, Števerjena	Al
Racuk	/		G
Ranfol	Štajerska belina, Štajerka, Urbanka, Vrbanka, Sremska lipovina, Svetla belina, Heunisch		Al
Rebula	Garganja, Ribolla gialla		L
Rečigla	/		G
Refošk	Teran, Refosco, Teranovka, Terrano, Refosco peduncolo rosso		L
Rožica	Rožca		G
Sladkočrn	/		Au
Števerjana	/		G
Teran	Terrano, Refošk	/	G, L
Teran Istra	Terrano, Refošk		G, L
Trevolina			Al
Vitovska grganja	Vitovska, Grganja, Beli Teran	Vitouska, Garganja, Beli Refošk, Racuk	Au
Volovnik	Volovna		Al, G
Vrtovka	/		G
Zelen	/		Au
Zelenika	/		G
Zunek	Cunek		G

Table 2. Names, designations and classifications of the less known cultivated grapevine varieties/accessions (*Vitis vinifera* L.) in Slovenia. Legend: Al – allochthon; Au – autochthon (landrace); L – local (domestic); G – germplasm (grown only in certain gene banks).

2.1. Genetic diversity and relations among varieties

SSR markers offer some advantages over other molecular markers, including their co-dominant inheritance, hypervariability, and high cross-species transferability [8,13]. Consequently a large number of markers has been developed for characterisation of grapevine by many research groups [8,12,19] – these markers have provided a very useful and convenient tool for analysing genetic diversity of grapevine. In order to efficiently manage these conserved local germplasm resources and to understand the genetic relationships among them, it is necessary to characterize the genetic diversity existing in the Slovenian collection and production vineyards. However, a majority of cultivated and grown grapevine accessions in the Sub-Mediterranean part of Slovenia and a subset of 6 widespread European varieties taken as reference have been genotyped with microsatellite loci in order to: (1) identify and/or differentiate varieties, especially those of similar morphological characteristics; (2) assess genetic diversity and relationships among them; (3) and compare these varieties to their synonyms [5,15-18] (Figures 2 and 3).

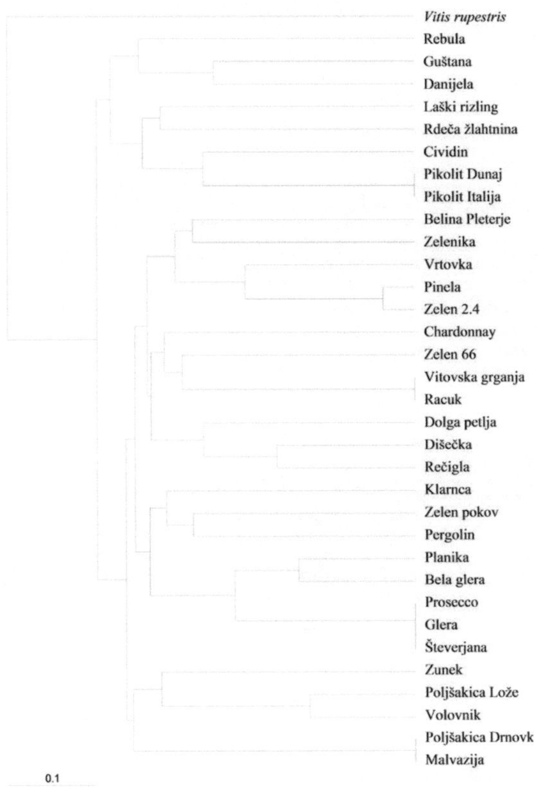

Figure 2. Dendrograms of genotyped grapevine accessions grown in Slovenia [5].

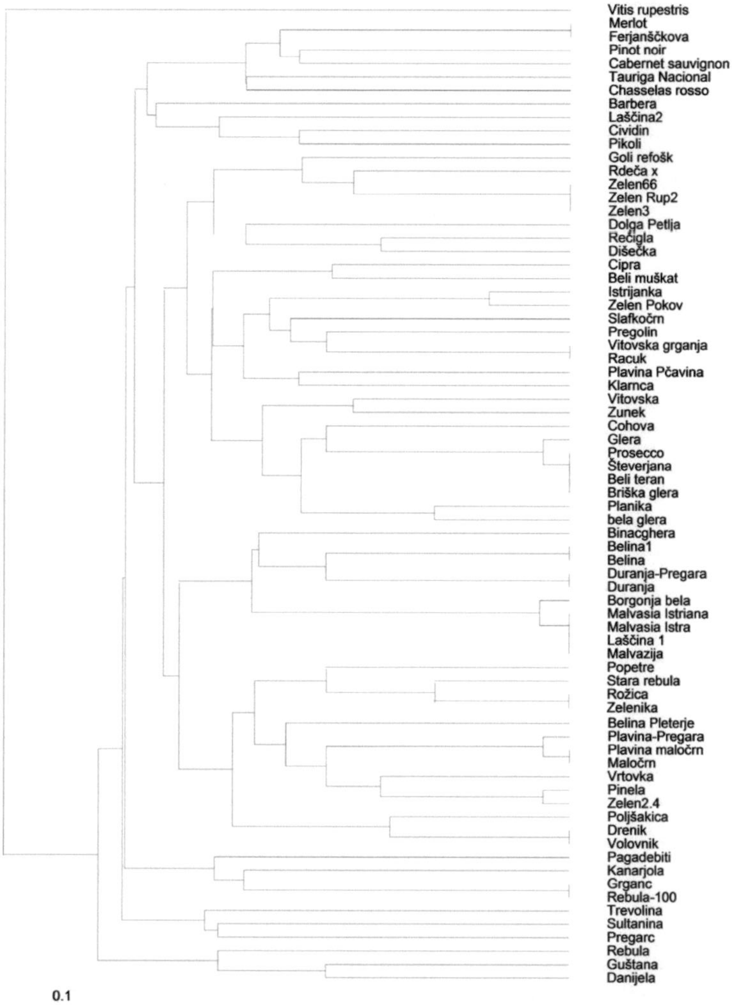

Figure 3. Dendrograms of genotyped grapevine accessions grown in Slovenia [17].

SSR profiles of groups of varieties with similar names but some different morphological characteristics were compared in order to assess their relationships and resolve existing doubts on their identity. In order to illustrate the population structure among Slovenian varieties/accessions dendrograms were constructed, which classify the varieties according to the proportion of shared alleles (Figures 2 and 3). The average similarity of all varieties is 34% of shared alleles (Figure 2, left), which is close to the average similarity observed for mid-European and Portuguese accessions [5]. Overall, two distinct clusters were obtained, with many sub-clusters. The accessions from the first cluster are related to 'Laški rizling' ('Welschriesling'), 'Rdeča žlahtnina' ('Roter Gutedel', 'Chasselas red') from the West European gene pool (*Proles occidentalis*), but also 'Rebula' ('Ribolla') known as *Proles pontica*. Another two varieties, 'Guštana' and 'Danijela', grouped with the first cluster, are both characterised by an early grape ripening. The name 'Guštana' may describe the ripening time of this variety, which is in August. The variety 'Danijela' is planted only in gene bank vineyards and its provenance is still uncertain. The variety 'Picolit' is cultivated in the western part of Slovenia (Northern Primorska) and in the north-east part of Italy – in Slovenia two distinct, morphologically different types of this variety/accession, 'Picolit Italia' and 'Picolit Vienna', are known. Microsatellite markers revealed no differences at 21 loci; matching also the same allelic profiles at all 7 SSR loci with a variety 'Picolit' from Italy [25]. In the second cluster, which includes 25 accessions, three groups of synonyms were discovered [5]: 'Vitovska grganja' = 'Racuk', 'Poljšakica Drnovk' = 'Istrska Malvazija' and 'Prosecco'= 'Briška Glera' = 'Števerjana'. 'Vitovska grganja' is an old grape variety cultivated in Slovenia in the winegrowing districts of the Vipavska dolina and Kras and also in the north-east part of Italy, where it played an important role in the past [1]. The synonymy found between 'Vitovska grganja' and 'Racuk' could not yet be confirmed despite obtaining identical allelic patterns, because an accurate morphological characterization of 'Racuk' is still lacking. The allelic profiles of our 'Vitovska grganja' have been further compared with the Italian variety 'Vitouska', recently published by [26] - dissimilarity was revealed at 14 out of 16 loci, and indicating that there is a compatible parent/progeny. The varieties 'Prosecco', 'Briška Glera' and 'Števerjana' form another group of synonyms. 'Prosecco' and 'Glera' have already been proved to be synonyms on the basis of morphological descriptors and isoenzyme analyses, while 'Števerjana' has not previously been considered to be a synonym. The variety 'Prosecco' takes its name from the village of Prosecco, in the Province of Trieste, where this variety is also known as 'Glera' [26]. The name for the variety 'Števerjana' may originate in Števerjan, a small village in North-East Italy (Collio), near the Slovenian border. The comparison of 16 SSR loci of our 'Prosecco' with Italian 'Prosecco tondo', which was recently analysed [26], revealed no differences - on the basis of this comparison 'Prosecco' = 'Prosecco tondo' = 'Glera' but according to [26] 'Glera' is related mainly to 'Prosecco lungo' and less frequently to 'Prosecco tondo' [5].

Comparison between the two accessions denominated 'Glera' ('Briška Glera' and 'Bela Glera') included in our analysis [5] revealed differences at 16 out of 21 loci, so they are considered homonyms. Two varieties, 'Poljšakica Drnovk' and 'Poljšakica Lože', which were expected to have the same genetic profile, were different at various analysed microsatellite loci. The synonymy of 'Heunisch' = 'Ranfol' = 'Belina', which was first mentioned by [2], was also analysed - discrepancies were discovered at 13 loci, but the two varieties share one common allele at all 13 loci [5]. Furthermore, the accession 'Belina Pleterje' was compared to the synonymic variety 'Ranfol Bijeli' from Croatia using microsatellite data [14] - 8 compared SSR loci data showed the same allelic profiles for all, except VVMD7, where a triallelic profile was observed. A mutation in the microsatellite sequence of locus VVMD7 was often responsible for identifying grapevine synonyms or related types [19].The SSR profiles of accessions linked to the designation 'Zelen' ('Zelen Pokov', 'Zelen 66' and 'Zelen 2.4') were compared as well. The differences at several microsatellite loci showed that 'Zelen' accessions are a heterogeneous group consisting of several genotypes. The differences were revealed also at the comparison of 'Zelen' to 'Verduzzo' and 'Verdicchio'andcan therefore not be considered/used synonyms.In pairwise comparison, excluding *Vitis rupestris* L., the greatest distances detected (86%) were between accesions 'Rebula' – 'Volovnik', 'Dolga petlja' – 'Guštana' and 'Klarnica' – 'Pikolit' (Figure 2).

Subsequent, more detailed study was done in 2011 [17], which included all accessions founded in the Sub-Mediterranean part of Slovenia. The observations are reported as follows [17] (Figure 3): The pairs/groups of vines denominated 'Duranja' and 'Duranja-Pregara', furthermore 'Belina' and 'Belina1', 'Zelen Rup2', Zelen66' and 'Zelen3' and 'Malvasia', 'Istrska Malvasia' and 'Malvasia Istriana' revealed identical genotypes in all SSR microsatellites analysed therefore they are regarded as synonyms. The designations related to 'Malvasia' very often comprise several types of grapevine varieties which are also morphologicaly heterogenous [20]. Very closely related to 'Malvasia Istra' was 'Borgonja bela', sharing 19 out of 20 alleles, what was also expected according to known ampelographic characteristics. Unexpectedly, vine denominated 'Laščina 1' also showed identical SSR profiles as 'Malvasia Istra' but completely different than accession 'Laščina 2', what is considered as a designation error or homonym. In the group of variety 'Plavina', the vines signed as 'Plavina maločrn' and 'Maločrn' showed same genotyping results and they are closely related to vine denominated 'Plavina Pregara', what cannot be affirmed for accession signed as 'Plavina-Pčavina'. 'Plavina-Pregara' in contrast to 'Plavina maločrn' and 'Maločrn' resulted in homozygous state (or heterozygous with a null allele) at locus VVMD7, but 'Plavina-Pčavina' was different from them in 8 loci out of 10 [17]. The difference in the stage of homozygosity/heterozygosity at locus VVMD7 was also obtained between accessions 'Glera' and 'Briška Glera' and these kind of mutations at locus VVMD7 have previously often been reported [5,9,21]. These results of close relatedness suggested that local varieties can be phenotypically partly different types or under-types and that one could have originated from the other through somatic mutations. Low relatedness (20% of similarity) was also found between Slovene 'Plavina' and Croatian 'Plavina' genotyped by [20]. 'Plavina' was reported the most common grape variety along the Dalmatian coast and was known as 'Pagadebiti' in 'Curzola', however in our study the vine denominated 'Pagadebiti' is more similar to variety 'Rebula' ('Ribolla') and 'Kanarjola' than to 'Plavina'; as pairwaise comparison revealed only 6% of genetic similarity. The obtained allelic profiles of 'Pagadebiti' are also different from the profiles of two 'Pagadebiti' accessions reported by [21] - 'Pagadebiti' translates as "pay the debt" and so the name may had been used for several very productive grapevine varieties, resulting in many homonyms.

To the group of synonyms related to the designation 'Glera' ('Glera', 'Prosecco', 'Briška Glera', 'Števerjana', 'Beli teran') the Croatian variety 'Teran bijeli' genotyped by [14] was compared. The comparison of data of 7 SSR loci resulted in identical SSR profile confirming their synonymy. A high genetic difference was revealed between the accessions denominated as 'Glera'/'Briška glera' and 'Bela glera' and could be explained with the fact that the term "Glera" was quite frequently used for white grapevine varieties in the Sub-Mediterranean part of Slovenia in the past [1,24], therefore the obtained homonymy is not a surprise. By comparing allele sizes of 7 SSR loci of the Croatian 'Muškat ruža Porečki' [14] with the Slovene 'Cipro', identical genotypes were found and therefore synonymy between these two varieties was confirmed. On the other hand, 'Beli muškat', which is clustered together with 'Cipro' in the dendrogram, is related to them with a similarity of 60%.

Unexpected the varieties 'Rožca' and 'Zelenika' showed the same genotype, what can be explained by the dissemination of scions of the same varieties throughout the winegrowing regions during the past, where the influence of dialects has also put its mark on the denomination. The variety 'Rožca' is found quite close to another white accession signed as 'Rebula stara'; they share 77% of alleles. Vines denominated 'Rebula stara' and 'Rebula-100' were expected to be closely related but share only 55% of alleles; in common at least one allele at each locus, which could imply their parent/offspring relation.A genotype of 'Rebula stara' (='Stara Rebula'='Rebula old') perfectly matched at 7 loci with variety 'Gouais blanc' ('Heunisch'). Another interesting finding was the similarity between the vines 'Grganc' and 'Rebula-100', which showed similar genotypes with exception at locus VVMD5 at which 'Rebula-100' revealed a triallelic pattern [17].

2.1.1. Grapevine accessions designated 'Rebula' and 'Vitovska'

2.1.1.1. Accessions related to designations 'Rebula' and 'Ribolla'

'Rebula' (*Vitis vinifera* L.) has been one of the most important white grapevine varieties from the ancient times down to the present days and has been mostly cultivated in the area of Goriška brda and Collio where it is still gaining in importance. The first mentions of the name 'Rebula' date back to 1299, later 1376,to a deed of sale »Notariorum Joppi« in the area of the Slovenian Collio [27], whereas the first ampelographic descriptions of the variety are found in Vinoreja [1] and in Ampelographie [2]. Six different types of the variety 'Rebula' are enumerated and described ("green rebula", "yellow rebula", "less fertile rebula", "unfertile rebula", "crazy rebula" and "rebula with smaller and dissected leaf") [1] what confirms its biodiversity which decreases with narrow clonal selection. For this variety many synonyms are used - 'Garganja', 'Glera', 'Ribolla gialla', 'Rebolla', 'Ribolla', 'Ribolla Bianca', 'Ribuèle', 'Rabuèle', 'Rosazzo', 'Ribollat', 'Raibola', 'Ràbola', 'Ribuole' and 'Gargania' [6,7,28]. DNA analyses regarding accessions of 'Rebula' have been already reported [15]. The variety 'Rebula' showed a very low similarity (16 %) with other analysed varieties, also with vines designed 'Rebula-100 years' and 'Rebula-old'. The accessions named 'Prosecco', 'Števerjana', 'Beli teran' and 'Briška Glera', however, revealed identical genotypes in all 11 SSR microsatellites analysed, and are therefore regarded as synonyms. The varieties 'Rebula' and 'Ribolla gialla' revealed an identical SSR profile at 8 out of 9 SSR loci (Figure 4).

Moreover, the genetic identity and relationship among 'Ribolla gialla', 'Rebula' and 'Robola' that have been traditionally cultivated in other European Countries (Goriška brda in Slovenia and Kefalonia Island in Greece) grown in private vineyards and in grape collections were additionally studied (Table 3)[29]. For this purpose, 35 SSR loci were analysed to fingerprint 19 accessions uniformly grown in the three cultivation areas (Friulian Collio, Goriška brda and Ionian Islands). 'Ribolla Gialla' and 'Rebula' accessions revealed identical genotype in all 35 analysed SSR markers, therefore are regarded as synonymous. Data proved the existence of full-sibling relationships between 'Ribolla gialla' and 'Robola' (Figure 5) [29].

Accession	Origin
'Bela Glera 5/6'	Ampelographical collection - University of Ljubljana – Slovenia
'Bela Glera 8/1'	
'Glera 1'	
'Glera 2'	
'Glera Medena'	
'Rebula 1'(clone B5)	Collection NEBLO – Slovenia
'Rebula 2'	Ampelographical collection – University of Ljubljana – Slovenia
'Ribolla 1'	Breeding ground GENRRJEVO of RADIKON STANISLAO – Italy
'Ribolla 2'	
'Ribolla 3'	Breeding ground AZ of RADIKON STANISLAO – Italy
'Ribolla 4'	Breeding ground CENO of RADIKON STANISLAO – Italy
'Ribolla 5'	Breeding ground of RADIKON STANISLAO – Italy
'Robola 1'	Ampelographical collection - University of Tessaloniki – Greece
'Robola 2'	Kefalonia – Greece
'Robola 3'	
'Robola 4'	
'Robola 5'	
'Robola 6'	
'Robola 7'	

Table 3. List of analyzed accessions and origin area of plant materials related to the designations 'Rebula', 'Ribolla' [29].

At least 9 distinct genotypes resulted from the 19 analysed accessions [29], due to several cases of detected synonyms (Figure 5) – for example the same genotype was founded among vines designated 'Robola 1 and 'Robola 2', than between 'Robola 6' and 'Robola 7', moreover among five Italian vines designated 'Ribolla' accessions (1-5) and three Slovenian accessions ('Glera 2', 'Rebula 1' and 'Rebula 2'), 'Bela Glera 5/6' with 'Glera 1'. Besides, several cases of homonyms revealed, two Slovenian accessions 'Bela Glera 5/6' and 'Bela Glera 8/1' showed different allelic profiles, sharing 48 % of alleles, so did the accessions 'Glera 1' and 'Glera 2'(34% of shared alleles). Five cases of homonyms of the group of Greek accessions were revealed [29]: 'Robola 1', 'Robola 2' and 'Robola 6', 'Robola 7' accessions showed allelic profiles different from other 'Robola' accessions. Vine designated 'Robola 1' and 'Robola 2' shared 81 % of alleles with 'Ribolla Gialla', 'Rebula' and 'Glera 2' accessions. In addition[29] report the SSR profiles of five accessions resulted to be unique: 'Bela Glera 8/1', 'Glera Medena', 'Robola 3', 'Robola 4' and 'Robola 5'. After the accurate consideration of the obtained results, the studied accessions are clustered into three major groups, but 'Glera Medena' and 'Robola 3' are presented as isolated samples. The cluster 2 presents a major number of analysed vines, what reveals high homogeneity of the vines cultivated in Italy, as well as in the comparison with vines cultivated in Slovenia under the name 'Rebula' and those from Greece named 'Robola'. All these designations from cluster 2 can be perceived as synonyms;

therefore the original name of this variety is still uncertain. It is demonstrated, that the groups of Italian and Slovenian vine from cluster 2 share 90% of alleles, suggesting the existence of a common ancestor[29]. In the case of the other Greek accessions the situation seems to be more complicated and variability is higher – the largest number of accessions is grouped in cluster 1 ('Robola 4', 'Robola 5', 'Robola 6'/'Robola 7' accessions) as they share about 75% of alleles, what suggests very close relationships. Grouped in cluster 3, we find the accessions' Bela Glera 8/1', 'Glera 1' and 'Bela Glera 5/6', which show a similarity of 68% in comparison with cluster 2 (Figure 5). No case of parent-offspring (PO) relationships was revealed among analysed accessions. True-to-type 'Rebula' and 'Ribolla Gialla' (cluster 2) showed possible full-sibling (FS) relationships with 'Robola 1'/'Robola 2' accessions (same cluster), but shared only 55/70 alleles. Among the Greek accessions, 'Robola 4', 'Robola 5' and 'Robola 6'/'Robola 7', belonging to a different cluster are related by a FS relationship; 'Robola 4' and 'Robola 5' accessions sharing 68 % of alleles, 'Robola 4' and 'Robola 6'/'Robola 7' sharing 80 % of alleles and 'Robola 5' and 'Robola 6'/'Robola 7' sharing 78 % of alleles. Accessions 'Bela Glera 5/6', 'Bela Glera 8/1' and 'Glera 1' shared 48 % of alleles and show a probable HS relationship [29].

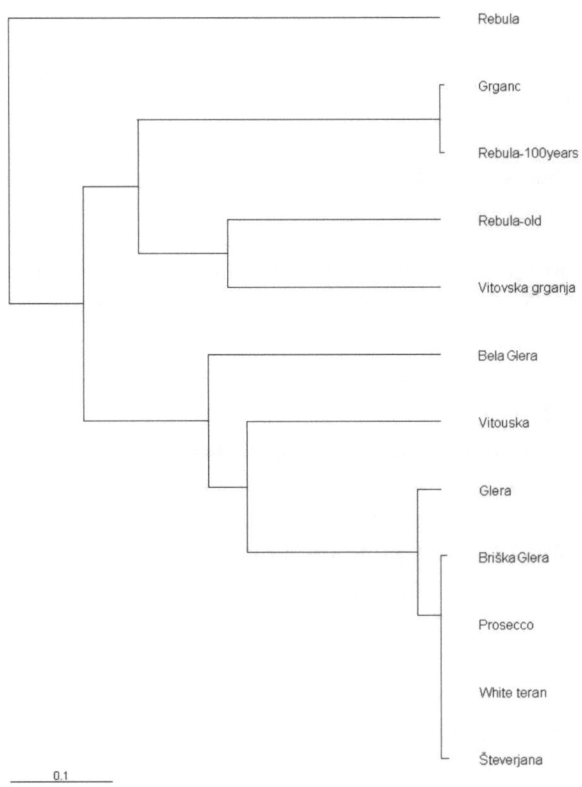

Figure 4. Dendrograms of grapevine accessions related to designations 'Rebula' and 'Ribolla' using D = 1-(proportion of shared alleles) as coefficient of distance and UPGMA as grouping method [15].

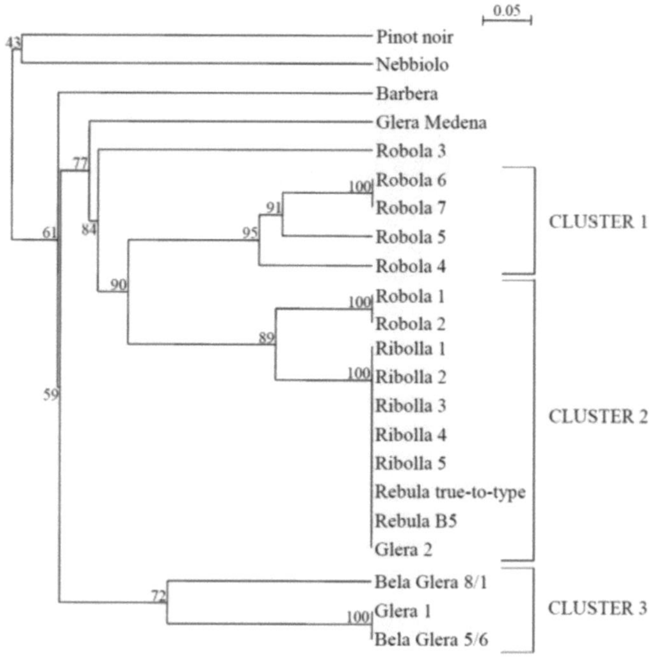

Figure 5. Dendrograms of grapevine accessions related to designations 'Rebula' and 'Ribolla' using D = 1-(proportion of shared alleles) as coefficient of distance and UPGMA as grouping method [29].

2.1.1.2. Accessions related to designations 'Vitovska' and 'Grganja'

The accessions related to 'Vitovska', 'Vitouska', 'Grganja' and 'Garganja' are cultivated in different winegrowing areas in Slovenia, Italy and Croatia; therefore their names (synonyms) still cause disorder, because they are often considered indigenous and ancient varieties in the area of cultivation. According to the accurate study [26], which included as much as 37 nuclear microsatellite locii confirmed the variety 'Vitouska' to be the progeny of 'Malvasia del Chianti' (syn. 'Malvasia bianca lunga', 'Malvasia lunga') and 'Prosecco tondo'. Following this, [30] found an original genotype of 'Malvasia del Chianti' with the varieties 'Pavlos' present in Greece and 'Maraština' cultivated in the south coastal region of Croatia (Dalmatia). Already in 1949 it was affirmed that 'Maraština' and 'Malvasia lunga' are one and the same variety [31]. In 2012 [18] reported genotyping and phenotyping of accessions (*Vitis vinifera* L.) cultivated in Slovenia, Croatia and Italia, mostly known as 'Vito(u/v)ska' and 'G(a)rgan(i)ja', while also referred as indigenous or landrace varieties [1,7,28]. The studied varieties, with an additional focus on variety 'Vitovska grganja', were mostly taken from germplasm collections: 'Vitovska grganja' from Slovenia, 'Vitouska', 'Ribolla gialla' and 'Prosecco' from Italy, 'Garganja' and 'Maraština' from Croatia.

Variety / Type / Vine / Abbreviation	Synonyms and homonyms (Translation)	Cultivation area and notes
'Vitovska grganja' 1 (VG)	Gerganja[1], Vitouska[2], Gargania[2], Ribolla gialla[2], Grganja[3], Garganja[3], Vitovška[3], Vitevška[3], Vitovka[3], Gorjanska[3], Malvazija s piko[3], Beli refošk[3], Vrbina[3], Vrbovna[3], Črna pika[3]	The vine is approx. 100 years old and grows in the village Sveto v (address Sveto 4), Kras (Slovenia) as an individual pergola-trained vine.
'Vitovska grganja' 2 (VG)	Gerganja[1], Vitouska[2], Gargania[2], Ribolla gialla[2], Grganja[3], Garganja[3], Vitovška[3], Vitevška[3], Vitovka[3], Gorjanska[3], Malvazija s piko[3], Beli refošk[3], Vrbina[3], Vrbovna[3], Črna pika[3]	The vine is over 110 years old and grows in the village Briščki (address Briščki 5), Carso, Italy cultivated as a pergola-trained individual vine.
'Vitouska' (V)	Vitovska grganja[2], Vitouska[2], Gargania[2], Ribolla gialla2	Sampled in the VCR (Rauscedo) selection of vines from vineyards planted near the Prosecco village, Italy. The variety is usually planted and cultivated in Carso region, north-east part of Italy.
'Prosecco' (P)	Gljera[6], Prosekar[6], Prosecco tondo[2], Prosecco bianco[2], Gargana[2], Brešanka[1,2] (Brescia)	Sampled in the VCR (Rauscedo) clone VCR101, Italy. The variety is usually planted and cultivated in Carso region, north-east part of Italy.
'Maraština' (M)	Pavlos[6], Malvasia del Chianti[6], Malvasia lunga[6], Rukatac[6], Marinkuša mala[6], Maraškin[6], Đerđevina[6], Kukuruz[6], Rukac[6], Krizol[6], Višana6	Sampled in the germplasm collection of the University of Zagreb, near Split (Dalmatia, Croatia). The variety is usually planted and cultivated in Dalmatia, coastal part of Croatia.
'Garganja' (G)	Rebula[6], Ribolla gialla[6]	Sampled in the germplasm collection of the Faculty of agriculture and tourism Poreč, near Poreč (Croatia). The variety is usually planted and cultivated in Croatian part of the Istria peninsula, near the village Buzet.
'Ribolla gialla' (RG)	Garganja[2], Glera[2], Refosco bianco[2], Teran bijeli[2], Ribola bijela[2], Erbula[2], Jerbula[2], Gorička ribola[2], Rebolla[2], Ribolla[2], Ribolla bianca[2], Ribuèle[2], Rabuèle[2], Rosazzo[2], Ribollat[2], Raibola[2], Ràbola[2], Ribuole[2], Gargania[2]	Sampled in the VCR (Rauscedo) clone VCR01, Italy. The variety is planted in Collio (Italy) presumably also in Goriška brda (Slovenian Collio), and Istria (Croatia). Sampled in the collection of the University of Ljubljana.

Table 4. List of the studied accessions related to the terms 'Vitovska', and 'Garganja' describing their already used synonyms, homonyms and cultivation areas[18]. Indexes: 1-[1]; 2-[28]; 3-[32]; 4-[7]; 5-[30]; 6-[31]

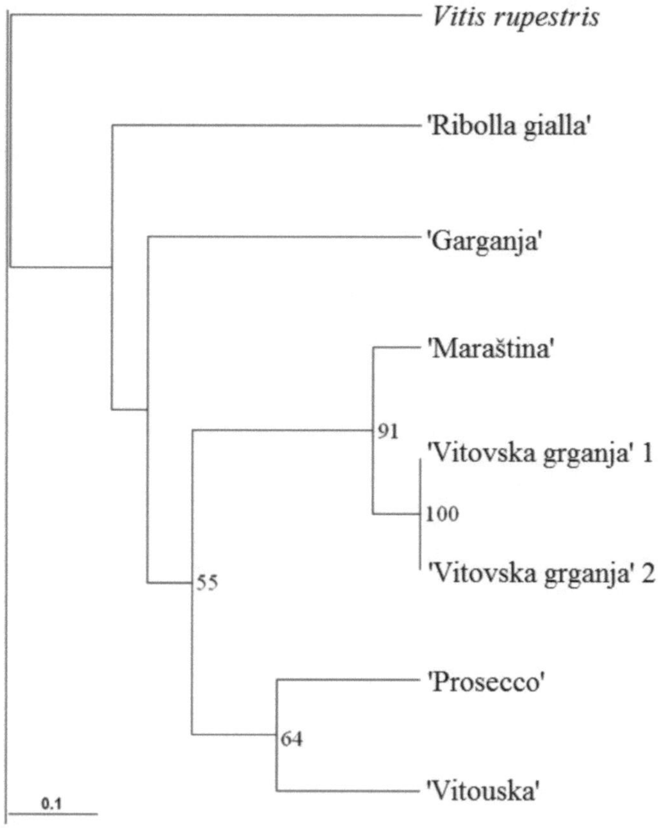

Figure 6. Dendrogram of studied grapevine (*Vitis vinifera* L.) accessions and of *Vitis rupestris* based on standard genetic distance [18].

Vines designated as 'Vitovska grganja 1' and 'Vitovska grganja 2', which have been grown for more than 100 years in the Slovenian Karst (Kras) revealed identical genotypes in all 11 analysed SSR microsatellites. The variety 'Maraština', nowadays known as an indigenous grapevine variety from Central and South Dalmatia (Croatia) surprisingly showed highly related SSR profile as both 'Vitovska grganja' cultivated in Slovenia. According to [30,31] the variety 'Maraština' coincides to the varieties 'Pavlos' and 'Malvasia del Chianti' (Malvasia lunga). In our study 'Maraština' coincided in seven of eleven SSR loci compared to 'Malvasia del Chianti' reported by [26], therefore we can conclude only that they are in strong relationship. Variety 'Vitouska' received from VCR Rauscedo (Italy), compared to 'Maraština' and 'Vitovska grganja' 1 and 2, differed in almost all used microsatellites, except at locus VVMD 5 and VVMD36, what additionally confirm the already reported affirmation of null (weak) relationship between both varieties [5,15,26]. Furthermore, considering that [26] affirmed the progeny of 'Vitouska' with 'Malvasia del Chianti' and 'Prosecco tondo', a quite weak match in their locii was found among them in our study.

2.2. Phenotyping experiences at grapevine varieties cultivated in Slovenia

2.2.1. Phyllometrical tolls for phenotyping

Morphological descriptions of different parts of the plant at different phenological stages, but also morphometry based on the measurements of parameters of plant organs are the oldest methods used in ampelography. Moreover, quantitative and qualitative analyses of biochemical compounds were also suggested at varietal description [33]. Despite the technical advances in biochemical and molecular approaches over the last decades, the descriptions of morphological characteristics remain an important methodology in the description and research of the diversity among species, varieties and clones. The most commune and used methodologies are descriptions according to the OIV codes [34] and ampelometry, especially phyllometry. The advances of morphological descriptions can be characterised as simple, cheap and applicable in the laboratory or directly in the field. Moreover, the present-day studies focus on finding new parameters in order to differentiate the varieties of vine precociously, quickly and efficiently [35]. The average vine leaves were reconstructed by measuring the proposed morphometrical parameters [36], furthermore a system for direct digitizing of phyllometric parameters from leaves was developed [37]. In grapevine collections approximately 5-10% of existing varieties are misnamed and even in commercial viticulture, misnaming and confusion related to synonyms and homonyms [38] is a frequent phenomena. We started to collect our allochthon, autochthone, local and endangered varieties systematically in 1980 and today we have few germplasm collections around Slovenia.

Additional attention has also been given to varietal phenotyping, especially at those varieties where some misunderstandings in nomination still cause disorders, for example at the varieties 'Vitovska grganja', 'Refošk', 'Rebula', 'Glera', 'Vitouska', 'Garganja' and others, which are frequently treated as an indigenous, autochthonous or landrace varieties in different countries [18]. Some new approaches to accelerate the phenotyping of leaves at different varieties were suggested; for example automatic measurements, as well as the possibility of typical-varietal mature leaf reconstruction [39,40].

2.2.1.1. Phyllometric measurements

Many studies in the last decades demonstrated the importance and contribution of phyllometry in grapevine description and distinction. The method provides around 82 criteria measurements of morphometrical characteristics on an individual leaf. According to a high variation in these characteristics among leaves of the same variety at least 10 mature, typically varietal leaves should be picked up, scanned or photocopied, placed in herbarium before being analysed. In our study we followed that procedure: Herbarium leaves were scanned (with the HP LaserJet M1005scanner) and analysed using AnalySIS image analysis program (Soft Imagining System GmbH). The quantitative base variables required the reconstruction of an average typical leaf of different varieties measured following the suggestions from [36] with some modifications. The following morphometrical parameters were measured in all leaves sampled per variety: leaf area (LA), length of nerves (L1, L2, L3, L4), distance from the petiole insertion to upper sinuses (OS), distance from the petiole insertion to lower sinuses (OI), total leaf height (H), total leaf width (W1), distance between the ends of left and right L2 (W2), distance between the basal tooth of the upper lobe (W3), angles between nerves measured from the petiole sinus to the first ramifications of vein (α, β, γ), angles between nerves from petiole sinus to apex of vein (α', β', τ), angle at the apex of upper lobe (δ1), angle between the apex of L1 and the apexes of L2 (δ2), angle between apex of L1 and apexes of L5 (δ3), distance between apex and basal tooth of upper lobes (D1), distance between the ends of L1 and L2 (D2), petiole length (lp), distance from the base of petiole sinus to the intersection of L3 vein to L4 vein (LO), tooth width at the end of L2 and L4 (b1, b2), tooth length at the end of L2 and L4 (h1, h2), distance between the ends of L5 (l), distance between the beginning of L5 (l') suggested by [41]. Moreover, [42] suggested the following characterizing equations, in which both parts of the same leaf are used separately: Rel.1 = Lp/L, Rel.2 (left side) = L1/L, Rel.3 (right side) = L1/L, Rel.4 (left side) = $\alpha+\beta+\gamma$, Rel.5 (right side) = $\alpha+\beta+\gamma$, Rel.6 (left side) = $\alpha'+\beta'+\tau$, Rel.7 (right side) = $\alpha'+\beta'+\tau$, Rel.8 (left side) = (OS+OI)/(L1+L2), Rel.9 (right side) = (OS+OI)/(L1+L2), Rel.10 (left side) = OS/L1, Rel.11 (right side)= OS/L1, Rel.12 (left side) = OI/L2, Rel.13 (right side) = OI/L2 (Figure 7).

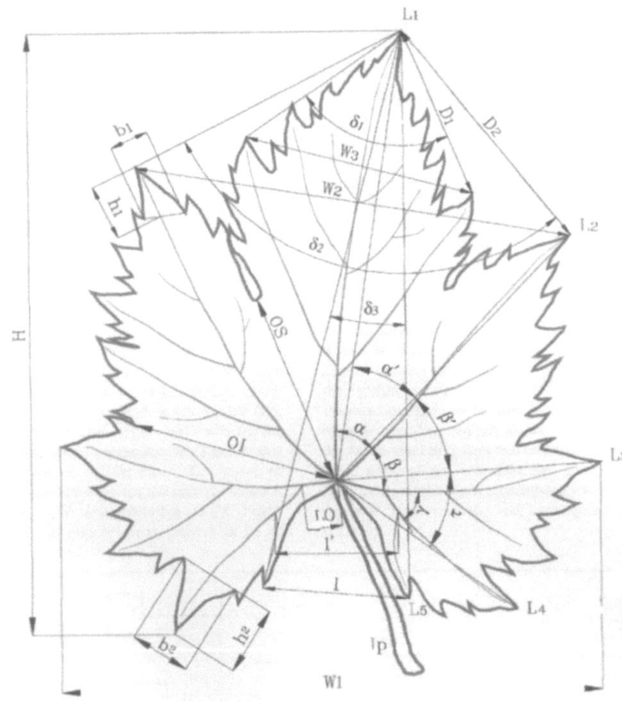

Figure 7. The morphometrical characteristics of mature leaves suggested by different authors [36,41,42].

The aim of the study was to investigate the efficiency of ampelographic and morphologic methods to evaluate the significance of O.I.V. code lists (Paragraph 2.2.2.2) and phyllometry regarding the varietal diversity. In addition, a graphic reconstruction method was applied and improved to form a typical varietal mature leaf (Figure 9). We evaluated 38 morphologic and 22 morphometric parameters per leaf. Statistical analysis was carried out, PCA (principal component analysis) was performed taking into account equations calculated from different leaf variables, and the qualitative variables proposed by the O.I.V. were preceded to Cluster Analysis. The combination of morphologic descriptors and phyllometric measurements proved to be complementary more than comparable methods. The results of the evaluation of O.I.V. descriptors showed a high level of varietal diversity, whereas accession 'Barbera type Bovcon' and 'Pokalca' as well as 'Barbera' and 'Syrah' showed some similarities. The most different variety in the group studied according to O.I.V. descriptors was 'Touriga national'. The PCA analysis showed the first three components responsible for more than 82% of the discriminating power. The reconstruction determinants were relations between the depth of the lateral sinuses and lateral veins (PC1), relation between first left lateral vein and central vein, relation between first right lateral vein and central vein (PC2), sum of angles between veins, relation between length of petiole and central vein and depth of the lateral sinuses (PC3). The enumerated relations enable and additionally improve the leaf reconstruction. At the varieties 'Sladkočrn' and 'Refošk' the shallowest lateral sinuses and at 'Tinta pinheira' the deepest ones were observed whereas 'Plovdina' and 'Touriga national'showed shorter lateral veins (L1d, L1g) in comparison with the main vein (L). The morphological descriptions and morphometric reconstructions of leaves gave a significant contribution to understanding the grapevine phenotypical biodiversity.

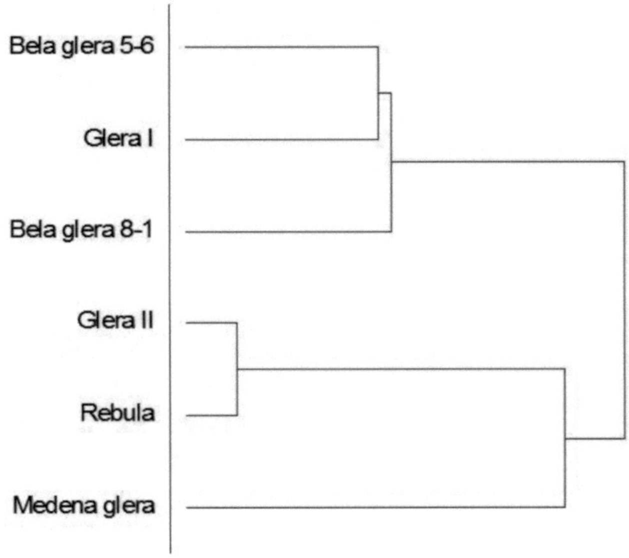

Figure 8. Dendrogram of grapevine accessions linked to the designations 'Glera' and 'Rebula' grouped on the basis of the 28 morphometric characteristics of mature leaves [40].

Furthermore, [40] studied 28 morphometric characteristics of leaves in detail and their contribution to the similarity among accessions with designations linked to 'Glera' and 'Rebula'. Phyllometric measurements affirmed the genotyping of same varieties studied by [29]. At the comparison of the groupings in Figure 5 and Figure 8, the measured 28 morphometric characteristics give suitable results what suggests that phyllometry grants an indispensable, cheap and credible methodology in the ampelography – varietal description and discrimination.

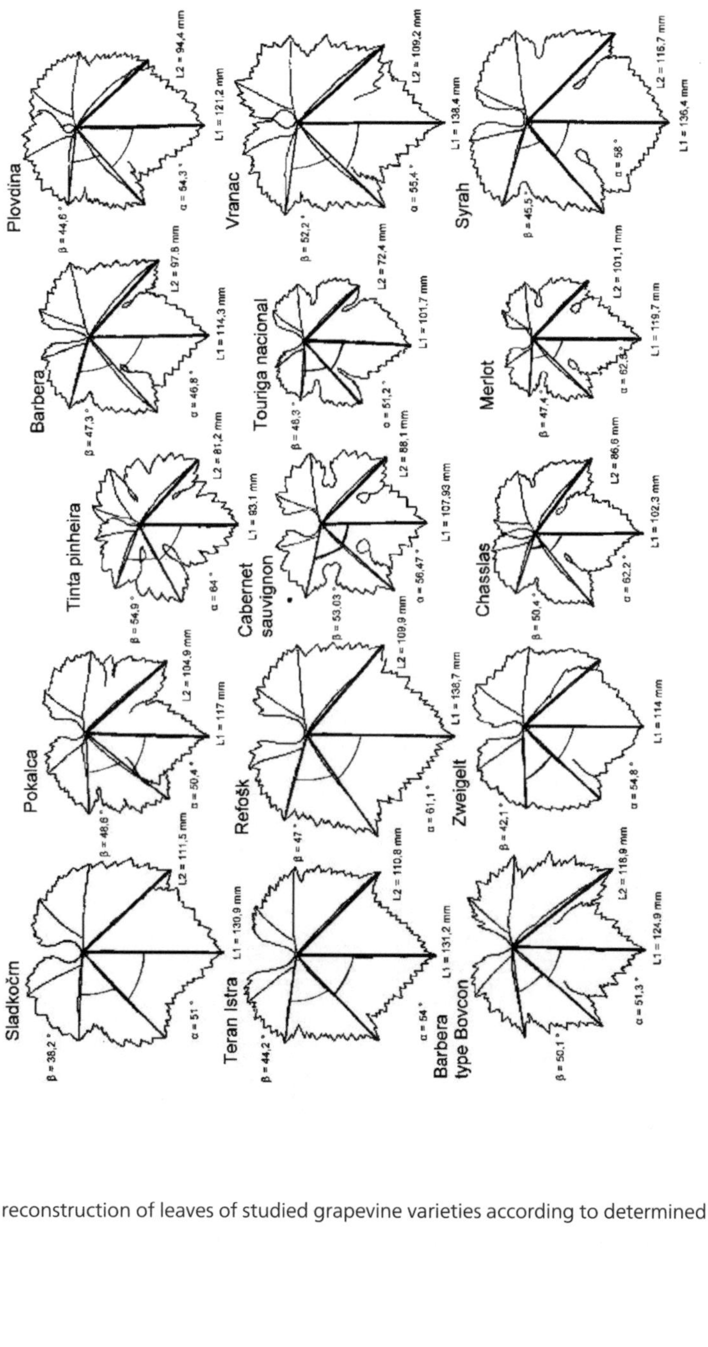

Figure 9. Graphical reconstruction of leaves of studied grapevine varieties according to determined phyllometric characteristics [39].

2.2.1.2. Morphological descriptions

The O.I.V. codes list still remains the most uses and suitable ampelographic tool in context of grapevine description [34]. There are approx. 147 descriptors; each regarding one significant characteristic, which offers a possibility of an accurate description of the main morphological parts of the vine, for example shoots tip, shoot, leaf, bunch, berry etc. [34]. For basic and relatively fast description of the varieties, the so called the Primary Descriptor list of OIV codes for Grapevine Varieties and Species *Vitis* was formed [34], which directs a description of 14 grapevine characteristics regarding shoots, leaves, bunches and berries. The OIV descriptors, for some leaf characteristics, overlap with phyllometric parameters: OIV 601 = L1, OIV 602 = L2, OIV 603 = L3, OIV 604 = L4, OIV 605 = OS, OIV 606 = OI, OIV 607 = α, OIV 608 = β, OIV 609 = γ, OIV 612 = h1, OIV 613 = b1, OIV 614 = h2, OIV 615 = b2, OIV 618 = l (Figure 7), therefore a collective use of both methodologies upgrades the varietal description. In few studies we used a different number of OIV codes, what depends especially on the condition of the grapevine, environmental conditions and the scope of the study.

The aim of the first study was the evaluation of selected OIV codes for grapevine description and discrimination – if the selected OIV code descriptorsare enough informative for a precise varietal discrimination? We sampled from 10 to 12 fully developed leaves from the sixth to ninth node of the shoot per certain variety. For the ampelographic description, shoot tips were sampled in the earlier phenological stages and berries were observed during the whole ripening period. The description and measurements of the particular parts of the vine were conducted according to the instructions of the individual descriptive codes.

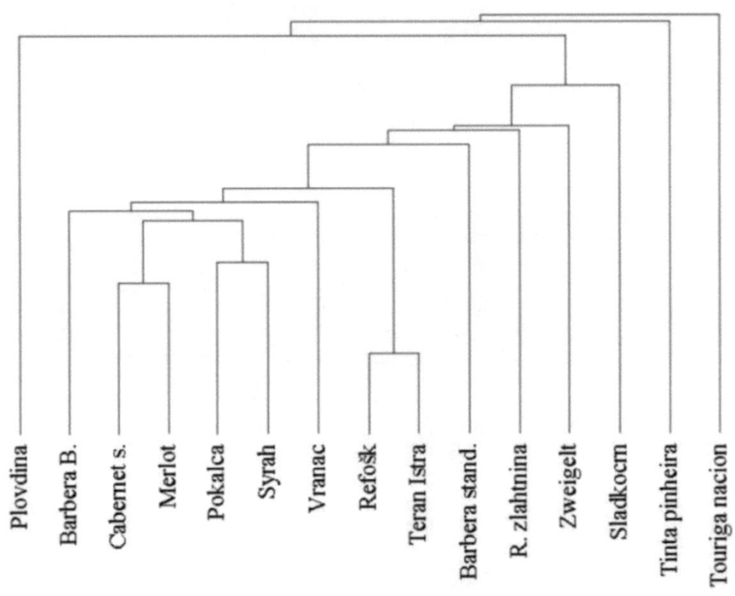

Figure 10. Dendrogram of grapevine varieties grouped according to the morphological evaluation, OIV codes [43].

According to the obtained data (Figure 10) and to the already acquired knowledge of the studied varieties some expected results revealed – for example similar OIV codes between accessions 'Refošk' and 'Teran Istra', frequently used as synonyms; or 'Pokalca' and 'Syrah' especially because of shape of the leaves berries. On the other hand the applied OIV codes revealed some imprecisions as well, for example similarities between varieties 'Merlot' and 'Cabernet sauvignon'. Therefore we suggest that the grouping according to the OIV code descriptors should be done and interpreted carefully, as many misunderstandingscould be provoked. For a comparison and interpretation of the studied accessions we suggest that the reference grapevine varieties should be included always, and as many as possible (also with greater differences in characteristics).

The further studies was focused on the variations, differences and possible mistakes linked to the subjective evaluations regarding OIV codes, where different descriptions of the same variety/accession were collected and compared. Therefore we selected accessions where many misunderstanding still provoke - accessions regarding the designations 'Vitovska' and 'Garganja' (Table 5) [18]. The misunderstandings in variety nomination are generally a result of false and misleading facts, which can partly be explained by neglect in observations and descriptions of variety characteristics, especially among those with a similar phenotype.

OIV code	Descriptor	VG	VGI*	V*	G*	M*,**
001	Young shoot: opening of the shoot tip	5	5	5	5	5
002	Young shoot: distribution of anthocyanin coloration on prostrate hairs of the shoot tip	2				2**
003	Young shoot: intensity of anthocyanin coloration on prostrate hairs of the shoot tip	5	6	1	1	5
004	Young shoot: density of prostrate hairs on the shoot tip	7	7	7	3	9, (7**)
005	Young shoot: density of erect hairs on the shoot tip	1				
006	Shoot: attitude (before tying)	3	1	1	1	3
007	Shoot: colour of the dorsal side of internodes	2	2	2	2	2
008	Shoot: colour of the ventral side of internodes	1	1	2	2	2
009	Shoot: colour of the dorsal side of nodes	3				
010	Shoot: colour of the ventral side of nodes	1				
011	Shoot: density of erect hairs on nodes	1				
012	Shoot: density of erect hairs on internodes	1				
013	Shoot: density of prostrate hairs on nodes	1				
014	Shoot: density of prostrate hairs on internodes	1				
015-1	Shoot: distribution of anthocyanin coloration of the bud scale	2	3			
015-2	Shoot: intensity of anthocyanin coloration of the bud scale	1	4			
016	Shoot: number of consecutive tendrils	1	1	1	1	1

OIV code	Descriptor	VG	VGI*	V*	G*	M*,**
017	Shoot: length of tendrils	5				5
051	Young leaf: colour of upper side of blade (4th leaf)	2,4[1]	3	1	3	2-3, (2-3**)
053	Young leaf: density of prostrate hairs between main veins on lower side of blade (4th leaf)	7	7	8	7	5-7
054	Young leaf: density of erect hairs between main veins on lower side of blade (4th leaf)	1				
055	Young leaf: density of prostrate hairs on main veins on lower side of blade (4th leaf)	7				
056	Young leaf: density of erect hairs on main veins on lower side of blade (4th leaf)	1				
065	Mature leaf: size of blade	7				7**
067	Mature leaf: shape of blade	3	3	4	4	4, (1,4**)
068	Mature leaf: number of lobes	3	2-3	3	2	5, (3**)
069	Mature leaf: colour of the upper side of blade	7				7**
070	Mature leaf: area of anthocyanin coloration of main veins on upper side of blade	2	1	2	2	1
071	Mature leaf: area of anthocyanin coloration of main veins on lower side of blade	2				
072	Mature leaf: goffering of blade	5	5	5	5	3, (7**)
073	Mature leaf: undulation of blade between main or lateral veins	1				
074	Mature leaf: profile of blade in cross section	4	5	5	5	1
075	Mature leaf: blistering of upper side of blade	5	3	7	3	3
076	Mature leaf: shape of teeth	3	2,4	4	5	3, (2**)
078	Mature leaf: length of teeth compared with their width	3				5**
079	Mature leaf: degree of opening / overlapping of petiole sinus	7	5	2	7	7
080	Mature leaf: shape of base of petiole sinus	1	3	1	3	1, (2**)
081-1	Mature leaf: teeth in the petiole sinus	1	1	1	1	1
081-2	Mature leaf: petiole sinus base limited by vein	1	1	1	1	3
082	Mature leaf: degree of opening/overlapping of upper lateral sinuses	2-3				
083-1	Mature leaf: shape of the base of upper lateral sinuses	1				
083-2	Mature leaf: teeth in the upper lateral sinus	1	1	1	1	1
084	Mature leaf: density of prostrate hairs between main veins on lower side of blade	7	1	6	5	7
086	Mature leaf: density of prostrate hairs on main veins on lower side of blade	7				

OIV code	Descriptor	VG	VGI*	V*	G*	M*,**
087	Mature leaf: density of erected hairs on main veins on lower side of blade	7	3	2	1	5
093	Mature leaf: length of petiole compared to length of middle vein	3				5**
094	Mature leaf: depth of upper lateral sinuses	5-7		4		3-5, (5**)
101	Woody shoot: cross section	1-2				2**
102	Woody shoot: structure of surface	1				1-2**
103	Woody shoot: main colour	1,3[2]				1,4**
104	Woody shoot: lenticels	1		1		1
105	Woody shoot: erect hair on nodes	1				
106	Woody shoot: erect hair on internodes	1				
151	Flower: sexual organs	3	3		3	3, (3**)
152	Inflorescence: insertion of 1st inflorescence	2				
153	Inflorescence: number of inflorescence per shoot	1				2**
155	Shoot: fertility of basal buds (buds 1-3)	5	1		5	5, (1**)
202	Bunch: length (peduncle excluded)	5-7	5	7	1	7, (7**)
203	Bunch: width	5-7				5**
204	Bunch: density	7	6	7	5	3, (5-7**)
206	Bunch: length of peduncle of primary bunch	3	4	5		5, (7**)
207	Bunch: lignifications of peduncle	5				5-7**
208	Bunch: shape	2	1	2	1	1-2, (1-2**)
209	Bunch: number of wings of the primary bunch	2	2-3	2	2	3, (2**)
220	Berry: length	3	5	5		3, (3**)
221	Berry: width	3	5	5		3, (3**)
222	Berry: uniformity of size	1				
223	Berry: shape	1,2	2	1	2	2, (2**)
225	Berry: colour of skin	1	1	1	1	1, (1**)
226	Berry: uniformity of skin colour	2				
227	Berry: bloom	5				7**
228	Berry: thickness of skin	3				7**
229	Berry: hilum	2				2**
231	Berry: intensity of flesh anthocyanin coloration	1	1	1	1	1
232	Berry: juiciness of flesh	2				2**
233	Berry: must yield	7				5**
235	Berry: firmness of flesh	2	1-2	1	2	3
236	Berry: particular flavour	1	5	1	1	1, (1**)
238	Berry: length of pedicel	3				7**
240	Berry: ease of detachment from pedicle	3				

OIV code	Descriptor	VG	VGI*	V*	G*	M*,**
241	Berry: formation of seed	3	3	3	3	3, (3**)
242	Berry: length of seeds	5,7				5,7**
243	Berry: weight of seeds	3				3-5**
244	Berry: transversal ridges on dorsal side of seeds	1		1		1
301	Time of bud burst	7	3			3
302	Time of full bloom	5-7				
303	Time of beginning of berry ripening (vérasion)	5	5			5
304	Time of full physiological maturity of the berry	5				7**
305	Time of beginning of wood maturity	7				
306	Autumn coloration of leaves	1				
351	Vigour of shoot growth	7	8			5, (5-7**)
352	Growth of lateral shoot	5				
353	Length of internodes	3-5				3**
354	Diameter of internodes	3				3**
401	Resistance to iron chlorosis	1				
402	Resistance to chlorides (salt)	1				
403	Resistance to drought	7				
452	Leaf: degree of resistance to *Plasmopara*	5				
453	Cluster: degree of resistance to *Plasmopara*	5				
455	Leaf: degree of resistance to *Oidium*	1,3				
456	Cluster: degree of resistance to *Oidium*	1,3				
458	Leaf: degree of resistance to *Botrytis*	5				
459	Cluster: degree of resistance to *Botrytis*	5				
501	Percentage of berry set	5				
502	Bunch: single bunch weight (g)	3	3			5, (3**)
503	Berry: single berry weight (g)	1-3	3			1, (1-3**)
504	Yield per m^2 (kg)	3-5[3]				5, (3**)
505	Sugar content of must	3	3		5	5, (5**)
506	Total acidity of must	3-5[4]	3		5	5, (5**)
508	Must specific pH	3,5[5]	3			7
601	Mature leaf: length of vein N1 (mm) = L_1***	5	7			5**
602	Mature leaf: length of vein N2 (mm) = L_2***	5	9			5**
603	Mature leaf: length of vein N3 (mm) = L_3***	5-7	7			5-7**
604	Mature leaf: length of vein N4 (mm) = L_4***	9	9			9**
607	Mature leaf: angle between N1 and N2 measured at the first ramification (°) = α***	5	7			7**
608	Mature leaf: angle between N2 and N3 measured at the first ramification (°) = β***	7	7			5**

OIV code	Descriptor	VG	VGI*	V*	G*	M*,**
609	Mature leaf: angle between N3 and N4 measured at the first ramification (°) = γ***	5	7			5**
612	Mature leaf: length of tooth of N2 = h1***	3	7			
613	Mature leaf: width of tooth of N2 = b1***	5	9			
614	Mature leaf: length of tooth of N4 = h2***	3	3-5			
615	Mature leaf: width of tooth of N4 = b2***	5	7			

Table 5. Ampelographic description according to O.I.V. descriptor codes and morphometric characteristics of 'Vitovska grganja' (VG), 'Vitovska grganija' (VGI), 'Vitouska' (V), 'Grganja' (G) and 'Maraština' (M) varieties [18]. *-[44]; **-[45]; ***-parameters of phyllometry [18]; [1]-copper-reddish between veins; yellow around veins; [2]-yellow and reddish; [3]-0.8-1 kg m^{-2}; [4]-total acidity 5.6 - 8.2 g L^{-1}; [5]-pH 3.10 - 3.45.

The study was carried out on grapevine (*Vitis vinifera* L.) varieties cultivated in Slovenia, Croatia and Italia, mostly known as 'Vitovska', 'Vitouska', 'Grganja' and 'Garganija' but also referred to as indigenous or landrace varieties from the mentioned countries [1,7,28]. The studied varieties, with an additional focus on the 'Vitovska grganja' variety, were mostly taken from germplasm collections: 'Vitovska grganja' from Slovenia, 'Vitouska', 'Ribolla gialla' and 'Prosecco' from Italy, 'Garganja' and 'Maraština' from Croatia.

The comparison of the ampelographic descriptions, morphometric and phyllometric characteristics among the 'Vitovska grganja' (VG), 'Vitovska grganija' (VGI), 'Vitouska' (V), 'Grganja' (G) and 'Maraština' (M) varieties is presented in Table 5. The ampelographic descriptions of the VGI, V and G varieties highly differed from the VG and M varieties [18]. Similarities between VG and M varieties were observed in the characteristics of young and mature shoots, the opening shoot tips showed similar anthocyanin coloration and a similar intensity and coloration of prostrate hairs. Similarities between VGI and M varieties have to be underlined, especially in the size of young and mature leaves, dark green coloured upper side of the blade, both convex shaped sides of the teeth. The latter descriptor was in contrast to the report of [45], where both sides were described as straight. Moreover, [45] described the shape of the petiole base sinus of the M variety as 'U'-shaped whereas in [44] it is shown brace-shaped, what is also in accordance with our observations on VG variety. Both M and VG varieties were characterized by high density prostrate hairiness between the main veins on the lower side of the blade; however, hair density differed at other blades/vein parts. Our observations regarding the depth of the upper lateral sinuses of VG variety coincided with those mentioned by [45] for the M variety. In our study, M variety had a lower fertility of basal buds (1-3) compared to VG variety and also to the results reported by [45].

A comparison of bunch and berry characteristics among the studied varieties were also reported [18], what partly explains the reasons of the misuse of the variety nominations due to the rather similar phenotypes. All studied varieties have a cylindrical or conical shape of the bunches, obloid and globose berries, and a bunch length ranging between 160 and 200 mm with the exception of M and VG varieties, where berries are distinctively shorter and narrower. The description of single bunch and berry weight, as well as yield of the M variety reported by [45], coincided with our description for the VG variety. On the other hand, many significant differences were observed among the studied varieties, also between VG and M variety. Compared to M variety, the VG variety blooms later, has thinner berry skin, higher must yield, less firm flesh, shorter length of pedicle, lower sugar and total acidity content and lower specific pH level of the must. Moreover VG also has later bud bursting, but reaches full physiological maturity of the berry earlier. [18] found also some similarities among the studied varieties, specifically in morphometrical characteristics of mature leaves. Significant similarities were observed between our measurements of the VG variety and data cited by [45] for the M variety, which only differed in 'α' and 'β' angles. The comparison of phenotypic characteristics of the studied varieties showed many similarities, which suggests that a consistent attention to the ampelographic distinguishing among varieties, especially among those with the same designations or synonyms, has to be emphasized in the future.

3. Conclusions

The grapevine genepool is particularly vulnerable in the marginal areas of its distribution range. Many grapevine varieties have been already described and their genotypes determined; but many local grapevine accessions remain unidentified and their ampelographic characteristics overlooked. Accurate ampelographic description of variety should be done with combined methodologies which involve morphological description and DNA analyses. In the last decade a great effort was given to the DNA analyses – genotyping of the grapevine varieties, and quite soon revealed that DNA analyses is simply not enough to answer all questions about varietal discrimination and description. Parallel to the development of genetic techniques, the morphological and morphometric evaluations have fall into oblivion. Although, microsatellites are very powerful means of identifying synonyms in germplasm collections, they have to be supported with morphological descriptions, what was also demonstrated in our researches. Only such works will allow a characterization of a large number of varieties/accessions and will contribute to improve the organization of grapevine collections and the possibility of exchanging true-to-type material. The presented works are a first step towards true-to-type identification, which will facilitate the registration of varieties and allow growers to be sure of the value of their products. Starting true-to-type variety identification by comparing results with neighbouring, historically linked areas, will significantly clarify the confusion in nomenclature and help determine the origin and relationships among varieties over the whole area. In the future all the obtained results from different studies should be upgraded to the same database for grapevine varieties, to obtain a total overview of World germplasm, what may be the key to stop the erosion of the many varieties.

Author details

Denis Rusjan*

Address all correspondence to: denis.rusjan@bf.uni-lj.si

Biotechnical Faculty, University of Ljubljana, Slovenia

References

[1] Vertovec, M. (1845). Vinoreja za Slovence. *Ajdovščina: Agroind Vipava.*

[2] Goethe, H. (1887). Berlin: Handbuch der Ampelographie. *Beschreibung und Klassifikation der bis jetzt kultivierten Rebenarten und Trauben-varietäten mit Angabe ihrer Synonyme,* Kulturverhältnisse und Verwendungsart. Berlin: P. Parey.

[3] Tolar, T., Jakše, J., & Korošec-Koruza, Z. (2008). The oldest macroremains of Vitis from Slovenia. *Vegetation history archaeobotany,* 17(1), 93-102.

[4] Tomažič, I., & Korošec, Koruza. Z. (2003). Validity of phyllometric parameters used to differentiate local Vitis vinifera L. cultivars. *Genetic resources and crop evolution,* 50(7), 779-787.

[5] Štajner, N., Korošec-Koruza, Z., Rusjan, D., & Javornik, B. (2008). Microsatellite genotyping of old Slovenian grapevine varieties (Vitis vinifera L.) of the Primorje (coastal) winegrowing region. *Vitis,* 47(4), 201-204.

[6] Plahuta, P., & Korošec-Koruza, Z. (2009). x sto vinskih trt na Slovenskem. Ljubljana: Prešernova družba

[7] Hrček, L., & Korošec-Koruza, Z. (1996). Sorte in podlage vinske trte: ilustrirani prikaz trsnega izbora za Slovenijo. Ptuj: Slovenska akademija Veritas

[8] Thomas, M. R., & Scott, N. S. (1993). Microsatellite repeats in grapevine reveal DNA polymorphisms when analysed as sequence-tagged sites (STSs). *Theoretical and Applied Genetics,* 86(8), 985-990.

[9] Crespan, M. (2004). Evidence on the evolution of polymorphism of microsatellite markers in varieties of Vitis vinifera L. *Theoretical and Applied Genetics,* 108(7), 231-237.

[10] Dettweiller, E. (1993). Evaluation of breeding characteristics in Vitis. Influence of climate on morphologic characteristics of grapevines. *Vitis,* 32, 249-252.

[11] Cipriani, G., Frazza, G., Peterlunger, E., & Testolin, R. (1994). Grapevine fingerprinting using microsatellite repeats. *Vitis,* 33, 211-215.

[12] Bowers, J. E., Dangl, G. S., Vignani, R., & Meredith, C. P. (1996). Isolation and characterization of new polymorphic simple sequence repeat loci in grape (Vitis vinifera L). *Genome*, 39(4), 628-633.

[13] Sefc, K. M., Lefort, F., Grando, M. S., Scott, K., Steinkellner, H., & Thomas, M. (2001). Microsatellite markers for grapevine: A state of the art. In: Roubelakis-Angelakis KA. (ed.) Molecular Biology & Biotechnology of the Grapevine. Dordrecht: Kluwer Academic Publishers , 433-466.

[14] Maletić, E., Sefc, M. K., Steinkellner, H., Karoglan, Kontić. J., & Pejić, I. (1999). Genetic characterization of Croatian grapevine cultivars and detection of synonymous cultivars in neighboring regions. *Vitis*, 38(2), 79-83.

[15] Rusjan, D., Jug, T., & Štajner, N. (2010). Evaluation of genetic diversity: which of the varieties can be named 'Rebula' (Vitis vinifera L.)? *Vitis*, 49(4), 189-192.

[16] Kozjak, P., Korošec-Koruza, Z., & Javornik, B. (2003). Characterisation of cv. Refošk (Vitis vinifera L.) by SSR markers. *Vitis*, 42(2), 83-86.

[17] Štajner, N., Rusjan, D., Korošec-Koruza, Z., & Javornik, B. (2011). Genetic characterization of old Slovenian grapevine varieties of Vitis vinifera L. by microsatellite genotyping. *American Journal of Enology and Viticulture*, 62(2), 250-255.

[18] Rusjan, D., Pipan, B., Pelengić, R., & Meglič, V. (2012). Genotypic and phenotypic discrimination of grapevine (Vitis vinifera L.) varieties of the 'Vitovska' and 'Garganja' Denominations. *European Journal of Horticultural Science*, 77(2), 84-94.

[19] Lefort, F., & Roubelakis-Angelakis, K. A. (2001). Genetic comparison of Greek cultivars of Vitis vinifera L. by nuclear microsatellite profiling. *American Journal of Enology and Viticulture*, 52(2), 101-108.

[20] Lacombe, T., Boursiquot, J. M., Laucou, V., Dechesne, F., Varès, D., & This, P. (2007). Relationships and Genetic Diversity within the Accessions Related to Malvasia Held in the Domaine de Vassal Grape Germplasm Repository. *American Journal of Enology and Viticulture*, 58(1), 124-131.

[21] Riaz, S., Garrison, K. E., Dangl, G. S., Boursiquot, J. M., & Meredith, C.P. (2002). Genetic divergence and chimerism within ancient asexually propagated winegrape cultivars. *Journal of the American Society for Horticultural Science*, 127(4), 508-514.

[22] Caló, A., Costacurta, A., Maras, V., Meneghetti, S., & Crespan, M. (2008). Molecular correlation of Zinfandel (Primitivo) with Austrian, Croatian, and Hungarian cultivars and Kratošija, an additional synonym. *American Journal of Enology and Viticulture*, 59(2), 205-209.

[23] Muganu, M., Dangl, G., Aradhya, M., Frediani, M., Scossa, A., & Stover, E. (2009). Ampelographic and DNA characterization of local grapevine accessions of the Tuscia area (Latium, Italy). *American Journal of Enology and Viticulture*, 60(1), 110-115.

[24] Zulini, L., Fabro, E., & Peterlunger, E. (2005). Characterisation of the grapevine cultivar Picolit by means of morphological descriptors and molecular markers. *Vitis,* 44(1), 35-38.

[25] Crespan, M., Crespan, G., Giannetto, S., Meneghetti, S., & Costacurta, A. (2007). Vitouska' is the progeny of 'Prosecco tondo' and 'Malvasia Bianca lunga'. *Vitis,* 46(4), 192-194.

[26] Crespan, M., Cancellier, S., Chies, R., Giannetto, S., Meneghetti, S., & Costacurta, A. (2009). Molecular contribution to the knowledge of two ancient varietal populations: 'Rabosi' and 'Glere'. *Acta Horticulturae ISHS,* 827(1), 217-220.

[27] Cosmo, I., & Polsinelli, M. (1957). Ribolla gialla. Roma: Annali della sperimentazione agraria.

[28] Caló, A., Scienza, A., & Costacurta, A. (2006). Vitigni d'Italia. Bolognia: Edagricole-Edizioni Agricole Srl.

[29] De Lorenzis, G., Failla, O., Scienza, A., Rusjan, D., Nikolaou, N., Vouillamoz, J., & Imazio, S. (2012). Study of genetic differences among 'Ribolla Gialla', 'Rebula' and 'Robola' (Vitis vinifera L.) respectively from North Eastern Italy, Northern Balkans and Ionian Islands (under review).

[30] Šimon, S., Maletić, E., Karoglan Kontić, J., Crespan, M., Schneider, A., & Pejić, I. (2007). Cv. Maraština a new member of the Malvasia group Mediterranean Malvasias: conference proceedings. October 3-6, 2007, Salina, Italija.

[31] Bulić, S. (1949). Dalmatinska ampelografija., *Zagreb: Poljoprivredni nakladni zavod.*

[32] Blažina, I., Štolfa, D., Korošec-Koruza, Z., & Cerkvenik, D. (1990). Vitovska grganja' (Vitis vinifera L. 'Vitovska grganja') in domača 'Grganja' (Vitis vinifera L. 'Domača grganja'). *Zbornik Biotehniške fakultete,* 1, 10-14.

[33] Di Stefano, R. (1996). Metodi chimici nella caratterizzazione varietale. *Rivista di Viticoltura é Enologia,* 49(1), 51-56.

[34] OIV. (2008). 2nd Edition of the OIV descriptor list for grape varieties and Vitis species. *Paris: Organisation Internationale de la Vigne at du Vin.*

[35] Schneider, A. (1996). Grape variety identification by means of ampelographic and biometric descriptors. *Rivista di Viticoltura é Enologia,* 49(1), 11-16.

[36] Martínez, M. C., & Grenan, S. (1999). A graphic reconstruction method of an average leaf of vine. *Agronomie,* 19, 491-507.

[37] Alessandri, S., Vignozzi, N., & Vignini, A. M. (1996). AmpeloCADs (Ampelographic Computer-Aided Digitizing system). An integrated system to digitize, file and process biometrical data from Vitis spp. Leaves. *American Journal of Enology and Viticulture,* 47(3), 257-267.

[38] Weihl, T., & Dettweiler, E. (2000). Differentiation and identification of 500 grapevine (Vitis vinifera L.) cultivars using notations and measured leaf parameters. *Acta Horticulturae*, 528, 35-41.

[39] Pelengić, R., & Rusjan, D. (2010). Efficacy of ampelographic and phyllometric tools for the validation of grapevine Vitis vinifera L. biodiversity in Slovenia. *International Journal Of Food, Agriculture & Environment*, 8(3&4), 563-568.

[40] Pelengić, R. (2012). Ampelografski model za identifikacijo sort žlahtne vinske trte (Vitis vinifera L.). PhD thesis. *University of Ljubljana; (under review).*

[41] Alleweldt, G., & Dettweiler, E. (1989). A model to differentiate grapevine cultivars with the aid of morphological characteristics. *Rivista di Viticoltura é Enologia*, 42(1), 59-63.

[42] Santiago, J. L., Boso, S., Martínez, M., Pinto-Carnide, O., & Ortiz, J. M. (2005). Ampelographic comparison of grape cultivars in Northwestern Spain and Northern Portugal. *American Journal for Enology and Viticulture*, 56(3), 287-290.

[43] Pelengić, R. (2006). Ampelographic and ampelometric description of red grape varieties (Vitis vinifera L.) from collection Ampelografski vrt. *Graduation thesis.University of Ljubljana.*

[44] The European Vitis Database. (2011). *Institute for Grapevine Breeding: Geilweilerhof.*, http://www.eu-vitis.de/index.php, accessed 10 November 2011.

[45] Maletić, E. (1993). Ampelographic investigation of cvs Maraština, Bogdanuša, Vugava and Pošip (Vitis vinifera L.) in conditions of Ravni Kotari, region Coastal Croatia. *MSc thesis.University of Ljubljana.*

Section 3

From Genotype to Product

Chapter 8

Italian National Database of Monovarietal Extra Virgin Olive Oils

Annalisa Rotondi, Massimiliano Magli,
Lucia Morrone, Barbara Alfei and Giorgio Pannelli

Additional information is available at the end of the chapter

http://dx.doi.org/10.5772/51772

1. Introduction

The abundance of indigenous Italian olive germplasm, numbering over 800 cultivars [1] and rising, guarantees the ongoing production of high quality extra virgin olive oils, thus contributing to the preservation of much of the ancient genetic biodiversity of the olive.

The *Olea Europea* species has maintained much of its genetic diversity as a result of limited genetic erosion. This is due to breeding programs of this species having begun relatively recently compared to those of other fruit species.

Knowledge and development of the characteristics of Italian monovarietal extra virgin olive oils will also lead to an improvement in knowledge of the areas where these oils are produced, in turn developing tourism, a crucial sector for the Italian economy.

In Italy, new regulation was recently introduced forcing virgin and extra virgin olive oil producers to indicate the location of both olive harvest and oil production. More recently the European Commission has established compulsory standards for the labelling of origin for extra virgin and virgin olive oils (Reg EC n.182/2009). The significant increase in demand for extra virgin olive oils is due not only to the health benefits it offers, but also to its organoleptic properties; the large number of Italian olive cultivars allows for the production of different monovarietal oils marked out by a wide range of pleasant flavours.

As the genotype of origin affects the chemical and sensory characteristics of extra virgin olive oil deeply, the preservation and characterization of authocthonous cultivars and clones play a key role in the marketing of high quality olive oils.

Conservation of genetic resources for olives has important implications for both adaptation of the cultivars to their local environment and their agronomical performance under specific conditions. This also implies that every initiative to promote olive cultivation ought to take into consideration the local varieties and also that every region should preserve its own plant material to safeguard olive adaptation and productivity and to maintain the intrinsic characteristics of its olive oil which represent a deep connection with the territory of origin.

In the EU olive oils can be linked to the cultivar of origin and in turn to its area of production under the rules of the Protected Denominations of Origin (PDO) or of the European Protected Geographical Indication (PGI).

Several typical Italian extra virgin olive oils have qualified for PDO and PGI status, as many as 42 PDO and 1 PGI. Generally, these products are blends of different varieties according to the different cultivar percentage reported in the Product specification; some Italian PDO oils are obtained from the transformation of a single cultivar (monovarietal oils) for example the PDO Nostrana di Brisighella.

Italian olive cultivation is marked out by its extremely rich and varied varietal heritage. An important objective being pursued by every region is the protection and preservation of autochthonous Italian olive cultivars. This can be seen in the spread of regional varietal catalogs and also in the ongoing rise in the number of monovarietal olive oils taking part in the Italian National Review of Monovarietal olive oils as organized by ASSAM Marche [2].

In Italy this review serves to characterize monovarietal oils in terms of both chemical and sensory profiles. The organization of events, courses and forums involving olive farmers, crushers, consumers and catering operators has contributed to an improvement in the visibility of the market for Italian monovarietal olive oils. Studies into the quality of monovarietal oils increase the value of the product while showcasing the region of origin and educating the consumer about their nutritional and organoleptic value.

In Italy the market for monovarietal and organic oils is growing due to consumers paying greater attention to both flavour and health benefits of the product.

2. The Italian National Database

ASSAM and IBIMET-CNR have created and are managing a database of chemical and sensory profiles of extra virgin oils participating in the Italian National Review of Monovarietal olive oils. This dynamic database includes a large number of observations for each monovarietal oil, and can allow for ongoing updates every year, thus providing more accurate chemical and sensory average data for the oils. For each monocultivar oil, chemical and sensory profiles were calculated and described, including a large number of oil samples from different regions.

The Review reached its ninth edition as of 2004, and the large number of oil samples has led to improvements in results, in turn diminishing the effect of the main variables which have a significant influence on the quality of the oil: seasonality, ripening and different milling technologies.

Sensory analysis was laid out by the "ASSAM – Marche Panel" as recognized by the IOOC (International Olive Oil Council) and the Italian Ministry for Agriculture, Food and Forestry Policy under the conditions described in EC Reg. 640/2008.

The 150 samples collected during the first and the second year (edition 2004-2005) were used to identify the specific descriptors for the sensory analyses of monovarietal extra virgin olive oil and to set up the relative profile sheet [3].

Each panellist smelled and tasted the oil, in order to analyse olfactory, gustatory, tactile and kinaesthetic characteristics. Thirteen attributes were evaluated: 9 during the olfactory phase (olive fruity, olive fresh leaf, grass, fresh almond, artichoke, tomato, apple, berries and aromatic herbs) and 4 during the gustatory phase (olive fruity, bitter, pungent and fluidity). Attributes were assessed on an oriented 10-cm line scale and quantified measuring the location of the mark from the origin. Data obtained for the 13 descriptors were used to define the sensory profile of each sample using the median values [4].

Fatty acid composition, determined according to Reg. EC Reg.796/2002 methodology [5], and total phenolic content determined according to the Folin-Ciocalteu spectrophotometric method expressed as milligrams of gallic acid per kilogram of oil, were determined by Centro Agrochimico ASSAM, Jesi (AN).

Chemical and sensory data were processed using SAS 9.1.3 (SAS Institute Inc., Cary, NC, USA). Explorative analysis and descriptive statistics were performed for each set of data in order to identify outliers, extreme observations and to obtain distributional properties of the data. Descriptive measures (moments, basic measures of location and variability, confidence intervals for the mean, standard deviation, and variance) of chemical and sensory variables were calculated for each monovarietal oil.

Currently, the database includes 2092 oils produced from 130 different cultivars from 18 Italian regions. Nutritional properties, expressed as fatty acid and total phenol content, and the sensory profiles of each Italian monovarietal oil were published in the Catalogue of Italian Monovarietal oils [6] and at http://www.olimonovarietali.it

Below are listed the average sensory profiles of the 16 most represented monovarietal oils. The number of samples belonging to each cultivar are indicated in brackets: Ascolana Tenera (36 samples), Bianchera (34), Biancolilla (28), Bosana (133), Casaliva (39), Coratina (80), Frantoio (122), Itrana (102), Leccino (105), Mignola (38), Moraiolo (83), Nocellara del Belice (49), Peranzana (47), Piantone di Mogliano (57), Raggia (54), Ravece (101).

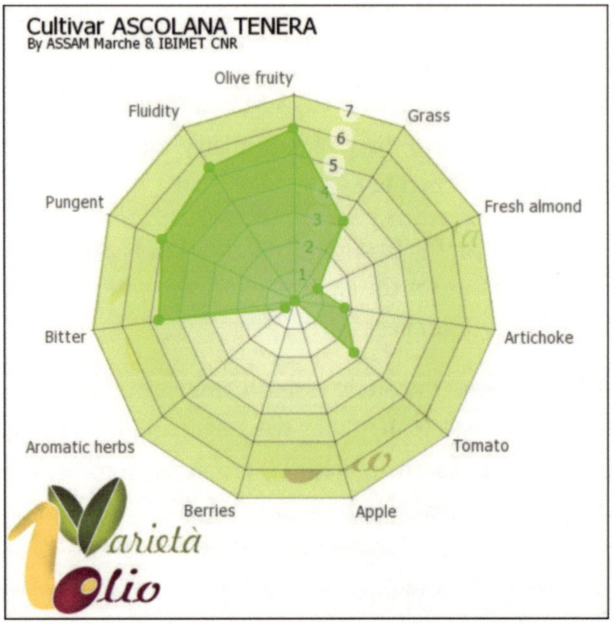

Figure 1. Ascolana Tenera – Marche region. Sensory profile: intense olive fruity, strongly grassy with hints of tomato and artichoke; balanced in taste, with medium intensity of bitter and pungent notes.

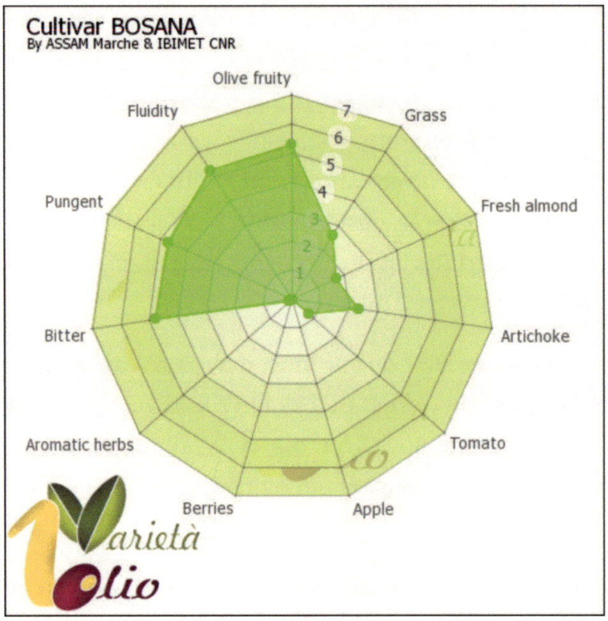

Figure 2. Bosana – Sardegna region. Sensory profile: medium olive fruity, grassy with prevalent scent of thistle and artichoke and hints of almond and tomato. Medium intensity of bitter and pungent notes.

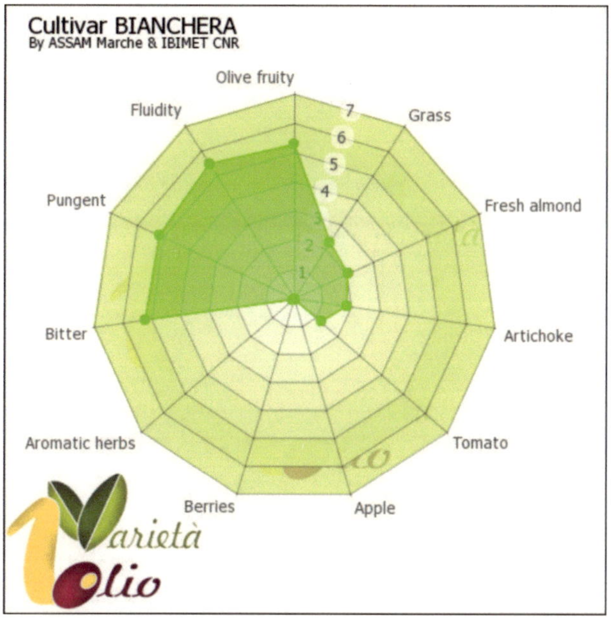

Figure 3. Bianchera – Friuli Venezia Giulia region. Sensory profile: medium-intense olive fruity, with grassy scent, artichoke, almond and tomato; medium bitter and pungent flavours.

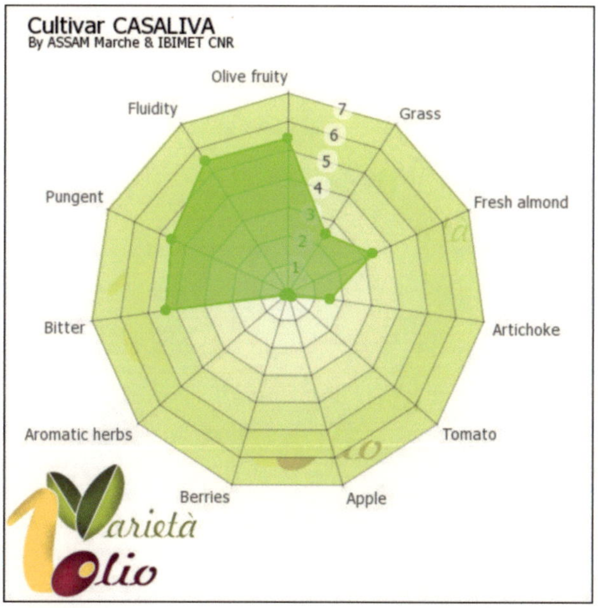

Figure 4. Casaliva – Lago di Garda area. Sensory profile: medium-intense olive fruity, with marked almond scent and light flavour of grass and artichoke; well balanced taste with medium intensity of bitter and pungent notes.

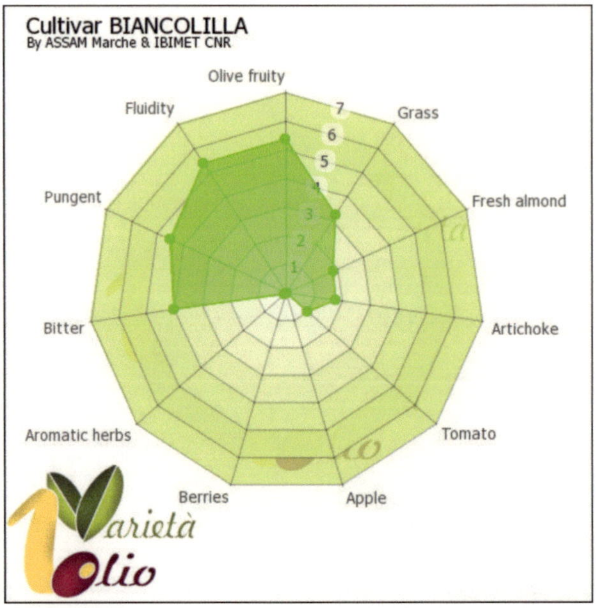

Figure 5. Biancolilla – Sicilia region. Sensory profile: medium-intense olive fruity, with a marked grass scent and light hint of almond, artichoke and tomato; bitter and pungent flavours are of medium-light intensity.

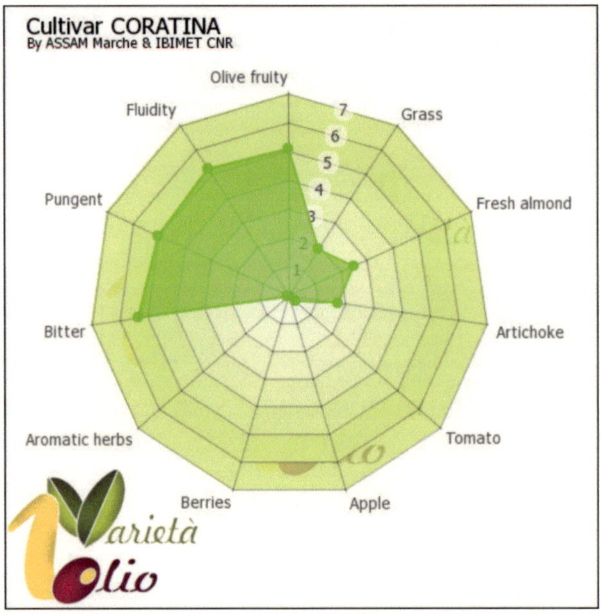

Figure 6. Coratina – Puglia region. Sensory profile: medium olive fruity, with a marked fresh almond scent together with notes of grass and artichoke; bitter and pungent flavours are of medium-high intensity.

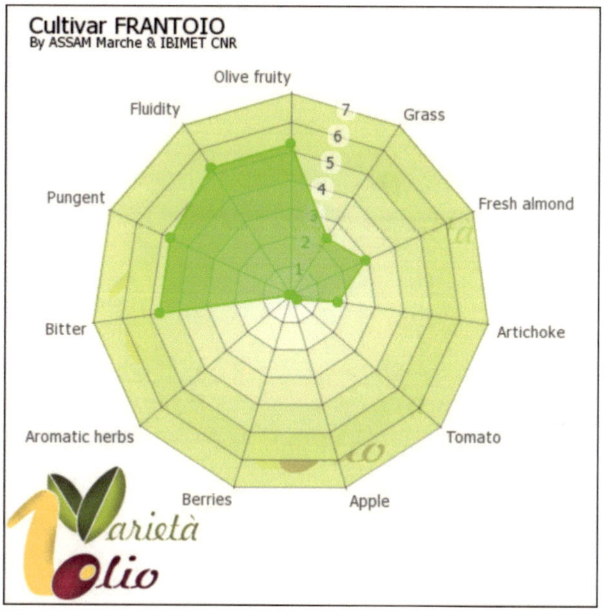

Figure 7. Frantoio – Central-North Italy. Sensory profile: medium-high olive fruity, with a marked fresh almond and and light flavour of grass and artichoke; bitter and pungent flavours are of medium intensity.

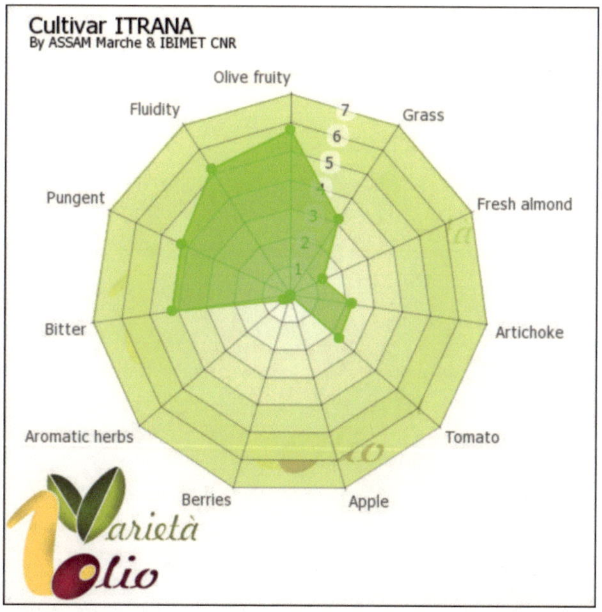

Figure 8. Itrana – Lazio region. Sensory profile: high olive fruity intensity, with grass, tomato and artichoke scent and light almond flavor; well balanced taste with a bitter and pungent medium-light intensity.

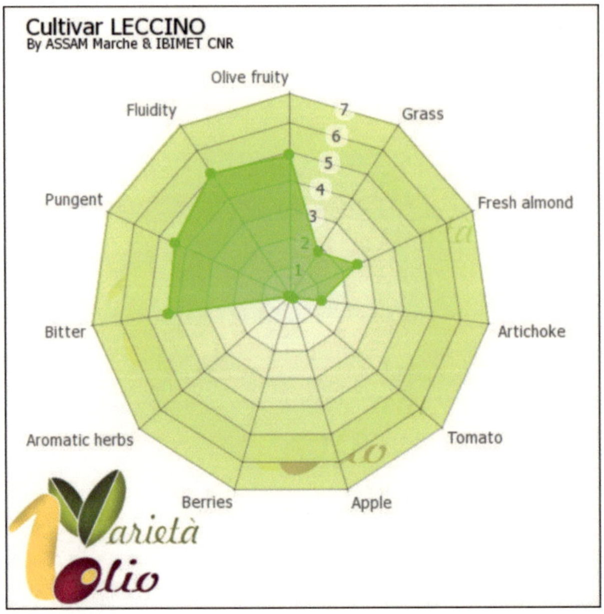

Figure 9. Leccino – North-central Italy. Sensory profile: medium olive fruity intensity, with almond scent and light grass and artichoke flavor; medium intensity of pungency and bitter taste

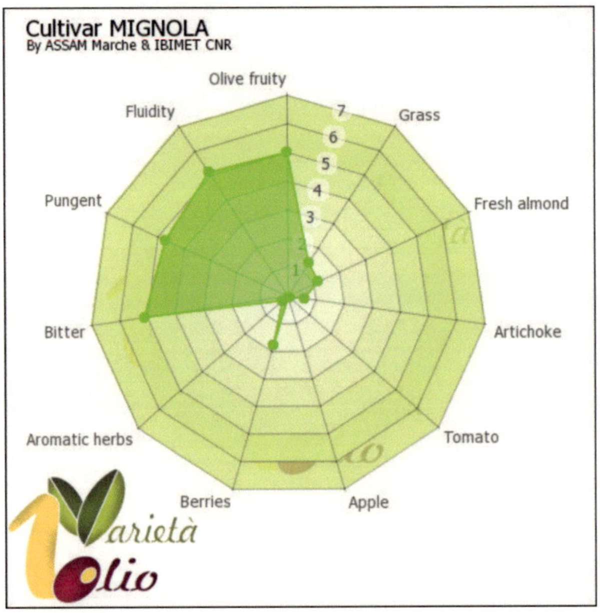

Figure 10. Mignola – Marche region. Sensory profile: medium olive fruity intensity, with a peculiar flavor of soft fruits; medium intensity of pungency notes and marked bitter taste.

http://dx.doi.org/10.5772/51772

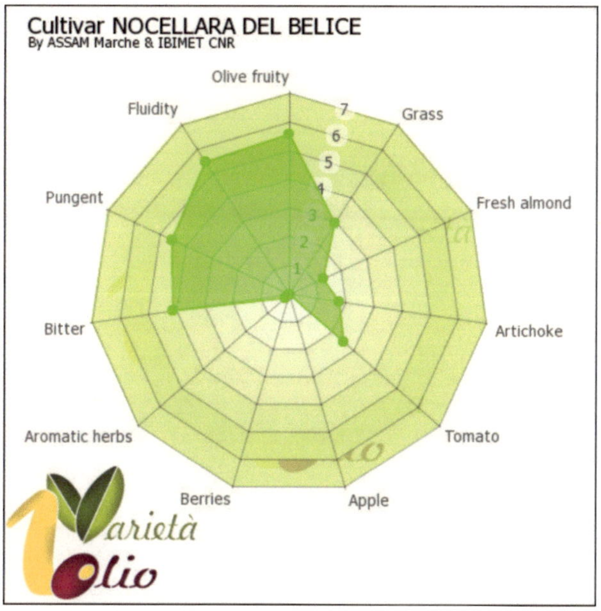

Figure 11. Nocellara del Belice – Sicilia region. Sensory profile: medium-high olive fruity intensity, with grassy and tomato notes and light scent of artichoke and almond; well balanced taste with medium intensity of bitter and pungency notes.

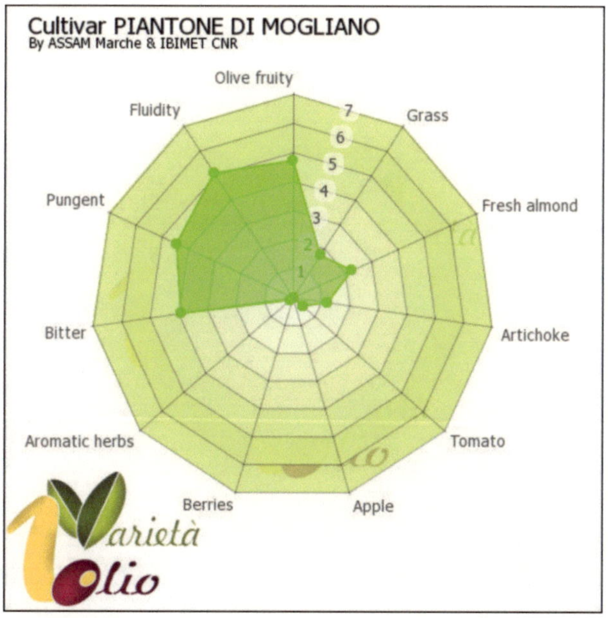

Figure 12. Piantone di Mogliano – Marche region. Sensory profile: medium olive fruity intensity, with almond scent; medium-light intensity of pungency and bitter taste.

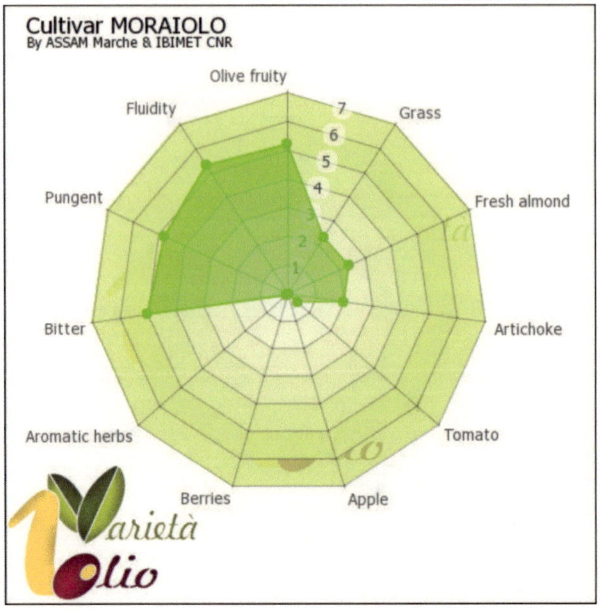

Figure 13. Moraiolo – Central Italy. Sensory profile: medium olive fruity intensity, with scents of grass, almond and artichoke; medium intensity of pungency and bitter taste.

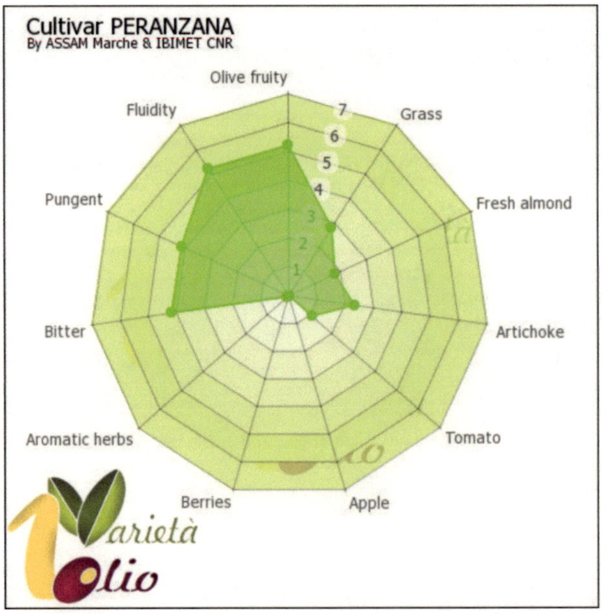

Figure 14. Peranzana – Puglia region. Sensory profile: medium olive fruity intensity, with scent of grass, artichoke, almond and tomato; medium intensity of pungency and bitter taste.

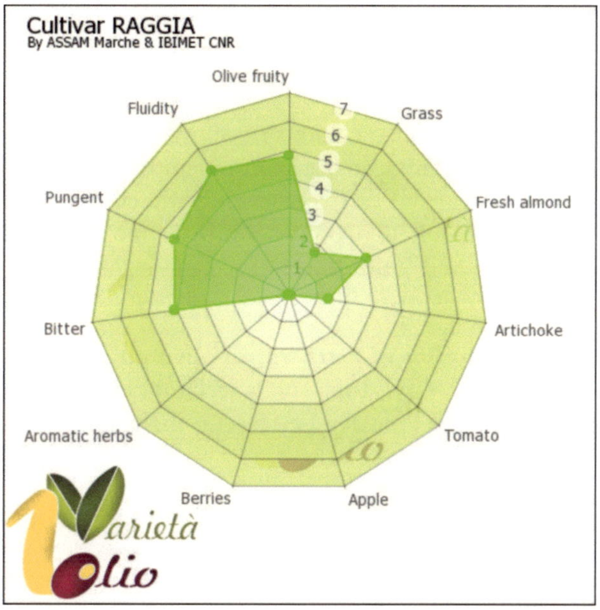

Figure 15. Raggia – Marche region. Sensory profile: medium olive fruity intensity, with strong green almond scent and light grass and artichoke flavor; well balanced to taste with medium intensity of pungency and bitter taste.

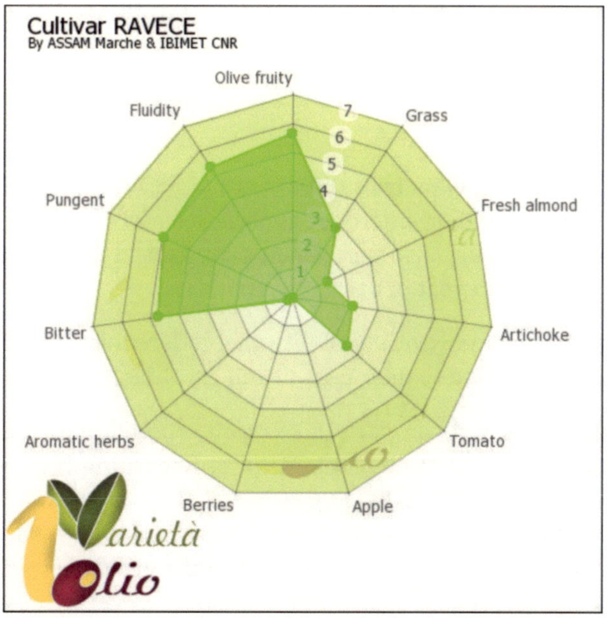

Figure 16. Ravece – Campania region. Sensory profile: medium-high olive fruity intensity, with grass, tomato and artichoke scent along with light almond scent; medium intensity of pungency and bitter taste.

3. The monovarietal olive oil quality and the influence of genetic matrix and of crop year on chemical and sensory profiles

The availability of this monovarietal oil database allows for a statistic elaboration of the data in order to meet different aims of the research into olive cultivation and olive oil quality. Some studies will be carried out considering the cultivars Frantoio and Leccino, which are widespread along the Italian peninsula, in order to evaluate the effect of the environment (climate, altitude and latitude) on the chemical and sensory profiles of monovarietal olive oils.

Moreover the quality and typicality of extra virgin olive oil are primarily determined by genetic, agronomical, environmental factors, and by technological parameters of oil processing [7,8,9]. Genetic matrix (cultivar) plays a key role in the chemical and sensory quality of the oil [10].

It is important to underline that a smaller number of studies has considered the seasonal effect on the chemical and sensory profile of olive oil. The seasonality, which is deeply related to the different climate events of the crop year, may influence the ripening process of olives, thus affecting the oil composition and the resulting quality of olive oil. This thesis is supported by a study carried out by IBIMET-CNR and ASSAM on the evaluation of the influence of the cultivar and seasonality, as well as their interaction on monovarietal oil composition.

The study was performed on 1108 monovarietal oils from the 16 most representative Italian cultivars.

Nutritional properties, expressed as fatty acid and total phenols contents, and the sensory profiles were considered.

The procedure was based on the analysis of variance (ANOVA) by a complete factorial design in order to examine treatment interdependencies (variety and crop year). A Principal Components Analysis (PCA) was also performed on chemical and sensory data separately, using mean values of each crop year of each cultivar collected.

Fatty acids with the highest index of variability (heptadecenoic, linoleic, oleic, stearic and palmitic) were selected according to their p-level and F-values and submitted to PCA.

Table 1 reports mean values and comparison of mean separation analysis of the fatty acids belonging to the 1108 monovarietal olive oil samples. Regarding oleic acid, the contents of which as reported to EC Official Reg. EC Reg.702/2007 [11] range from 55 to 83%, the range of oleic acid of oils considered in this study varied from 71% to 77% showing their high nutritional level, some cultivars such as Coratina and Itrana are characterized by the production with the highest amount of acid oleic (above 77%).

Considering organoleptic quality, Table 2 shows mean values and comparison of mean separation analysis of the sensory attributes of the monovarietal oils. All monovarietal oils have presented significant intensities of grass attributes with the highest levels of 3.2 noted in Ascolana Tenera and Biancolilla. Considering the peculiar attributes, which are heavily cultivar–dependent, such as fresh almond, artichoke, tomato, aromatic herbs and berries [12, 13], oils produced by Coratina, Frantoio, Leccino, Moraiolo and Piantone di Mogliano are distinguished for their high intensity of fresh almond, a typical pleasant flavour which characterized these cultivars.

	Heptadecenoic	Linoleic	Oleic	Palmitic	Stearic	Total phenols
ASCOLANA T.	0.20c	6.10gh	75.57cd	13.42cd	1.98de	394def
BIANCHERA	0.10de	5.98h	76.26bc	12.70ef	2.52b	646a
BIANCOLILLA	0.25a	9.38b	71.92i	13.69bc	2.22c	327f
BOSANA	0.10de	10.07a	72.83h	12.71ef	2.09cd	440bcd
CASALIVA	0.10de	6.74e	76.84b	12.38fg	1.70fg	411cde
CORATINA	0.08f	7.07de	77.74a	11.24h	1.80efg	588a
FRANTOIO	0.10de	6.99de	76.09bc	12.77ef	1.81efg	495b
ITRANA	0.09def	6.17fgh	77.66a	12.06g	1.80efg	329f
LECCINO	0.11d	6.69ef	75.00ed	14.01b	1.74fg	414cde
MIGNOLA	0.10de	8.71c	71.42i	14.63a	1.84efg	503b
MORAIOLO	0.09def	7.41d	75.59cd	13.00de	1.68g	504b
NOCELLARA B.	0.11d	8.20c	73.73fg	12.85ef	2.63ab	358ef
PERANZANA	0.08ef	9.41b	73.30gh	13.00de	1.87efg	375def
PIANTONE M.	0.22b	6.55efg	76.58b	12.14g	1.97de	395def
RAGGIA	0.10de	7.49d	74.45ef	13.40cd	1.89ef	414cde
RAVECE	0.08ef	9.23b	73.28gh	12.51efg	2.78a	473bc

Table 1. Mean values and comparison of mean separation analysis (ANOVA) of the fatty acids relative to the 1108 monovarietal olive oil samples.

All monovarietal oils considered in this study presented a significant intensity of artichoke flavours.

Bosana, Peranzana, Itrana and Ravece oils exhibited the highest intensity, while in Piantone di Mogliano and Leccino oils, slight artichoke attributes were noted. With regard to this last attribute, oils of Ravece, Ascolana Tenera and Itrana are distinguished also for their high intensity of tomato flavour. Berries flavour characterized the oil produced by the Mignola cultivar.

All monovarietal extra virgin olive oil considered in this study were characterized by a significant level of bitterness showing a range from 3.9 to 5.3. In particular Piantone di Mogliano, and Biancolilla oils presented the lowest intensity of bitterness. It is interesting to underline that the same oils were also characterized by a lower total phenol content (see tab. 1). By contrast, the monovarietal oils of Coratina, Bianchera and Mignola which exhibited the highest intensity of bitterness, also showed the highest phenolic content.

	Olive fruity	Grass	Fresh almond	Artichoke	Tomato
ASCOLANA T.	5.9a	3.2a	0.9h	1.8bcde	2.7a
BIANCHERA	5.3cde	2.2def	2.1ef	1.6cdef	1.0c
BIANCOLILLA	5.4cd	3.2a	1.8f	1.8bcde	1.0c
BOSANA	5.3de	2.6bcd	1.7fg	2.3a	0.8cd
CASALIVA	5.4cd	2.4cd	3.3a	1.5defg	0.2e
CORATINA	5.1efg	1.9efg	2.5cde	1.7bcde	0.3e
FRANTOIO	5.2def	2.3de	2.9abc	1.7cdef	0.3e
ITRANA	5.7ab	3.1a	1.3gh	2.2ab	2.3b
LECCINO	4.8gh	1.8fgh	2.6bcd	1.1g	0.2e
MIGNOLA	5.0fg	1.4h	1.2h	0.6h	0.1e
MORAIOLO	5.2def	2.3de	2.4de	2.0abcd	0.4de
NOCELLARA B.	5.6bc	2.9ab	1.3gh	1.8bcde	2.5ab
PERANZANA	5.2def	2.8abc	1.8f	2.3a	1.1c
PIANTONE M.	4.7h	1.7gh	2.2de	1.2fg	0.4de
RAGGIA	4.8gh	1.7gh	3.0ab	1.4efg	0.1e
RAVECE	5.7ab	2.8abc	1.3gh	2.1abc	2.5Ab
	Berries	**Aromatic herbs**	**Bitter**	**Pungent**	
ASCOLANA T.	0.0b	0.4a	4.7cd	5.0ab	
BIANCHERA	0.0b	0.1c	5.1ab	5.2a	
BIANCOLILLA	0.0b	0.1c	4.0f	4.5cdef	
BOSANA	0.0b	0.1c	4.8bc	4.7bcd	
CASALIVA	0.0b	0.1c	4.3def	4.5cdef	
CORATINA	0.0b	0.1c	5.3a	5.0ab	
FRANTOIO	0.1b	0.1c	4.7cde	4.7bcd	
ITRANA	0.0b	0.3ab	4.2f	4.2ef	
LECCINO	0.0b	0.1c	4.3ef	4.4def	
MIGNOLA	1.8a	0.3abc	5.1ab	4.7bcd	
MORAIOLO	0.0b	0.1c	5.0abc	4.8bc	
NOCELLARA B.	0.0b	0.2bc	4.1f	4.5cde	
PERANZANA	0.0b	0.1c	4.2f	4.1f	
PIANTONE M.	0.0b	0.1c	3.9f	4.4cdef	
RAGGIA	0.0b	0.1c	4.1f	4.5cdef	
RAVECE	0.0b	0.2bc	4.7bcd	4.9ab	

Table 2. Mean values and comparison of mean separation analysis (ANOVA) of the sensory attributes relative to the 1108 monovarietal olive oil samples.

Regarding the influence of the genetic matrix and the crop year in Table 3 we see that in all fatty acids analysed, the effects of the cultivar and crop year are highly significant. The effect of the interaction between the two factors is also highly significant on the content of fatty acid, with the exception of palmitoleic, heptadecenoic and heptadecanoic acids. Also the total phenolic contents are heavily influenced by both the cultivar and the year, as well as the interaction between the two factors. It is interesting to underline that both factors (cultivar and crop year) have similarly significant influence on the contents of the most important fatty acids as linoleic, linolenic, oleic, palmitic and palmitoleic.

For oleic and palmitoleic the main factor was, however, the cultivar, in fact ANOVA procedure explains the 68.30% and 70.32% of its variation respectively. The cultivar did not represent a great source of variability for linolenic acid: only 7.88%, while the crop year shows a variation of 85.95%.

Parametr	Cultivar	Crop year	Cultivar x crop year
Eicosanoic	24.34 ***	10.49 ***	65.17 ***
Eicosenoic	10.98 ***	80.78 ***	8.24 ***
Heptadecanoic	56.17 ***	20.08 ***	23.75 *
Heptadecenoic	82.95 ***	8.11 ***	8.94 ns
Linoleic	69.61 ***	20.07 ***	10.32 ***
Linolenic	7.88 ***	85.95 ***	6.17 ***
Oleic	68.30 ***	17.58 ***	14.12 ***
Palmitic	55.37 ***	27.72 ***	16.91 ***
Palmitoleic	70.32 ***	9.44 ***	20.24 *
Stearic	52.43 ***	39.64 ***	7.93 ***
Total phenols	51.32 ***	26.64 ***	22.04 ***

Table 3. Variability expressed as percent of the Total Sum of the Squares for fatty acid composition and total phenols. *, **, *** Significant F-values the 0.05 (*), 0.01 (**) or 0.001 (***) level, respectively; ns = nonsignificant.

The sensory profiles of the 1108 oil samples were submitted to the ANOVA procedure by a complete factorial design. The effect of the cultivar factor is highly significant on the sensory attributes. Olive fruity, grass, fresh almond, tomato and berries were strongly influenced by the cultivar. For these attributes the variability, expressed as percentage of the total sum of the squares, the cultivar factor is characterized by a range from 71.77% to 90.75%. In general the crop year factor for all the sensory attributes remains limited and the interaction with the cultivar is not significant with the exception of berries, bitter and pungent (table 4).

Parametr	Cultivar	Crop year	Cultivar x crop year
Olive fruity	71.77 ***	5.93 ***	22.30 ns
Grass	75.92 ***	2.54 ns	21.54 ns
Fresh almond	73.59 ***	9.81 ***	16.60 ns
Artichoke	56.96 **	15.38 ***	27.66 ns
Tomato	90.75 ***	1.48 *	7.77 ns
Berries	75.71 ***	0.99 **	23.30 ***
Aromatic herbs	26.86 **	10.17 *	62.97 ns
Bitter	49.02 ***	17.04 ***	33.94 ***
Pungent	26.35 ***	42.63 ***	31.02 ***

Table 4. Variability expressed as percent of the Total Sum of the Squares for sensory attributes. *, **, *** Significant F-values the 0.05 (*), 0.01 (**) or 0.001 (***) level, respectively; ns =not significant.

4. The Italian olive oil typology

The decision to carry out a study of a large number of labelled commercial extra virgin olive oils was taken in order to provide the consumer with information about the chemical and sensory properties of extra virgin olive oils which are currently available on the Italian market. The commercial potential of the monovarietal oils can be exploited either in terms of purity, relying on the specific characteristics of the single cultivar, or mixing the monovarietal oils from each cultivars as a "blend" based on the different typologies of Italian olive oil.

For this purpose these oil typologies were assessed by clustering the collected olive oil data according to different sensory profiles. Descriptive analysis and hierarchical cluster analysis of sensory characters were performed. Monovarietal oils were clustered in six different sensory typologies emphasising the variability and the depth of aromas characterising Italian Monovarietal oils.

Such classification of monovarietal oils typologies may help the consumer in making an informed choice, and in matching more easily with the wide range of flavours found in Italian cuisine.

These monovarietal oils were classified as belonging to typology 1:

Caninese, Carboncella, Carpellese, Cornetta, Dolce Agogia, Dolce di Rossano, Dritta, Gentile di Chieti, Gentile di Larino, Leccino, Limoncella, Nebbio, Ogliarola, Ogliarola del Bradano, Paesana Bianca, Piantone di Mogliano, Raggia, Raggiola, Rajo, Razzola, Rosciola, Salviana, Sargano di Fermo, Taggiasca.

http://dx.doi.org/10.5772/51772

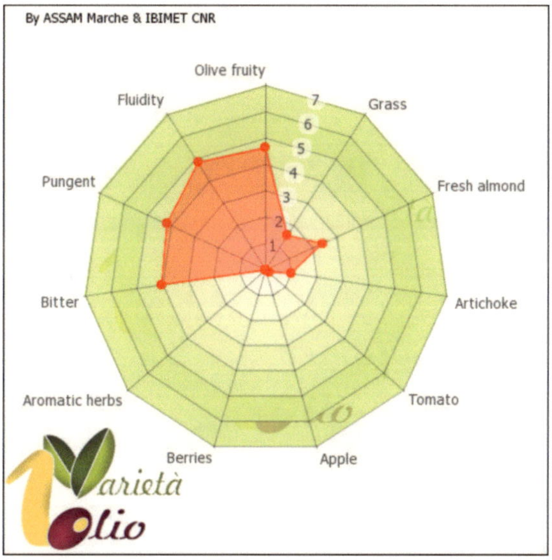

Figure 17. Typology 1 Sensory profile: medium olive fruity intensity, with prevalent almond scent and light notes of grass/leaf and artichoke; pungent and bitter taste of medium-light intensity.

These monovarietal oils were classified as belonging to typology 2:

Casaliva, Coratina, Correggiolo, Frantoio, Moraiolo, Ogliarola Garganica, Oliva Nera di Colletorto, Olivastra Seggianese, Pendolino, Raggiolo, Razzo, San Felice, Sargano di Ascoli.

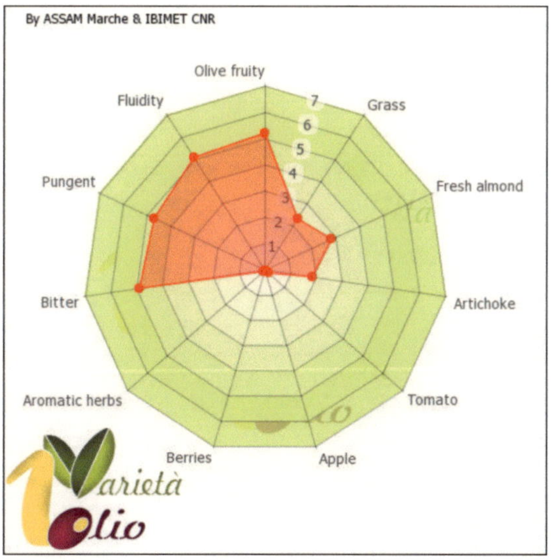

Figure 18. Typology 2 : Sensory profile: medium olive fruity intensity, with prevalent almond scent and light notes of grass/leaf and artichoke; pungent and bitter taste of medium intensity.

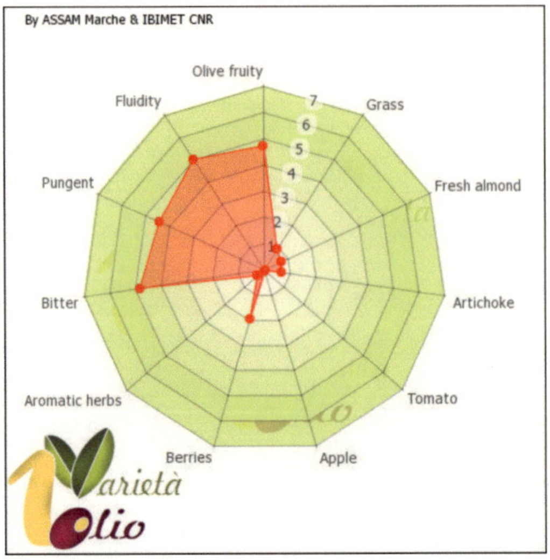

Figure 19. Typology 3 Sensory profile: medium olive fruity intensity, with peculiar soft fruits scent; pungency and bitter taste of medium intensity.

These monovarietal oils were classified as belonging to typology 3: Cellina di Nardò, Mignola, Ogliarola Salentina

These monovarietal oils were classified as belonging to typology 4: Biancolilla, Bosana, Carolea, Coroncina, I77, Majatica di Ferrandina, Maurino, Orbetana, Peranzana, Prempesa, Salella, Semidana, Tonda del Matese.

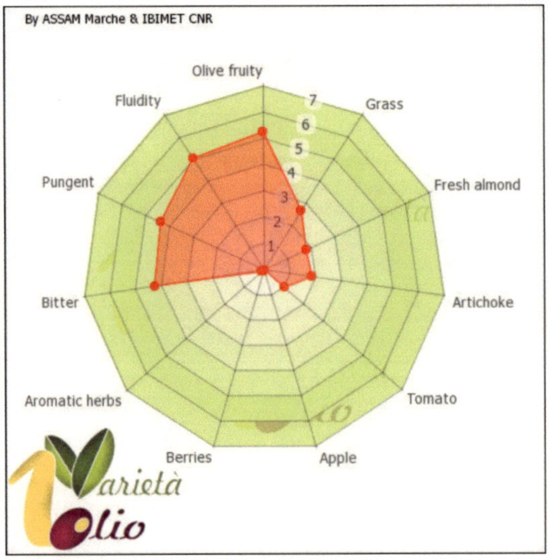

Figure 20. Typology 4 Sensory profile: medium olive fruity intensity, with scent of grass, artichoke fresh almond and tomato; pungency and bitter taste of medium light intensity.

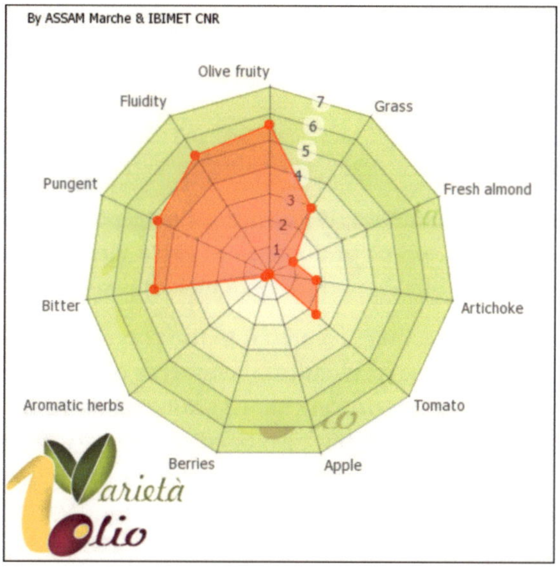

Figure 21. Typology 5 Sensory profile: medium high olive fruity with grassy notes, tomato and artichoke scent and light flavour of fresh almond; pungency and bitter taste of medium intensity.

These monovarietal oils were classified as belonging to typology 5. Ascolana Tenera, Cerasuola, Ghiacciolo, Itrana, Nera di Oliena, Nocellara del Belice, Nocellara Etnea, Nocellara Messinese, Ortice, Ravece, Tonda Iblea

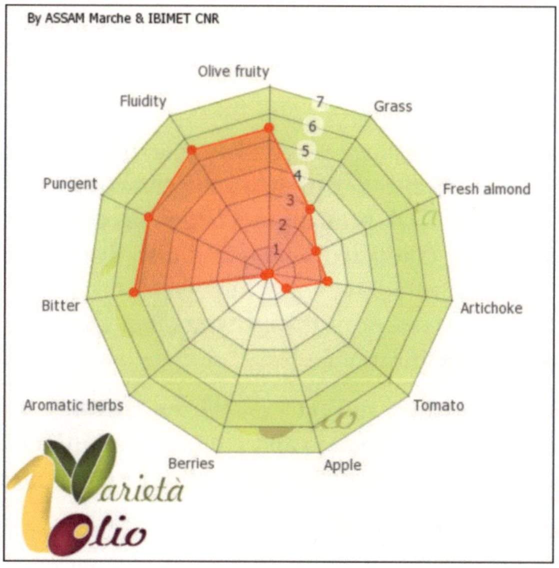

Figure 22. Typology 6 Sensory profile: medium high olive fruity with grass/leaf and artichoke notes, light flavor of fresh almond and tomato, medium intensity of bitter and pungency taste.

These monovarietal oils were classified as belonging to typology 6: Bianchera, FS17, Intosso, Lantesca, Leccio del Corno, Nostrana di Brisighella, Piantone di Falerone, Picholene

5. Conclusions

At a national level the varietal biodiversity culture is being promoted ever more heavily, resulting in increasing diversification of production of extra virgin olive oil which constitutes the necessary basis for creating blends and PDOs which appeal to consumers.

Various Italian regions are conducting research into the promotion of the genetic heritage of the olive cultures, drawing on social and cultural elements of olive culture. For a strong and healthy olive culture, the cultivation process should not only fulfil the demands of intensification and optimization of production, but also balance this with respect for the ancient traditions and heritage- traditions and heritage which we see throughout Italy in the form of monumental trees, archaeological exhibits, ancient tools and gastronomic traditions which have extra virgin olive oil at their very heart.

Development of olive production with care being taken to respect biodiversity and cultural traditions- and, as a consequence, the different autochthonous genotypes- is key to ensuring sustainable and environmentally friendly olive production processes.

The will to proceed with recovery and exploitation of Italian germplasm, encourages the development of marginal areas, but also allows for the protection of biodiversity and ecological systems in specific areas where the olive tree plays an important role for buoyancy and hydro-geological protection for the characterization of the landscape.

The exploitation of monovarietal oil results in the propagation of many native Italian varieties involving research institutes, University and nurseries called upon to halt the erosion of genetic heritage of Italian olive.

The unique Italian monovarietal heritage plays a key role in the "diversification" culture that should guide Italian production in order to avoid the standardization of products recently observed in the GDO market. The promotion of the "diversification" culture has to be reached both by increasing the cultivated land – and also discouraging the substitution of pre-existing cultivar or the implantation of new "universal" ones with the sole aim of greater production - and by the reinforcement of the elements that characterize GDO territoriality: organoleptic and sensorial diversification.

The quality oil market is expanding and Italy is still the reference point at an international level. The operators of the production chain and national institutions have the task of developing appropriate strategies to strengthen the position of production, sales and marketing.

As has happened in the case of wine, the varieties typical to these regions may become a symbol of high quality product and find a better place in the market.

This approach can be a first step toward traceability and authenticity of these particular productions in order to protect the interest of both consumer and producer.

The authors are aware of the numerous variables: mill typology, olive ripening index and agronomic practice, which influence the overall olive oil quality. These variables are not usually known for commercial oils.

Knowledge of the chemical and sensory profiles of the Italian monovarietal olive oils could potentially start a certification process for these oils, thus leading to greater guarantees of origin and consequently greater guarantees of quality for the consumer.

Author details

Annalisa Rotondi[1*], Massimiliano Magli[1], Lucia Morrone[1], Barbara Alfei[2] and Giorgio Pannelli[3]

*Address all correspondence to: a.rotondi@ibimet.cnr.it

1 Institute of Biometeorology, National Research Council, Italy

2 Agri-food Service Agency of Marche Region, Italy

3 CRA, Sperimental Institute for Olive Cultivation, Italy

References

[1] Bartolini, G. (2008). Olive germplasm. *Cultivar and World Wide Collections.*, http.// www.oleadb.eu/.

[2] Alfei, B., Magli, M., Rotondi, A., & Pannelli, G. (2008). Chemical and organoleptic characterization of Italian Monovarietal olive oils. *The Sixth International Symposium on Olive Growing*, September 9-13 Evora Portugal.

[3] Alfei, B., Magli, M., Rotondi, A., & Pannelli, G. (2006). Statistical analyses of sensory prperties of Italian Monovarietal olive oils. *Olivebioteq- Second International Seminar*, November 5-10 ,Marsala Italy.

[4] Rotondi, A., Alfei, B., Magli, M., & Pannelli, G. (2010). Influence of genetic matrix on chemical and sensory profiles of italian monovarietal olive oils. *28th International Horticultural Congress*, August 22-27Lisboa Portugal.

[5] European Commission regulation No. 640/2008 (2008). Official Journal of European Community, L 178, July 4th, 11-16.

[6] Alfei, B., Magli, M., Rotondi, A., & Pannelli, G. (2012). La varietà da l'impronta all'olio, ma anche la stagione può influire sulle caratteristiche chimiche e sensoriali. *Catalogo degli oli monovarietali Olivo&Olio* [6], 11-15.

[7] Cerretani, L., Bendini, A., Rotondi, A., Lercker, G.., & Gallina, Toschi. T. (2005). Analytical comparison of monovarietal virgin olive oils obtained by both a continuous industrial plant and a low-scale mill. *European Journal Lipid Science Technology*, 107, 93-100.

[8] Lazzez, A., Perri, E., Caravita, M. A., Khlif, M., & Cossentini, M. (2008). Influence of olive maturity stage and geographical origin on some minor components in virgin olive oil of the Chemlali variety. *Journal of Agricultural and Food Chemistry*, 56, 982-988.

[9] Angerosa, F., Mostallino, R., Basti, C., & Vito, R. (2001). Influence of malaxation temperature and time on the quality of virgin olive oils. *Food Chemistry*, 72(1), 19-28.

[10] Rotondi, A., Alfei, B., & Magli, M. (2011). Pannelli Influence of genetic matrix and of crop year on chemical and sensory profiles of Italian monovarietal olive oils. *Journal of Science Food Agriculture*, 90, 2641-2648.

[11] European Commission regulation No. 702/2007. (2007). *Official Journal of European Community*, L 161, June 21th, 11-27.

[12] Angerosa, F. (2002). Influence of volatile compounds on virgin olive oil quality evaluate by analytical approaches and sensory panels. *European Journal Lipid Science Technology*, 104, 639-660.

[13] Aparicio, R., Morales, M. T., & Alonso, V. (1997). Authentication of European virgin olive oils by their chemical compounds, sensory attributes and consumer attitudes. *Journal of Agricultural and Food Chemistry*, 45, 1076-1083.

Chapter 9

Challenges for Genetic Identification of Olive Oil

Sattar Tahmasebi Enferadi and Zohreh Rabiei

Additional information is available at the end of the chapter

http://dx.doi.org/10.5772/53546

1. Introduction

Olive oil is the oil extracted exclusively from fruit of *Olea europaea* L. only by means of mechanical methods or other physical procedures that do not cause any alteration of the glyceric structure of the oil thus preserving its characteristics and properties. The healthy properties of olive oil are well known in the Mediterranean diet, in which virgin olive oil is the main source of fat [1].

In comparison to commonly used vegetable oils, the cost of olive oil is higher. As such, olive oil comes up against adulteration with other cheaper oils as well as the use of unapproved production methods [2]. Blending premium olive oil with low quality oils (mostly pomace) or with other plant oils such as hazelnut (*Corylus avellana*), soya (*Glycine max*), almond (*Prunus dulcis*), maize (*Zea mays*), sunflower (*Helianthus annuus*) and sesame (*Sesamum indicum*) has a great negative effect on olive oil trade [3]. This defines an urgent action in confirming olive oil authenticity

In the recent years, the European Union (EU) has introduced sever regulations about virgin olive oil origin confirmation with the aim of governing its label and protecting its producers and consumers from fraudulent activities [3]. Besides, the concept of certified brands such as PDO (Protected Designation of Origin) and PGI (Protected Geographical Indication) has been recently introduced to provide more tools to protect olive oil within both EU and non EU countries. Although aforementioned tools confirms oil origin (geographical cultivation origin or place of processing), it is still to decide if olive oil PDOs and PGIs are safeguarded from fraudulent labeling [4].

The authenticity of olive oil, and particularly that of a virgin olive oil, is conventionally assessed by monitoring of several components such as sterols, phenols, fatty acids, triacylglycerols, volatile compounds and tocopherols. However, the analytical analyses have their limits and chemical composition of virgin olive oil is influenced by genetic (variety) and environmental factors (climatological and edaphologic conditions). As the composition of extra virgin olive oil is the result of a complex interaction among olive variety, environmental conditions, fruit ripening and oil extraction technology, two main approaches, botanical and geographical origin identification, are focused to trace olive oil. However, in both cases, the selection of appropriate markers is sophisticated and needs more attention [5]. This has promoted a growing interest towards the application of DNA-based markers since they are independent from environmental conditions. The evaluation of DNA nucleotide sequences can provide more precise information, which can be obtained through traditional morphological markers or chemical composition analysis. Thus, specific protocols for DNA isolation from olive oil have been developed [6-9]. However, the application of DNA-based methods requests the knowledge on nucleotide sequences of olive. This information for olive is back to 1994, when the first sequence of *Olea europaea* L. has deposited in NCBI (National Center for Biotechnology Information) [10].

On the other hand, the advent of molecular markers offers a powerful tool to uncover synonymy and parentages between olive cultivars, and reveals a large admixture amongst varieties of different geographic origin of olive that consequently improves PDO olive oil recognition [11]. Individual fingerprinting based on molecular markers has become a popular tool for studies of population genetics and analysis of genetic diversity in germplasm collections, including the solution of synonymy/homonymy and analysis of paternity and kinship.

In this chapter, researches for genetic identification of olive oil in introducing unequivocal identifiers for authentication and traceability of olive oil will emphasis as a crucial concept to be overcome for international olive oil trade.

2. An overview on olive oil

Olive tree (*Olea europaea* L.) represents the most important oil producing crop in the Mediterranean basin. Olive tree is a diploid species (2n = 46) that is able to survive for a long time [12-13], is outcrossing and sometimes self-incompatible which implies that seeds are produced by cross-pollination [13-15].

This species belongs to monophyletic *oleaceae* family. *Oleaceae* comprises about 600 species and 24 genera [16, 17]. Within this family, *Olea* and ten other (extant) genera constitute the subtribe *Oleinae* within the tribe *Oleeae* [16]. Thirty-three species and nine subspecies of evergreen shrubs and trees have been circumscribed in *Olea* based on morphological characters [18]. In addition, these taxa are classified in three subgenera, *Olea, Paniculatae* and *Tetrapilus,* the first of which has two sections (*Olea* and *Ligustroides*). Section *Olea* is formed exclusively by the olive complex (*Olea europaea*), in which six subspecies are recognized [18, 19]. This subgenus is distributed from South Africa to China, across the Saharan mountains, Macaronesia and the Mediterranean basin. The cultivated one is *Olea europaea* subsp. *europaea* var. *europaea.* that found outside of its native range as a result of human-mediated dispersal; it has been repeatedly introduced in the New World and has become naturalized and has invaded numerous areas in Australia, New Zealand and the Pacific islands [18, 20].

Olive is the second most important oil fruit crop cultivated worldwide after oil palm. Its cultivation covers over eight million hectares of land, predominantly concentrated in the Mediterranean basin, where 70% of the olive oil produced is consumed [21]. The olive tree is a glycophytic species that shows a high tolerance to drought and salt stresses, if compared with other fruit trees that are generally salt sensitive [22]. Olive oil is produced solely from the fruit of the olive tree (*Olea europaea* L.) and differs from most of the other vegetable oils in the method of extraction, allowing it to be consumed in crude form, hence conserving its vitamins and other natural healthy high-value compounds [13].

The olive oil is known for its beneficial effects on health, such as ability to reduce blood pressure and low-density lipoprotein (LDL) cholesterol, as well as for its cancer prevention, antimicrobial and antioxidant virtues [23, 24]. The international olive council Resolution no. res-3/89-iv/03 [25, 26] catagorised virgin olive oil as: (i) "Extra virgin olive oil": virgin olive oil that has a free acidity, expressed as oleic acid, of not more than 0.8 g per 100 g; (ii) "Virgin olive oil": virgin olive oil that has a free acidity, expressed as oleic acid, of not more than 2 g per 100 g; (iii) "Ordinary virgin olive oil": virgin olive oil that has a free acidity, expressed as oleic acid, of not more than 3.3 g per 100 g.

The increase in the demand for high-quality olive oils has led to the appearance in the market of olive oils, elaborated olive oil with specific characteristics. They include oils of certain regions possessing well-known characteristics, that is, olive oils with a denomination of origin, or with specific olive variety composition. Olive oils obtained from one genetic variety of olive or from several different varieties are called monovarietal or coupage, respectively. Monovarietal olive oils have certain specific characteristics related to the olive variety from which they are elaborated [5]. However, coupage olive oils are obtained from several olive varieties to achieve a special flavor or aroma [13].

3. Olive oil adulteration

Olive oil is one of the most valuable single products of the agro-food industry. It is made from diverse cultivars either mixed or single. Those ensure different tastes and typicity, and these may be also enhanced by the region of production of cultivars [8].

Recently PDO and PGI olive oils appeared that requiring precise definition of several parameters such as cultivar, geographical origin, agronomic practice, production technology, and organoleptic qualities. The quality of these monovarietal oils is associated with superior taste, consistency and colour and is directly related to the olive cultivar. Therefore, the authenticity efforts concentrated on the identification of their varietal origin as well as their adulteration with lower grade, processed olive oils [27].

That is why a well-documented traceability system has become a requirement for quality control in the olive oil chain.

DNA, being not environmentally labile, considered a great potential to be used as a means of identifying the varieties of the trees purporting to be the source of a given sample of olive oil [3].

Genetic traceability implies the control of the entire chain of food production and marketing, allowing the food to be traced through every step of its production back to its origin. The verification of olive oil traceability is necessary for the prevention of deliberate or accidental mixing or mislabeling, which is very important in the international trade.

4. Conventional identification tools of olive oil: Chemical analysis

As the quality of an olive oil depends on the olive variety from which it is elaborated, the production of olive oils from certain varieties with appreciated quality has increased [28]. The olive variety selection is mainly based on its adaptation to different climatic conditions and soils.

Several analytical determination has improved during years, because of improving both of knowledge on olive oil composition and of analytical techniques and instrumentation. Unfortunately, adulterations, too, improved and became more sophisticated; for example, it is recognized that by careful blending high oleic oils (such as sunflower oil) the obtained product well fits the fatty acid composition of olive oil. However, they usually have a "wrong" sterol composition and those who perform frauds were able to eliminate sterols, because of this act, the absolute amount of sterols was enclosed in the standard as well as the measurement of sterol dehydration products (such as stigmastadienes).

To relate the fatty acid composition of olive oils with the cultivar, Mannina *et al.*, (2003) studied olive oil in a well-limited geographical region, with no consideration of the pedoclimatic factor (soil characteristics such as temperature and humidity) [29]. A relationship between the fatty acid composition and some specific cultivars has been observed [5].

The volatile fraction in olive oils, which represents one of the most important qualitative aspects of this oil, consists of a complex mixture of more than 100 compounds, but the most important substances useful for olive cultivar differentiation are the products of the lipoxygenase pathway (LOX). Only a subset of volatile compounds and a combination among them could provide valuable information for olive cultivar differentiation [5].

In fact, genetic and geographic factors influence the volatile compound production of the olive fruits and affect the differentiation of olive oils according to their olive variety [30, 31]. The volatile compound contents allowed differentiation among monovarietal olive oils and even identification of the technique used for olive oil production [32].

The colour of a virgin olive oil is due to the solubilization of the lipophilic chlorophyll and carotenoid pigments present in the fruit. The green-yellowish colour is due to various pigments, that is, chlorophylls, pheophytins, and carotenoids [33]. Several researchers reported the same qualitative composition in chlorophyll and carotenoid pigments, independent of the olive variety and the time of picking [34, 35]. Cerretani *et al.*, (2006) showed that the carotenoid and chlorophyll content determination using Uv-vis spectrophotometry was not useful to discriminate oils produced from different olive varieties [36]. Lutein/β-carotene ratio has been reported as a tool to differentiate oils from a single cultivar [5].

Tocopherols and hydrocarbons are the compositional markers less studied to date to differentiate olive oils. An important common aspect is that the content and composition of these markers are highly affected by the environmental conditions, the fruit ripening, and the extraction technology [5].

5. Need for cultivar identification, search for new approaches: Molecular markers (DNA) vs. biochemical descriptors

Olive oils labeled with their region of origin are sold at a premium price. This premium is greatest for oil from those regions associated with superior taste, consistency or colour. For cold-pressed oils (extra virgin and virgin), these properties are associated with the cultivar and the environment.

A lot of research has been carried out to assure the authenticity of olive oil through chemical analysis [37-40]. However, several difficulties have been encountered in distinguishing olives and olive-oils from different cultivars because their characteristics are strongly influenced by environmental conditions. On the other side, accurate and rapid identification of cultivars is especially important to obtain a reliable label of origin.

In such a case, some DNA-based technologies can help in revealing either the authenticity or the different origin of lots that have contributed to the olive oil. This action discourage from the adulteration with extraneous material of lower cost and value.

6. Olive oil genetic traceability

The genomic analysis of olive oil involves two main obstacles; extraction of DNA from an oily matrix and selection of appropriate molecular markers that can provide a trustable result. Overcoming these two important limits make possible the assessment of genetic traceability of olive oil.

DNA analysis offers an alternative approach, relative to other macromolecules and metabolites, due to it is less influenced by environmental and processing conditions, enables genome fingerprinting with consequent identification of variety/type composition.

Significant amounts of DNA are present in olive oil obtained by cold pressing [6]. However, the filtration process lowers DNA concentration, which tends to disappear due to nuclease degradation [7, 41]. Application of molecular markers to trace foods brought new benefit to consumers.

Several works described the application of molecular markers to genetic recognition of the cultivar composition of monovarietal olive oils. However, these works describe the application of multilocus markers, such as RAPDs or AFLPs [9, 41] or microsatellites [8, 42].

6.1. Achievements in olive oil genetic traceability between 2000 – 2006

During the years between 2000 -2006, many important events regarding genetic traceability of olive oil have been recorded. The most outstanding achievement was defining a protocol for DNA extraction from olive oil [8-9, 41]. The other success in this context refers to the deposition of several sequences of olive genome on NCBI database, including 458 nucleotides and 24 ESTs (Expressed Sequence Tags). These sequences facilitated the access to more DNA markers. For example, Cipriani *et al.*, (2002) isolated and sequenced 52 microsatellites or simple sequence repeats (SSRs) from nearly 60 positive clones obtained from two 'Frantoio' olive genomic libraries enriched in (AC/GT) and (AG/CT) repeats, respectively [43]. Furthermore, many authors have reported on SSR development in olive and several of them are currently available for DNA analysis [43-46].

These approaches offered new opportunities for researchers to perform such an analysis involving olive cultivar identification and olive oil authentication.

6.1.1. A big obstacle: DNA Extraction

The greatest challenges one faces while using DNA technology is the low quality and highly degraded DNA recovered from the fatty matrices and the impact of oil extraction processing on the size of the recovered DNA. DNA of low, difficult to determine content and of unknown, variable quality would potentially lead to inconsistent and consequently inconclusive results. Although, the concentration of DNA did not appear to be limiting; rather, successful PCR amplification likely depended on the ability of the DNA extraction method to free DNA from inhibitors of PCR present in the olive oil.

In olive oil, once the barrier of DNA extraction has been overcome, several markers could be used to identify olive cultivars that made up a certain olive oil [6].

Muzzalupo *& Perri* (2002) tested some enzymatic mixtures to prevent DNA damage that occurs during crushing and malaxation [41]. They emphasized positive effect of proteinase-K treatment during the malaxation process to provide DNA amenable to random amplified polymorphic DNA (RAPD)-PCR amplification.

However, Busconi *et al.*, (2003) defined a reliable DNA extraction method via CTAB method from 50-100 mL lab-made monovarietal oil and commercial extra virgin olive oil [9]. The suggested method concerned both quantity and quality of DNA.

Breton *et al.*, (2004) have used several supports to retain DNA checking different techniques (silica extraction, hydroxyapatite, magnetic beads, and spun column) [8]. The method using magnetic beads has been introduced as the most efficient method and they claimed that the running protocol is usable in routine labs to control virgin or crude oil samples and may be used for refined oil, as well.

DNA has been extracted from cell residues recovered by oil centrifugation as reported by Pasqualone *et al.,* (2004) [42]. Doing successful DNA extraction and SSR analysis, the electrophoretic patterns showed an adequate level of amplification and were identical to those obtained from leaves and drupes of the same cultivar [42].

Pafundo *et al.,* (2005) reported optimization of AFLPs for the characterization of olive oil DNA, to obtain highly reproducible, and high quality fingerprints [47]. Her group found that correspondence of fingerprinting by comparing results in oils and in plants was close to 70% and that the DNA extraction from olive oil was the limiting step for the reliability of AFLP profiles, due to the complex matrix analyzed [47].

6.1.2. Molecular markers selection and PCR analysis

AFLP, RAPD, SCAR and SSR have been employed as genetic marker for various cultivar / genotype identification. Muzzalupo & Perri (2002) reached the first unambiguous and reproducible RAPD-PCR amplification of DNA recovered from virgin olive oil [41]. The presence of additional alleles in RAPD profiles deriving from monovarietal oil, missing in the leaves of original varieties were interpreted as signature of pollen DNA, but no data were provided as support of this hypothesis [41].

However, Busconi *et al.,* (2003) showed the correspondence between profiles of the DNA purified from monovarietal oil with that from the leaves of the same cultivar [9]. Although Pafundo *et al.,* (2005) found the aforementioned correspondence close to 70% [47]. Based on their findings, DNA extraction from olive oil was the limiting step for the reliability of AFLP profiles. Their results also suggest that increasing the DNA amount above 200 ng does not improve significantly the quality of AFLPs. However, below this concentration of DNA, the quality of AFLP profiles was reduced.

Breton *et al.,* (2004) and Pasqualone *et al.,* (2004) used SSR to identify olive cultivars contributed in commercial olive oil samples and virgin olive oils [8, 42]. The electrophoretic patterns showed an adequate level of amplification and were identical to those obtained from leaves and drupes of the same cultivar.

6.2. Improvements in olive oil genetic traceability between 2007 – 2012

By 2007 the main obstacle in genetic traceability of olive oil, DNA extraction, has been overcome successfully; however, there are still some publications which introducing DNA extraction protocols specified for olive oil. In this period, the main activities were focused on using different molecular approaches; meanwhile the volume of sequences deposited on NCBI has being increased exponentially which enabled the researchers in profiting other molecular markers such as designing new and more effective primer pairs, on different regions over nuclear regions, such as chloroplast, mitochondrial, and plastomal sequences (NCBI database, including 1405 nucleotides, 7865 ESTs and 26 GSS (Genome Survey Sequences) hints by 18/09/2012). Also, there have been many attempts to establish a better understanding of cultivar differentiation, genetic diversity [48] and identification of new polymorphic regions.

6.2.1. A glance on DNA extraction

Until 2007 a wide range of protocols on DNA extraction from olive oil, either cold press or refined, had been introduced. Then on, the researchers have focused on the application of current molecular tools or searching for new and more appropriate ones. The general idea that has been accepted and reported by several researchers is that olive oil provides very low yields of DNA and has variable degrees of degradation which may limit the applicability of molecular markers [49, 6]. Furthermore, it has been shown that DNA is damaged by oxidation reactions, which may cause DNA lesions and base transitions with production of dangerous adducts [50, 51]. If the DNA is damaged, it could be not properly accessible to the DNA polymerase, which stalls at the sites of damage and the reaction may be interrupted; this being able to influence the length and significance of the synthesized amplicons [52].

6.2.2. Molecular markers selection

Many studies in recent years have employed AFLP, RAPD, ISSR, SSR, LDR/UA, qRT-PCR, SNPs, CE-SSCP, and DNA barcode as genetic tools to produce a reliable platform to identify the cultivars contributing to an olive oil. Pafundo *et al.*, (2007) suggested developing of sequence characterized amplified region (SCAR) markers [53] derived from AFLP profile of olive oil can be instrumental to simplify the determination of varietal composition of an oil sample. A procedure to visualize AFLPs of oil in agarose gel was developed to avoid the usual procedure for SCAR isolation from polyacrylamide gel electrophoresis, expensive, time-consuming, and requiring the use of radioactive isotopes [53]. Finally, Pafundo *et al.*, (2007) mentioned the high correspondence between the profiles obtained with agarose and capillary electrophoresis as an index of the reliability of the used method [53]. In addition, Montemurro *et al.*, (2008) reported a good quality of AFLP profile in oil [49]. It has been possible to improve the sensitivity of AFLP with the optimisation of DNA extraction and restriction/ligation condition.

On the other hand, the effect of olive oil storage time on the quality and quantity of DNA as an analyte for molecular traceability has been assessed via AFLP, which was used due to its multilocus nature, which allows a better discrimination of DNA composition [51]. Comparison of AFLPs was made among profiles of leaves and monovarietal olive oil stored at different times. Montemurro *et al.*, (2008) it has been detected that for some cultivars such as Taggiasca cv. although the AFLP profiles of leaf and of the oil DNA extracted after one and three weeks of storage were similar, the same profiles were quite different for DNA extracted in the one year stored oils [49]. For Carolea, some homologies were found between leaves and oil, whereas for other cultivars such as Leccino and Ogliarola Leccese these homologies were not detectable. Finally, it has been declared that nine months after production; profiles of oil DNA were highly different from the leaf profiles in all examined cultivars [49].

Another study to assess DNA stability in olive oil during storage time proposed the use of lambda DNA as a marker. In this study the progress of DNA fragmentation in olive oil has been monitored throughout a 12-month storage period. *Lambda* DNA was introduced into filtered olive oil samples in three different concentrations as a DNA marker. It has been revealed that the inhibitory effect of olive DNA extracts was increased partially and gradually with the storage period of the olive oil samples used for the DNA extraction [54].

Martines-Lopes *et al.,* (2008) evaluated the efficiency of RAPD, ISSR, and SSR molecular markers for olive oil varietal identification and their possible use in certification purposes [55]. Among eleven RAPD primers tested only two produced reproducible bands in all olive oil samples, which demonstrate low efficiency of this tool. However, it has been shown that ISSR marker system in olive oil is more informative than RAPD. Finally, Martinez-Lopes confirmed that SSR amplification was satisfactory only when water phase DNA was used in the reaction. SSR analysis used to compare the profile of DNA isolated from monovarietal oil with that from leaves of the same cultivar [55].

SSRs have a high discrimination power and so far they are the most employed markers. Several authors [25, 42, 56-59] reported a good match between olive oil and leaf profiles but they did not report any data about repeatability of results. In addition the SSR sequence dramatically influences the efficiency of analysis, as well as the kind of oil [58]. The reproducibility of results was low confirming that the choice of SSR loci and primers is relevant for an efficient analysis. Furthermore, assigning the true size of alleles and resolution of conflicts considers as another obstacle in interpretation of SSR results.

All the authors agree that differences in size and allele drop out in oil may be due to components interfering with PCR reaction, or to the lower quantity and quality of DNA, which makes difficult the selective amplification of DNA for any allele pairs [3].

The appearance of extra alleles detected in oil addresses either the mixing with traces of other oils present on the machinery during milling process or the accidental mixing with other cultivars during harvesting, transportation and processing. In addition, DNA from the pollinators present in the genome of the seed embryo, could potentially contain alleles not present in the genome fruit pulp, invalidating the molecular traceability of olive oil [60].

Pasqualone *et al.,* (2007) demonstrated that microsatellites are useful in checking the presence of a specific cultivar in PDO oil, thus verifying the identity of the product [61]. However, they obtained only the marker profile of the main cultivar in the oil: no signal was detected for the secondary varieties [62]. Specifically, Pasqualone *et al.,* 2007 confirmed the sufficiency of a single microsatellite, GAPU103A, to distinguish Leucocarpa oil from the other samples [61]. An identification key based on the amplification profile of this microsatellite was set up to distinguish the oils from different cultivars [57]. Rotondi *et al.,* (2011) performed a comparison between genetic results, chemical and sensory properties of monovarietal olive oils and demonstrated a very good correspondence between the clustering obtained by SSR analysis and the clustering based on selected fatty acids composition [63].

Single nucleotide polymorphisms (SNPs) are molecular markers which require short DNA amplicons for genotyping [27]. In addition, they are the most abundant markers in the genome; they are stably inherited; bi-allelic in most cases and co-dominant [64]. The most significant comparative advantage of SNPs among all molecular markers is the requirement for short, even shorter than SSRs, approximately 100 nucleotides PCR templates as analytical targets. This advantage can be considered highly critical for heavily processed food matrices such as olive oils and other plant oils due to the highly degraded nature of DNA present in these matrices. Moreover, although the identification of SNPs was considered an expensive task few years ago, the recent developments on high throughput sequencing technological platforms allows the cost- and time-efficient identification of SNPs in every plant species [64].

Consolandi *et al.*, (2008) gave precise and accurate genotype results using ligation detection reaction (LDR)/universal array (UA) on a SNP-containing DNA sequences for the genotyping of olive cultivars. In this assay, alleles are distinguished by a ligation detection reaction and, subsequently, detected by hybridization onto a universal array [6].

Bazakos *et al.*, (2012) reported a successful use of SNPs in tracing olive oil and mentioned neither paternal contribution of embryos was detected in olive oil samples nor did additional peaks in leaf samples [27].

Kumar *et al.*, (2011) proved SNPs variation in noncoding spacer region between psb-trnH and partial coding region of matK of plastid genome. This procedure enabled to discriminate the mixing of canola and sunflower oil into olive oil. This plastid based molecular DNA technology proposed to be used for rapid detection of adulteration easily up to 5% in olive oil [65]. The development of this kind of marker requires a high level of genome sequence information: it is therefore not surprising if only a few SNPs have been reported in olive, where only a small amount of sequence data was available before the year 2009.

To overcome the lack in sequence knowledge, Reale *et al.*, (2006) used both a sequence-based and an arbitrary approach to identify eight SNPs in olive [66].

As conventional PCR technique is not optimal for authentication purposes when quantification is needed, qRT-PCR has been introduced as an efficient tool allowing discarding primers with low PCR efficiency [67]. Wu *et al.*, (2008) employed a sensitive real-time PCR method using the novel fluorescence stain Evagreen (Evagreen intercalates in a sequence independent way in DNA duplexes) established for detection of olive oil, which successfully distinguished olive oil from inferior plant oils [68].

A more recent study using a CE-SSCP (Capillary Electrophoresis-Single Strand Conformation Polymorphism) method based on PCR technique was established to trace olive oil authenticity from adulteration with other vegetable oils [69]. SSCP is based on the dependence of electrophoretic mobility of a single-stranded DNA on its folded conformation, which is dependent on the nucleotide sequence. Even single base change in a sequence is likely to result in different conformations, which results in slight difference of molecular mobility [70]. The method developed was very suitable for the determination of modeled and of unknown adulterants [71].

Another novel method for identification of different species of vegetable oils based on suspension bead array has been reported by Li *et al.*, (2012) [72]. The suspension bead array as a rapid, sensitive, and high-throughput technology has a great potential to identify more species of vegetable oils with increased species of probes [72].

7. Future challenges and prospective

Future research in the concept of olive oil genetic traceability will concern the application of high-throughput platforms including functional genes, non-nuclear genes, transcriptome analysis, and developing more sophisticated SNPs detection.

Understanding the function of genes and other parts of the genome is known as functional genomics that describes the relationship between an organism's genome and its phenotype. This approach provides a more complete picture of how biological function arises from the information encoded in an organism's genome and such information will contribute in intra-species determination especially with PDO oils.

Chloroplast DNA considers as a most important non-nuclear genes and has been investigated for cultivar identification in olive oil [73]. One advantage of chloroplast DNA is the high copy number of chloroplast per cell (about 50), which is especially beneficial for refined oil sample. It is to develop suitable markers on this region and a compositional test able to identify a cultivar in a monovarietal olive oil. The designed markers will be applied in a high-throughput platform to assess and quantify the contribution of a single cultivar in commercial multivarietal oils.

Olive transcriptome will address the identification of genes differentially expressed during fruit development, with particular attention to those involved in lipid and phenolic metabolism. The provided information will discuss the case of olive oil PGI.

Improving SNPs detection using high resolution melting (HRM) RT-PCR analysis allows olive cultivar genotyping, results in an informative, easy, and low-cost method able to greatly reduce the operating time is also recommended.

Finally, Zhang *et al.*, (2012) proposed an alternative strategy would be using fast and less accurate sensor technology, such as electronic nose, as screening method and verifying suspected samples by DNA method [71].

8. Conclusions

Appropriate method, DNA-based analysis, has been developed to verify the authenticity of olive oil and detect possible adulteration to protect the consumer against any fraud practices. DNA analysis represents an attractive and alternative choice to the more classical analytical methods, because DNA, rather than the macromolecules and metabolites, is less influenced by environmental and processing conditions [74].

Although significant progress has been made in the last decade on DNA extraction from olive oil and the choice of molecular markers, still a consensus protocol, by which the mystery behind olive oil authentication reveals, has not been accepted in trade markets, yet.

At present, DNA extraction from oil is no longer problematical, and the critical point is the choice of markers. Basic criteria to evaluate the suitability of molecular markers at this purpose are: i) the discrimination power; ii) the correspondence between leaf and oil-DNA profiles; iii) reproducibility and repeatability of results; iv) simplicity of analysis [3].

RAPDs and AFLPs give complex profiles that can be applicable to monovarietal oils but not to mixtures of three to four cultivars, such as those usually adopted in PDO oils [53]. Single-locus microsatellites are more effective to this aim, but they are not applicable to high-throughput screening such as microarray.

Single locus (SSRs and SNPs) are preferred to multilocus (AFLP, RAPD) markers because they are simpler to perform, more easily interpretable and can be combined in high-throughput platforms.

Since most olive cultivars are auto-incompatible (pollen could not germinate on an ovary from the same tree) the DNA extracted from oil contains alleles of the tree (fruit pulp somatic tissues) as well as alleles of the seed embryo which may contain exogenous alleles from the pollinator. Thus, Bracci *et al.,* (2011) concluded that care needs to be taken in the interpretation of DNA profiles obtained from DNA extracted from oil for resolving provenance and authenticity issues [62].

Besides, using capillary electrophoresis permits to differentiate alleles with very small differences in molecular weight and to detect a very low or partially degraded DNA, which is the case of extracted DNA from olive oil [75].

The availability of other approaches such as semi-automated SNP genotyping assay proposed to verify the origin and authenticity of monovarietal extra virgin olive oils [6].

Finally, generating an "Identity Card" which, can be used for the unequivocal identification of highly prized oil, has been considered as a potential for olive oil DNA fingerprinting [3].

Acknowledgment

We would like to acknowledge National Institute of Genetics and Biotechnology (NIGEB) and Ministry of Jihad-e-Agriculture for supporting "the national Project of Iranian Olive Germplasm, ..."and this study.

Author details

Sattar Tahmasebi Enferadi* and Zohreh Rabiei

*Address all correspondence to: tahmasebi@nigeb.ac.ir

National Institute of Genetic Engineering and Biotechnology, Tehran, Iran

References

[1] Alonso-Salces RM, Holland MV, Guillou C, Héberger K. Quality Assessment of Olive Oil by ^{1}H-NMR Fingerprinting In Boskou D. (ed.) Olive Oil – Constituents, Quality, Health Properties and Bioconversions; Rijeka: InTech; 2011. p185-210. http://www.intechopen.com/books/olive-oil-constituents-quality-health-properties-and-bioconversions (accessed 1 February 2012).

[2] Kiritsakis K, Christie WW. Analysis of edible oils. In L. John, & R. Aparicio (Eds.), Handbook of olive oil. Gaithersburg (USA): Aspen Publishers; 2000. p129-151.

[3] Agrimonti C, Vietina M, Pafundo S, Marmiroli N. The use of food genomics to ensure the traceability of olive oil. Trends in Food Science & Technology 2011;22 237-244.

[4] Garcia-Gonzalez Dl, Aparicio R. Research in Olive Oil: Challenges for the Near Future J. Agric. Food Chem. 2010;58 12569–12577.

[5] Montealegre C, Alegre MLM, Garcia-Ruiz C. Traceability markers to the botanical origin in olive oils. Journal of Agricultural and Food Chemistry 2010;58(1) 28–38.

[6] Consolandi C, Palmieri L, Severgnini M, Maestri E, Marmiroli N, Agrimonti C, Baldoni L, Donini P, De Bellis G, Castiglioni B. A procedure for olive oil traceability and authenticity: DNA extraction, multiplex PCR and LDR-universal array analysis. European Food Research and Technology 2008;227 1429-1438.

[7] De la Torre F, Bautista R, Canovas FM, Claros MG. Isolation of DNA from olive oil and oil sediments: application in oil fingerprinting. J Food Agric Environ 2004;2 84–86.

[8] Breton C, Claux D, Metton I, Skorski G, Berville A. Comparative study of methods for DNA preparation from olive oil samples to identify cultivar SSR alleles in commercial oil samples: possible forensic applications. Journal of Agricultural and Food Chemistry 2004;52(3) 531-537.

[9] Busconi M, Foroni C, Corradi M, Bongiorni C, Cattapan F, Fogher C. DNA extraction from olive oil and its use in the identification of the production cultivar. Food Chemistry 2003;83 127–134.

[10] http://www.ncbi.nlm.nih.gov/

[11] Araghipour N, Colineau J, Koot A, Akkermans W, Moreno Rojas JM, Beauchamp J, Wisthaler A, Mark TD, Downey G, Guillou C, Mannina L, van Ruth S. Geographical origin classification of olive oils by PTR-MS. Food Chem. 2008;108 374–383.

[12] Minelli S, Maggini F, Gelati MT, Angiolillo A, Cionini PG. The chromosome complement of Olea europaea L.: characterization by differential staining of the chromatin and in situ hybridization of highly repeated DNA sequences. Chromosom Res 2000;8 615–619.

[13] Rabiei Z, Tahmasebi Enferadi S. Traceability of Origin and Authenticity of Olive Oil. In: Boskou D. (ed.) Olive Oil - Constituents, Quality, Health Properties and Bioconversions. Rijeka: InTech; 2011. p163-184. http://www.intechopen.com/books/olive-oil-constituents-quality-health-properties-and-bioconversions (accessed 1 February 2012)

[14] Besnard G, Khadari B, Villemur P, Berville′ A. Cytoplasmic male sterility in the olive (Olea europaea L.). Theor Appl Genet 2000;100 1018–1024.

[15] Diaz A, Martin A, Rallo P, Barranco D, de la Rosa R. Selfincompatibility of 'Arbequina' and 'Picual' olive assessed by SSR markers. J Am Soc Hortic Sci 2006;131 250–255.

[16] Wallander E, Albert VA. Phylogeny and classification of Oleaceae based on rps16 and trnL-F sequence data. American Journal of Botany 2000;87 1827–1841.

[17] Green PS. Oleaceae. In: Kubitzki K, Kadereit JW. (eds.) The families and genera of vascular plants. Vol. VII: Flowering plants, dicotyledons. New York: Springer; 2004. p 296–306.

[18] Green PS. A revision of Olea L. Kew Bulletin 2002;57 91-140.

[19] Vargas P, Kadereit JW. Molecular fingerprinting evidence (ISSR inter-simple sequence repeats) for a wild status of Olea europaea L. (Oleaceae) in the Eurosiberian North of the Iberian Peninsula. Flora 2001;196 142–152.

[20] Besnard G, Rubio de Casas R, Vargas P. Plastid and nuclear DNA polymorphism reveals historical processes of isolation and reticulation in the olive tree complex (Olea europaea L.). J Biogeogr 2007; 34 736–752.

[21] Baldoni L, Belaj A. Olive. In: Vollmann J, Rajean I (eds.) Oil crops. Handbook of plant breeding, vol 4: Springer Science Business Media, New York;2009. p97–421.

[22] Gucci R, Tattini M. Salinity tolerance in olive. Hortic Rev. 1997;21 177–213.

[23] Roche HM, Gibney MJ, Kafatos A, Zampelas A, Williams CM. Beneficial properties of olive oil. Food Res Int 2000;33 227–231.

[24] Ben Ayed R, Grati-Kamoun N, Moreau F, Rebaï A. Comparative study of microsatellite profiles of DNA from oil and leaves of two Tunisian olive cultivars. Eur Food Res Technol 2009;229 757–762.

[25] Doveri S, Donal M, Lee D. Non concordance between genetics profiles of olive oil and fruit: a cautionary note to the use of DNA markers for provenance testing. J Agric Food Chem 2006;54(24) 9221–9226.

[26] http://www.internationaloliveoil.org/web/aa-ingles/oliveWorld/aceite.html and http://www.oliveoil.org/eng/doc/RES-1-94e.pdf

[27] Bazakos C, Dulger AO, Uncu AT, Spaniolas S, Spano T, Kalaitzis P. A SNP-based PCR–RFLP capillary electrophoresis analysis for the identification of the varietal origin of olive oils. Food Chemistry 2012;134 2411–2418.

[28] Sanz-Cortes F, Parfitt DE, Romero C, Struss D, Llacer G, Badenes ML. Intraspecific olive diversity assessed with AFLP. Plant Breeding 2003;122 173-177.

[29] Mannina L, Dugo G, Salvo F, Cicero L, Ansanelli G, Calcagni C, Segre A. Study of the cultivar-composition relationship in Sicilian olive oils by GC, NMR, and statistical methods. J. Agric. Food Chem. 2003;51 120–127.

[30] Tura D, Failla O, Bassi D, Pedo S, Serraiocco A. Cultivar influence on virgin olive (Olea europaea L.) oil flavor based on aromatic compounds and sensorial profile. Sci. Hortic. 2008;118 139–148.

[31] Mahjoub-Haddada F, Manai H, Daoud D, Fernandez X, Lizzani-Cuvelier L, Zarrouk M. Profiles of volatile compounds from some monovarietal Tunisian virgin olive oils. Comparison with French PDO. Food Chem. 2007;103 467–476.

[32] Torres Vaz-Freire L, Gomes da Silva MDR, Costa Freitas AM. Comprehensive two-dimensional gas chromatography for fingerprint pattern recognition in olive oils produced by two different techniques in Portuguese olive varieties Galega Vulgar, Cobranc-osa and Carrasquenha. Anal. Chim. Acta 2009;633 263–270.

[33] Cichelli A, Pertesana GP. High-performance liquid chromatographic analysis of chlorophylls, pheophytins and carotenoides in virgin olive oils: chemometric approach to variety classification. J. Chromatogr A 2004;1046 141–146.

[34] Giuffrida D, Salvo F, Salvo A, La Pera L, Dugo G. Pigments composition in monovarietal virgin olive oils from various Sicilian olive varieties. Food Chem. 2007;101 833–837.

[35] Roca M, Gandul-Rojas B, Gallardo-Guerrero L, Mı´nguez- Mosquera MI. Pigment parameters determining Spanish virgin olive oil authenticity: stability during storage. J. Am. Oil Chem. Soc 2003;80 1237–1240.

[36] Cerretani L, Motilva MJ, Romero MP, Bendini A, Lercker G. Pigment profile and chromatic parameters of monovarietal virgin olive oils from different Italian cultivars. Eur. Food Res. Technol. 2008;226 1251–1258.

[37] Bianchi G, Giansante L, Shaw A, Kell DB. Chemometric criteria for the characterisation of Italian protected denomination of origin (DOP) olive oils from their metabolic profiles. Eur. J. Lipid Sci. Technol. 2001;103 141-150.

[38] Caponio F, Gomes T, Pasqualone A. Phenolic compounds in virgin olive oils: influence of the degree of olive ripeness on organoleptic characteristics and shelf-life. Eur. Food Res. Technol. 2001;212 329-333.

[39] Pasqualone A, Catalano M. Free and total sterols in olive oils. Effects of neutralization. Grasas Aceites 2000;51 177-182.

[40] Guillén MD, Ruiz A. High resolution 1H nuclear mag-netic resonance in the study of edible oils and fats. Trends Food Sci. Technol. 2001;12 328-338.

[41] Muzzalupo I, Perri M. Recovery and characterization of DNA from virgin olive oil. Eur. Food Res. Technol. 2002;214 528–531.

[42] Pasqualone A, Montemurro C, Caponio F, Blanco A. Identification of virgin olive oil from different cultivars by analysis of DNA microsatellites. J. Agric. Food Chem. 2004;52 1068-1071.

[43] Cipriani G, Marrazzo MT, Marconi R, Cimato A, Testolin R. Microsatellite markers isolated in olive (Olea europaea L.) are suitable for individual fingerprinting and reveal polymorphism within ancient cultivars. Theoretical and Applied Genetics, 2002;104 223–228.

[44] De la Rosa R, James C, Tobutt KR. Isolation and characterization of polymorphic microsatellite in olive (Olea europaea L.) and their transferability to other genera in the Oleaceae. Mol Ecol 2002;2 265–267

[45] Sabino-Gil F, Busconi M, Da Câmara-Machado A, Fogher C. Microsatellite markers are powerful tools for discriminating among olive cultivars ans assigning them to geographically defined populations. Genome 2006;49 1606–1615.

[46] Sefc KM, Lopes MS, Mendoncxa D, Machado M, Da Camara Machado A. Identification of microsatellite loci in olive (Olea europaea) and their characterization in Italian and Iberian olive trees. Mol Ecol 2000;9 1171–1173.

[47] Pafundo S, Agrimonti C, Marmiroli N. Traceability of Plant Contribution in Olive Oil by Amplified Fragment Length Polymorphisms. J. Agric. Food Chem. 2005;53 6995-7002.

[48] Diez1 CM, Trujillo I, Barrio E, Belaj A, Barranco D, Rallo L. Centennial olive trees as a reservoir of genetic diversity. Annals of Botany 2011;108 797–807.

[49] Montemurro C, Pasqualone A, Simeone R, Sabetta W, Blanco A. AFLP molecular markers to identify virgin olive oils from single Italian cultivars. European Food Research and Technology, 2008;226 1439-1444.

[50] Bjelland S, Seeberg E. Mutagenity, toxicity and repair of DNA base damage induced by oxidation. Mutation Research, 2003;531 37–80.

[51] Pafundo S, Busconi M, Agrimonti C, Fogher C, Marmiroli N. Storagetime effects on olive oil DNA assessed by amplified fragments length polymorphisms. Food Chemistry, 2010;123 787-793.

[52] d'Abbadie M, Hofreiter M, Vaisman A, Loakes D, Gasparutto D, Cadet J, Woodgate R, Paabo S, Holliger P. Molecular breeding of polymerases for amplification of ancient DNA. Nature Biotechnology, 2007;25 939–943.

[53] Pafundo S, Agrimonti C, Maestri E, Marmiroli N. Applicability of SCAR markers to food genomics: olive oil traceability. Journal of Agricultural and Food Chemistry, 2007;55 6052-6059.

[54] Spaniolas S, Bazakos C, Ntourou T, Bihmidine S, Georgousakis A, Kalaitzis P. Use of lambda DNA as a marker to assess DNA stability in olive oil during storage. European Food Research and Technology, 2008;227 175-179.

[55] Martins-Lopes P, Gomes S, Santos E, Guedes-Pinto H. DNA markers for Portuguese olive oil fingerprinting. Journal of Agricultural and Food Chemistry, 2008;56 11786-11791.

[56] Alba V, Sabetta W, Blanco A, Pasqualone A, Montemurro C. Microsatellite markers to identify specific alleles in DNA extracted from monovarietal virgin olive oils. European Food Research and Technology, 2009;229 375–382.

[57] Pasqualone A, Di Rienzo V, Blanco A, Summo C, Caponio F, Montemurro C. Characterization of virgin olive oil from Leucocarpa cultivar by chemical and DNA analysis Food Research International 2012;47 188–193.

[58] Marmiroli N, Maestri E, Pafundo S, Vietina M. Molecular traceability of olive oil: from plant genomics to food genomics. In L. Berti, & J. Maury (Eds.), Advances in olive resources. Kerala (India): Transworld Research Network. ; 2009 p. 157-172.

[59] Corrado G, Imperato A, La Mura M, Perri E, Rao R. Genetic diversity among olive varieties of Southern Italy and the traceability of olive oil using SSR markers Journal of Horticultural Science & Biotechnology 2011;86(5) 461–466.

[60] Muzzalupo I, Pellegrino M, Perri E. Detection of DNA in virgin olive oils extracted from destoned fruits. European Food Research and Technology, 2007;224 469-475.

[61] Pasqualone A, Montemurro C, Summo C, Sabetta W, Caponio F, & Blanco A. Effectiveness of microsatellite DNA markers in checking the identity of PDO extra virgin olive oil. Journal of Agricultural and Food Chemistry 2007; 55 3857–3862.

[62] Bracci T, BusconiM, Fogher C, Sebastiani L. Molecular studies in olive (Olea europaea L.): overview on DNA markers applications and recent advances in genome analysis. Plant Cell Rep 2011;30 449–462.

[63] Rotondi A, Beghè D, Fabbri A, Ganino T. Olive oil traceability by means of chemical and sensory analyses: A comparison with SSR biomolecular profiles. Food Chemistry 2011;129 1825–1831.

[64] Brookes AJ. The essence of SNPs. Gene, 1999;234(2) 177–186.

[65] Kumar S, Kahlon T, Chaudhary S. A rapid screening for adulterants in olive oil using DNA barcodes. Food Chemistry 2011;127 1335–1341.

[66] Reale S, Doveri S, Dıaz A, Angiolillo A, Lucentini L, Pilla F, Martin A, Donini P, Lee D. SNP-based markers for discriminating olive (Olea europaea L.) cultivars. Genome, 2006;49 1193-1205.

[67] Giménez MJ, Pistón F, Martín A, Atienza SG. Application of real-time PCR on the development of molecular markers and to evaluate critical aspects for olive oil authentication. Food Chemistry, 2010;118 482-487.

[68] Wu Y, Chen Y, Ge Y, Wang J, Xu B, Huang W, Yuan F. Detection of olive oil using the Evagreen real-time PCR method. European Food Research and Technology, 2008;227 1117-1124.

[69] Wu Y, Zhang H, Han J, Wang B, Wang W, Ju X, Chen Y. PCR-CE-SSCP applied to detect cheap oil blended in olive oil Eur Food Res Technol 2011;233 313–324.

[70] Hayashi K. PCR-SSCP: a simple and sensitive method for detection of mutations in the genomic DNA. Genome Research, 1991;1 34-38.

[71] Zhang H, Wu Y, Li Y, Wang B, Han J, Ju X, Chen Y. PCR-CE-SSCP used to authenticate edible oils. Food Control 2012;27 322-329.

[72] Li Y, Wu Y, Han J, Wang B, Ge Y, Chen Y. Species-Specific Identification of Seven Vegetable Oils Based on Suspension Bead Array. J Agric Food Chem 2012;7 60(9) 2362-2367.

[73] Intrieri MC, M R, Buiatti M: Chloroplast DNA polymorphisms as molecular markers to identify cultivars of Olea europaea L. Journal of Horticultural Science & Biotechnology 2007;82(1) 109-113.

[74] Woolfe M, Primrose S. Food forensics: using DNA technology to combat misdescription and fraud. Trends in Biotechnology,2004;22 222-226.

[75] Ben Ayed R, Grati-Kamoun N, Sans-Grout S, Moreau F, Rebai A. Characterization and authenticity of virgin olive oil (Olea europaea L.) cultivars by microsatellite markers. Eur Food Res Technol 2012;234 263–271.

Section 4

And All Begins with Genetics

Chapter 10

Adaptation of Local Grapevine Germplasm: Exploitation of Natural Defence Mechanisms to Biotic Stresses

Massimo Muganu and Marco Paolocci

Additional information is available at the end of the chapter

http://dx.doi.org/10.5772/51976

1. Introduction

The history of human civilization is closely intertwined with the development of viticulture, considering the consumption of grapevine fresh fruits and their use for wine-making since Neolithic Age. Evidences are given by the archeobotanical discovers of ancient seeds belonging to *Vitis vinifera* subsp. *sativa* (domestic grape) and *Vitis vinifera* subsp. *sylvestris* (wild grape) and by the discovery of wine jars and primitive stone wine presses, known locally as "pestarole" [1-2]. These evidences suggest that wine production was initially dependent from the collection of wild fruits, and subsequently from the cultivation of plants derived from the domestication of the Eurasian grapevine. Probably grapevine domestication occurred in different areas, from most ancient Anatolian and Caucasian centers to recent west Mediterranean basin and Central-Europe centers. The study of grapevine domestication is complicated by the presence of large para-domestication areas where wild plants were protected for fruit utilization [3]. Grapevine domestication was significant for the development of Mediterranean agriculture, based on cereal-olive-grape cultivations, typical of Greek and Roman civilizations. During the Middle-Age grape cultivation was maintained by monks as wine is bound to Christian liturgy. During late Middle-Age a description of grape varieties used for the production of high value wines was drawn [4] and among listed varieties, some are currently cultivated. The diffusion through the Europe of several grape pathogens during XIX century, such as downy and powdery mildew, caused the end of the ancient vitiviniculture, the erosion of grape genetic variability and the increase of chemicals used for plant disease protection.

During last years climate changes are causing the increase of favorable environmental conditions for the development of grapevine diseases, with the reduction of suitable areas for traditional crops particularly in some Mediterranean regions [5]. At the same time the large use of copper compounds to control grape diseases lead to accumulation of the toxic heavy metal in soil and groundwater [6]. Considering economical losses, the limits to the use of chemicals for plants protection and the wide market requirement of high quality wines able to express terroir characteristics, the interest of researchers and viticulturist for grape local varieties increased. In many countries with viticultural historical tradition the recovery and description of local varieties are undertaken, due both to vines adaptation to local environment and grape ability to determine typical qualitative characteristics of wines. Even thought the short coevolution period with invasive pathogens and the use of agamic propagation reduced the probability to have disease resistant grapevine genotypes, *V. vinifera* local germplasm can includes minor varieties which characteristics are often unknown with respect to genetic profile, viticultural and oenological potential [7-9] and to the degree of resistance to pathogen, that was already observed during XIX century [10-12]. Different responses to biotic stresses were described among grapevine varieties, particularly concerning the widespread pathogens *Plasmopara viticola, Erysiphe necator* and *Botrytis cinerea,* the causal agents of downy mildew, powdery mildew and of gray mold respectively [13-19]. It is well-known that grapevine genetic resources for pathogen resistance are mainly found in American and Asian wild *Vitis* species [20-23], but at present breeding programs involving *V. vinifera* and aimed to obtain resistant genotypes, released disease resistant interspecifics hybrids able to produce wines suitable only for local markets. For these reasons the study of natural defence mechanisms to biotic stresses of *V. vinifera* varieties has a scientific and applicative interest to improve both the management of grape genetic resources than wine quality.

The aim of the present paper is to review most studied constitutive and inducible defence mechanisms in *V. vinifera*. Among constitutive defences, anatomical and morphological features of leaf, bunch and berry were described. Furthermore induced defence mechanisms, including callose synthesis, stilbenes production and pathogenesis related (PR) proteins induction were discussed. The analysis of different *V. vinifera* varieties indicate that in many cases grapevine varieties have or activate defence responses to biotic stresses.

2. Plant defence mechanisms

The relationships between plant and pathogen start with the initial contact phase between infective propagules and the plant tissue surfaces. As response plants are able to activate defence mechanisms that may be referred to constitutive or inducible defences.

2.1. Constitutive defences

Constitutive defences are active in the plant before pathogen challenge. They are considered able to limit the entry phase of parasite in host tissues through direct penetration or pre-existing tissues opening and to contrast the infection during the first phases. Constitutive defences are generally referred to morpho-anatomical characteristics of leaf, bunch and berry, developed independently from fungal attack [13] or include constitutive compounds that can have antimicrobial activity. The synthesis of some antimicrobial constitutive compounds may be also enhanced as plant response to stresses [24-25].

• Leaf Hairs

Grapevine leaf hairs (trichomes and bristles) are morphological characters with ampelographic value. The density of leaf prostrate and erect hairs is included in the OIV descriptors list for grape varieties and *Vitis* species [26]. The number and length of leaf hairs differ according to *Vitis* species and varieties and may be influenced by environmental conditions (Figure 1). The hairs of abaxial leaf surface can constitute a hydrophobic barrier able to reduce the contact area among water droplets and leaf lamina, with the reduction of wettability of epidermal tissues. The presence of very dense leaf hairs leads to a reduction of water retention capacity of the leaf surface [27-28], that are decisive during the infection process [29-31]. The density of abaxial leaf hairs has been related to the different degree of tolerance of *Vitis* species to pathogens [32-33]. In *V. doaniana* and *V. davidii*, downy mildew resistant species, the reduction of leaf wettability prevent zoospores emission from zoosporangia and the pathogen is hampered to reach host tissues. In these species the use of wetters to reduce water surface tension increases the infection and lead to the regular sporulation of the pathogen. A similar behavior was demonstrated in *V. cinerea* e *V. labrusca*, which downy mildew infections were enhanced by wetter use, but the subsequently pathogen growth was blocked, supporting the hypothesis of the presence of further defence mechanisms in these two species. In *V. vinifera*, even though any significative correlation was demonstrated between the hair density of abaxial leaf surface and plant resistance to downy mildew [34-35], furthers investigations might be useful, according to the great variability of the character among varieties and clones.

• Stomata

Stomata are plant natural openings bordered by two guard cells, that exert a control over plant water and carbon cycles by variation in both size and number. In grapevine leaf they occupy a small percentage of the surface and are mainly located in the abaxial side. In *V. vinifera* cultivars, stomatal leaf density (number of openings per leaf surface unit) varies according to environmental conditions, including CO_2 concentration, light intensity, air temperature, photoperiod [36] and genotype [37-39] (Figure 2). Stomata are one of the most important way for pathogen entry [40-42]. The penetration of the grapevine obligate biotrophic parasite *P. viticola* occurred exclusively through stomata, while sporulation can rarely occurs also through other tissue openings [43]. During sporulation stomatal density can affect *P. viticola* secondary infections [43]. The mobility of pathogen zoospores to health stomata was related to a chemotaxis process that is regulated by chemical compounds as aminoacids, isoflavons, pectins and cell wall fragments, which production might be influenced by stomata opening [44]. Other hypothesis have been evaluated to explain a functional relationship between stomata and zoospore, including the presence of electrical fields produced by stomata [45-46]. Infected stomata are preferential sites of attraction for zoospores. This process, known as adelphotaxis, is the cause the accumulation of more than one zoospore on the same stoma [47-48]. Studies carried out on *V. vinifera* varieties showed a no clear relation between leaf stomatal density and susceptibility to downy mildew [35], even though the lower percentage of infected stomata occurred in *V. vinifera* varieties with the lower number of stomata per surface unit (Paolocci and Muganu, unpublished data) (Figure 3). Functional stomata found on the berry surface are possible entries for pathogens [49]. After berry set and under the influence of climate, stomata are quickly covered by wax layers and originates lenticels, that are often surrounded by cuticle tears [50-51]. These morphological transformation was correlated to the acquisition of berry ontogenetic resistance to downy mildew, even though the occurrence of berry infection during this phase remains still possible through berry pedicel [49]. Starting from veraison lenticels and peristomatic tissues represent the main entry sites for *B. cinerea* infection, but a significative correlation between the number of lenticels and the degree of berry susceptibility to gray mold was not demonstrated [52, 13]. *V. vinifera* stomatal opening/closure is influenced also by plant health, considering that in downy mildew infected leaves stomata are open in darkness and during water stress, leading to an increase of transpiration. This functional relationships is not systemic, being restricted to the infected area, and could be related to non-systemic compounds affecting stomatal activity and produced by pathogen or by the infected plant [53].

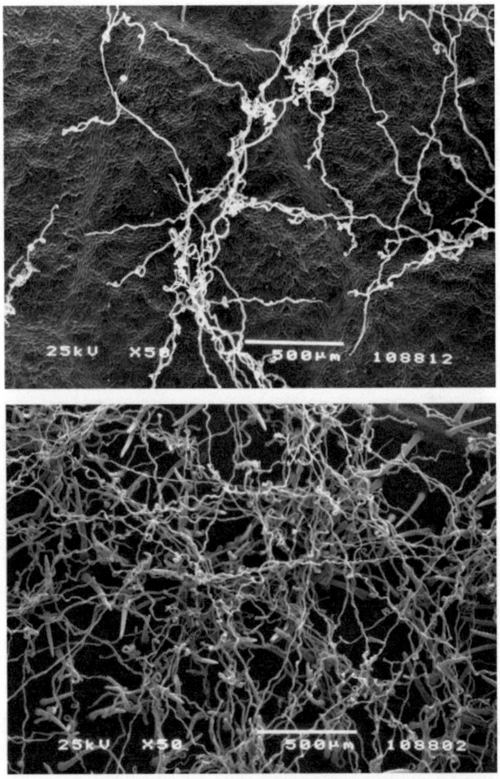

Figure 1. Scanning electron micrographs of hair density assessed on abaxial leaf surface on the two V. vinifera local varieties Romanesco (above) and Trebbiano giallo (below) (pictures by Muganu and Paolocci)

http://dx.doi.org/10.5772/51976

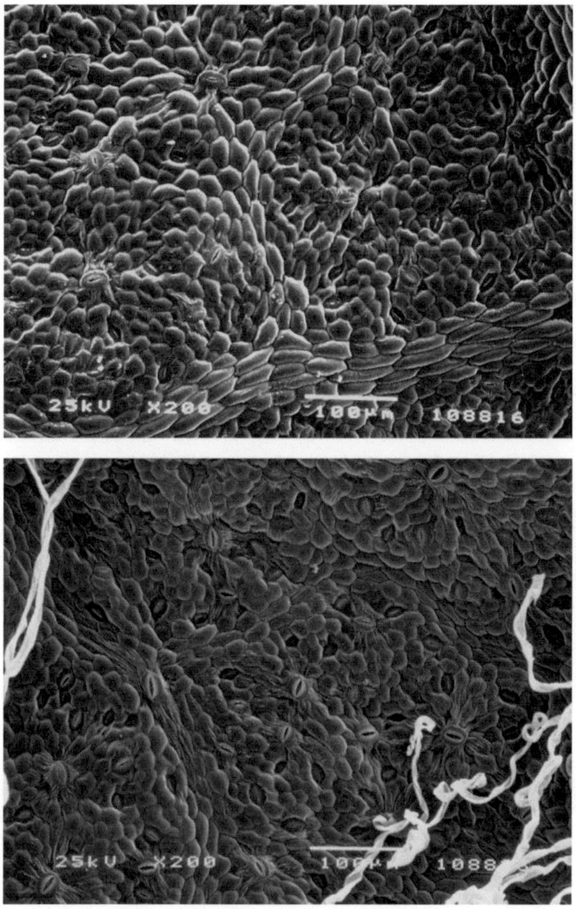

Figure 2. Scanning electron micrographs of stomata density assessed on abaxial leaf surface on the two V. vinifera local varieties Romanesco (above) and Trebbiano giallo (below) (pictures by Muganu and Paolocci)

- Cuticular membrane

The cuticle is a protective membrane of aerial plant tissues able to maintain a stable tissue form, to reduce water loss and to control gas exchanges [54-55, 51]. The cuticle is formed by an insoluble cutin layer and a soluble epicuticular wax layer. Quantitative differences in cuticle content among varieties have a genetic control even though wax amount and plate-like structure are influenced by environmental factors [13, 56-57]. The cuticle membrane is the first defence barrier that many plant pathogens must overcome to infect plant tissues and its variation in thickness, structure and composition have been analyzed to study its protective role against several grapevine diseases. The thickness of leaf cuticle of different grape varieties was positively correlated to their susceptibility to *E. necator* [58-59]. Nevertheless the increase of cuticle thickness during berry growth was not related to the acquisition of ontogenetic resistance to *E. necator* of mature berries [60].

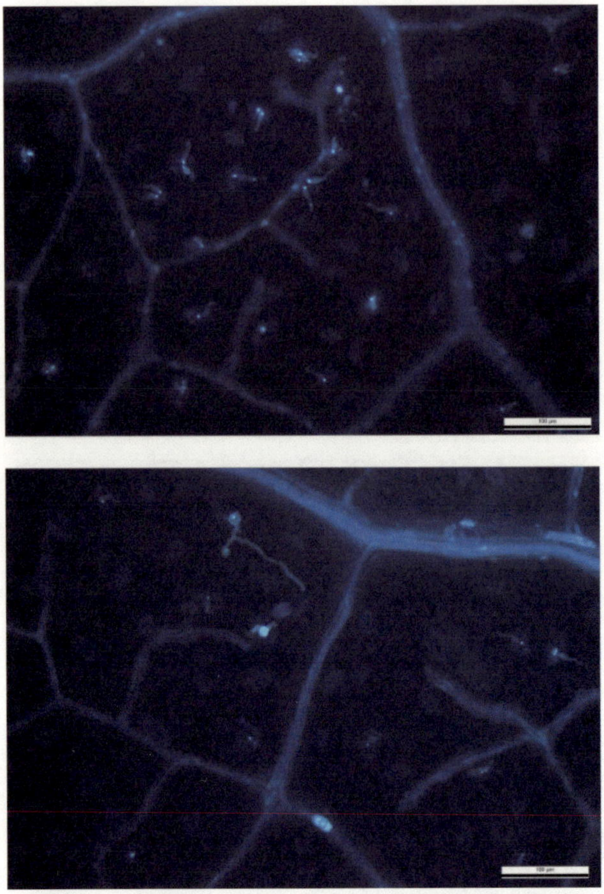

Figure 3. Epifluorescence micrographs of leaf stomata infected by P. viticola assessed at 24 hours post inoculation on the two V. vinifera varieties Aleatico (above) and Trebbiano toscano (below) (pictures by Muganu and Paolocci)

The amount of berry epicuticular wax positively affected the level of resistance to *B. cinerea* of different *V. vinifera* varieties as the influence of wax content on berry skin hydrophobicity and reduction of pathogen adhesion [13]. The removal of berry epicuticular wax increases the susceptibility to *B. cinerea,* indicating a role of wax layer on the infection phase [61]. The significant decrease of cutin content per surface unit of berry skin from berry set to veraison influenced the susceptibility to gray mold of three different clones of Pinot noir [62].

Anyway mere dimensional or quantitative variations of the cuticle membrane seem not explain the changes of grapevine degree of resistance to pathogen during annual vine cycle [63]. For this reason the presence of morphological and/or chemical differences occurred at cuticle level during berry growth and able to influence the development of infection, must be considered. Several studies analyzed the chemical composition of berry epicuticular wax from berry set to harvest. Variations of lipidic and alchoolic composition of the cuticle were shown in the transition from bunch closure to veraison phase and the presence of compounds with an inhibitory effect on the germination of *B. cinerea* conidia were detected [64-65].

- Bunch and berry features

Morphological characteristics of bunch and berry anatomy can affect grape resistance to pathogens. The evaluation of bunch density allows us to distinguish loose bunches, with movable berries, and dense or very dense bunches with not movable and sometimes deformed berries, as consequence of the contact with each other. Tight bunches determine the presence of micro-environmental conditions in the fruit zone, such as the increase of air temperature, low ventilation and relative humidity, that could promote pathogen growth, as showed for *B. cinerea,* which occurrence could be enhanced [62]. A part the frequency of berry skin cracks, that cause the release of free water required in the germination of *B. cinerea* conidia, the increase of physical contact among berries during growth leads to the development of flattened areas on berry contact surfaces and affects the structure of epicuticular wax. The contact surfaces show the larger areas of amorphous and thinner wax and the higher number of gray mold infections compared to non-contact berry skin surfaces [61, 54]. Recent studies about ampelographic characters described a negative correlation between bunch density and berry degree of resistance to *E. necator,* whereas any relation was obtained among powdery mildew infection and bunch length, width and shape [18].

Some morphological and anatomical characteristics of the berry were related to the susceptibility to *B. cinerea.* A positive correlation was found among berry resistance to gray mold and berry skin thickness and number of epidermal cell layers [13]. The intravarietal evaluation of Spanish Albariño variety showed that clones with small berries and short pedicels were low susceptible to gray mold [66].

- Constitutive compounds

Constitutive compounds with antimicrobial activity are preformed in plant tissues before host-parasite interaction. Many of these compounds may be related to the group of phytoanticipins according to the following definition "phytoanticipins are low molecular weight antimicrobial compounds that are present in plants before challenge by microorganisms or are produced after infection solely from preexisting constituents" [67]. These metabolites are complementary to phytoalexins, antimicrobial metabolites which synthesis occurs after plant-parasite contact [68]. Preformed phenolic compounds were demonstrated to have antimicrobial activity [69-70] such as constitutive pterostilbene that showed antifungal properties against *B. cinerea.* Pterostilbene was detected in low concentration in gray mold resistant young berries, but its toxic activity against pathogen was enhanced by the high content of glycolic acid during berry set [71]. Other constitutive phenols, such as cathechin, epicatechin-3-*O*-gallate, caftaric acid and cutaric acid are able to inhibit fungal stilbene oxidase activity between flowering and veraison, and a high content of catechin was detectectd in *B. cinerea* resistant grape varieties after veraison [72]. The non-specific inhibition of *B. cinerea* lytic enzymes was related to the detection of proantocyanidins, polymeric flavonoids which are considered inhibitors of the oxidative fungal enzyme laccase, responsible of pterostilbene detoxification [73]. These results suggest that the resistance of young berries to gray mold depends on both catechins and proantocyanidins contents which contribute in maintaining the pathogen in a quiescent state [72, 74]. Considering that the decrease of proantocyanidins content during berry ripening lead to the increase of gray mold susceptibility, proantocyanidins are considered as markers of grapevine *B. cinerea* resistance [75].

It is well known that light exposure affect the synthesis of phenolic compounds. For this reason several studies evaluate the relationships between the intensity of tissue sun-light exposure and grapevine susceptibility to pathogens [76-78]. Shaded *P. viticola* infected leaves, besides showing the lower content of flavonoids compared to full light exposed ones, also displayed the highest disease severity [69]. A similar result was obtained with the artificial inoculation of detached berries with *E. necator:* in this case shaded berries showed the highest susceptibility to the pathogen [79].

Insects use of plants as a source of nutrients causes tissue mechanical damages and, in many cases, compromises plant health as consequence of virus or phytoplasma transmissions. Phytophagous find host plant using mainly olfactory signals produced by the host plant itself. These chemical signals, known as volatile organic compounds (VOCs), are plant secondary metabolites including alcohols, aldehydes, terpenoids and aromatic phenols, that showed a different role in plant-insect relationships [80]. Each plant species releases specific bouquets, which blend is influenced by plant phenology and health conditions [81]. Grape berries and leaves release hundred of volatiles compounds among which α-farnesene, (E)-β-farnesene and (Z)-3-hexenyl acetate. These compounds detected in Chardonnay varieties between pre-flowering and green berry developmental phases, significantly elicited female attraction of *Lobesia botrana*, the most important insect of *V. vinifera* in Europe [82]. *Lobesia botrana* feeds on all *V. vinifera* cultivars but a different susceptibility to the insect among grape varieties was shown [83].

In the study of Grapevine Yellows the involvement of VOCs in the ecology of *Scaphoideus titanus* Ball (the causal agent of Flavescence Doreé) and of *Hyalesthes obsoletus* Signoret (the causal agent of Bois noir), are under investigation. Observation on preference of *H. obsoletus* for different plant species were made testing different plant extracts among which *V. vinifera* [84].

2.2. Induced defences

The induced defences are the result of plant reaction to pathogen attack and require the perception of plant-tissues signals resulting from pathogen infection.

Plants have evolved different active defence strategies aimed at the protection against biotic stresses. A first strategy is founded on the recognition, by host extra-cellular receptors, of pathogen associated molecular patterns (PAMPs) which are microbial products, among which chitin [85-87]. This recognition triggers active plant defence mechanisms (PAMP-Triggered Immunity PTI), including the synthesis of pathogenesis related (PR) proteins, and the strengthening of plant tissue cell walls [88]. PTI strategy is considered a plant basal immunity against non-host specific pathogens and can be overcame from host specific pathogens, which developed the ability to produce effectors, molecules able to suppress PTI resistance. As consequence plants evolved effector-triggered immunity (ETI) defence mechanism, which enable plant recognition of the PTI-suppressing effectors [86-87]. This strategy, which involves the activation of specific resistance (*R*) genes, lead to a hypersensitive response (HR) which is one of the most efficient mechanism used by plants to arrest biotrophic pathogen infections. HR involves the massive production and accumulation of reactive oxygen species (ROS), among which hydrogen peroxide (H_2O_2), that can modulate localized plant cell death (PCD) of infected tissues which prevents pathogen nutrition and growth. It must be considered that H_2O_2 can act also as diffusible signals for the induction of different plant defence reactions, among which the production of phytoalexins, of PR proteins and of cell wall polymers. HR mechanism was described in the american species *V. rotundifolia*, resistant to *E. necator* and in some of their hybrids with *V. vinifera* [89]. Recently PCD activity was also proved for the two grape varieties Kishmish vatkana e Dzhandzhal kara, belonging to *V. vinifera* subsp. *sativa* proles *orientalis* subproles *antasiatica*, native of Uzbekistan [90]. It may be useful to highlight that effector-triggered immunity is dependent on the activation of single, dominant genes and can be overcome by the deletion or mutation of a single effector [86-87, 91].

Besides to tissue-localized defence activities, plant pathogen recognition also induce plant systemic reactions, known as systemic acquired resistance (SAR). SAR enhances defence responses against a wide range of biotrophic pathogens in plant organs remotely located from the initial site of infection [92-94]. It has long been thought that salicylic acid (SA) is a key signaling molecule in plant defence resistance against biotrophic pathogens and it is required for activation of SAR [95- 96]. Endogenous salicylic acid level was higher in powdery mildew resistant *V. aestivalis* than in susceptible *V. vinifera* varieties, which salicylic acid content increased only at 120 hours after infection, being inadequate to limit disease progression [97]. Evidence of the involvement of salicylic acid in SAR is showed by the exogenous application of SA that increases the synthesis of stilbenes [98] and of PR proteins [99-100].

As above described the different grapevine defence mechanisms trigger the production of physical barrier or the synthesis of anti-microbial compounds that are involved in grapevine pathogen resistance strategies. Among which:

- Callose synthesis

The synthesis and accumulation of callose, a sugar polymer of (1-3)-β-D-glucose, occurs in phloematic tissues, root hairs, epidermal cells and in parenchimatic tissues as a consequence of fungal infections. Callose synthesis is considered a grapevine induced defence response to powdery and downy mildew [101-102]. Callose deposition on stomata as response to *P. viticola* infection is able to block the penetration of zoospores to the substomatal cavity 7 hours post infection (hpi) and at 24 hpi infected stomata are surrounded by necrotic areas, showing a HR-like reactions. The deposit of callose was also detected at 120 hpi in stomata close to infections sites, even though the presence of necrotic areas did not occur. The nature of signals that affect neighboring health stomata are unknown, but their callose deposition is able to prevent secondary infection and could be referred to a systemic acquired resistance process (SAR) against *P.viticola* [43]. The percentage of infected stomata that showed callose deposition at 48 hpi is used as a histological marker to evaluate the degree of resistance to downy mildew of grape varieties [103]. Late callose deposition was detected in the mesophyll both in susceptible than in resistant *Vitis* varieties. At this time the pathogen block, occurred at 3-4 days after inoculation, was observed only in resistant varieties and was related with the presence of further defence mechanisms [102]. The presence of callose deposit was observed as a consequence of the grape leaf infection of *E. necator*. In this case the penetration of the haustorium in epidermal cells was stopped by the formation of a papilla, a structure formed by different layers containing carbohydrates, silica and phenolic compounds and callose deposits could be observed around the haustorial neck and papilla [101].

The role of callose in grape defence mechanisms was validated by the increase of the number of sporangia produced in leaf tissues infected with *P. viticola* treated with 2-deoxy-D-glucose (DDG), an inhibitor of callose synthesis. The increase of sporangia was observed also in *P. viticola* resistant variety Solaris, even thought Solaris treated tissues showed higher resistance compared to the basal resistance of susceptible Chasselas variety, indicating the involvement of further resistance factors, besides callose synthesis, in Solaris [104].

- Stilbenes synthesis

Stilbenes are low molecular weight phenolic compounds found in several plant genera, included many *Vitis* species. Stilbenes show low solubility in water and high solubility in organic solvents. In *V. vinifera* they are costitutive compounds of the berry and of woody tissues [105]. Grapevine stilbenes include several compounds among which resveratrol, with *cis* and *trans* isomers, piceid and resveratroloside, two glucosides of resveratrol [106] and different molecules derived from resveratrol, that include pterostilbene and viniferins [71, 107-108]. Resveratrol was the first described stilbenic compound and its activity is studied since the first half of XX century. Resveratrol content in grape berries is influenced by environmental conditions, vineyard agronomic management and genotype characteristics [105-106,109-110]; it is included among wine components [105-106] and recently its regular presence in human diet was positively correlated with the protection from cancer and other cardiovascular diseases [110-112]. Stilbenes production has been related to plant response to abiotic elicitors among which UV-irradiation, ozone, fosetyl-Al, metyl-jasmonate, benzothiadiazole, chitosan olygomers, ciclodextrins and salicylic acid [113-114]. Stilbenes are induced in non-woody vine tissues, such as flowers, leaves and berries, by different pathogen infections among which *B. cinerea,* one of the first studied elicitors [107], and by *P. viticola, E. necator, Phomopsis viticola, Rhizopus stolonifer, Aspergillus* spp., *Trichoderma viride* [106, 114-115]. Induced stilbenes are considered phytoalexins, compounds with antimicrobial activity and the involvement of stilbenic phytoalexins in grapevine induced defences against *B. cinerea* was observed for a long time [113]. Stilbenes are able to inhibit some fungal ATPases and fungal cells respiration [73, 116], and their effectiveness is related to the rapidity of their synthesis. Stilbene-sinthase is the key enzyme in resveratrol synthesis. The decrease of berry resveratrol content during berry ripening and sugar accumulation goes with the increase of berry susceptibility to *B.cinerea.* Resveratrol reduction in ripe berries was related to the decline of stilbene-sinthase gene expression and to the contemporary increase of chalcone-sinthase enzyme which is bound with flavonoyds synthesis [73, 116]. Among stilbenes,

http://dx.doi.org/10.5772/51976

resveratrol did not show an instant antimicrobial activity [117], even though the long term incubation of the pathogen together with resveratrol can inhibit conidia germination and the growth of germ tubes [118]. Also the production of resveratrol in micropropagated grape explants was correlated with the severity of gray mold infection [119]. *B. cinerea* evolved the ability to detoxify grape berry phytoalexins by stilbene oxidase activity [72,74] and the accumulation of resveratrol in leaf tissues of *in vitro* transgenic plants for stilbene synthase gene was related to the reduction of disease severity [120]. Among other stilbenes the role of pterostilbene against gray mold infection remains unclear, considering that its content did not increase after berry inoculation [73].

Several studies analyzed stilbene production during downy mildew infection. The toxicity of pterostilbene and of the two resveratrol dimers δ-viniferin and ε-viniferin against *P. viticola* was demonstrated, while piceid, a resveratrol derived compound, did not show antimicrobial activity as its high synthesis and accumulation was showed in infected leaf tissues of the susceptible Chasselas variety [103, 121]. The involvement of stilbenes in induced defence mechanisms against *P. viticola* was shown in *V. rotundifolia,* which infection with the oomycete lead both to the extrusion of pathogen cells from stomata and to the accumulation in infected tissues of one hundred fold of stilbenic molecules compared to the stilbene content detected in infected tissues of resistant hybrids [122]. Pterostilbene is considered the most toxic stilbene compound against downy mildew. Its inhibition of the mobility of *P.viticola* zoospores was shown in laboratory tests, whereas resveratrol and piceide did not influence pathogen propagules activeness [121]. In the resistant grape hybrid IRAC 2091 pterostilbene was one of the most synthetized stilbenes in infected tissues, and its toxic activity caused the reduction of pathogen growth and development [122]. Anyway the average constitutive content of pterostilbene in *V. vinifera* varieties is very low: less than 5 μg/g in leaves and fruit [73] and its concentration still remains very low in infected leaves and berries. As consequence its role in defense mechanisms is difficult to study [121]. Among stilbenes also viniferins showed a toxic activity agaist *P. viticola* zoospores, particularly δ-viniferin that has higher toxicity compared to ε-viniferina. Both compounds were identified as the major stilbenes synthesized in grape leaves infected with *P. viticola*, playing an important role in grapevine resistance to downy mildew [121]. Stilbenes have been proposed as early selection markers for resistance in grapevine breeding programs aimed to obtain downy mildew resistant genotypes. The analysis of contents of viniferins in the leaves of seedlings at 48 hpi can predict the degree of resistance to downy mildew in the selection of resistant hybrids [103].

Plant stilbene synthesis was related to the grapevine disease powdery mildew [123]. The exogenous application of methyl-jasmonate on susceptible Cabernet-Sauvignon variety increased its resistance to *E. necator* and the content of resveratrol, piceide, ε and δ- viniferins and of pterostilbene in the epidermis of leaves, suggesting a role of stilbenes in plant defence mechanisms against powdery mildew [124]. The determination of viniferins content as marker of resistance to powdery mildew has been proposed to carry out genetic selection programs. Considering that *E. necator* infections are restricted to the first layer of epidermis the amount of viniferins must be related to the number of fungal appressoria [119].

- Other phenolic compounds

Plant phenolic compounds are a very heterogeneous group of metabolites which presence in plant tissues is considered an adaptive response to adverse environmental conditions. The role of these metabolites may be physiologically important as a means of storing carbon in presence of plant nutritional deficiencies [126] and the abundance of different phenolic compounds in plant tissues has been explained as an evolutive strategy of protection against plant tissues photodamages [25]. Anyway many evidences suggest that phenolic compounds accumulation may be related to plant defence responses induced by pathogen infection [25]. The analysis of plant responses showed that the accumulation of polyphenols in cell wall of infected tissues and non-infected neighbouring tissues is related to plant HR response induced by pathogen penetration [59]. The accumulation of electrondense deposits referable to phenolic compounds was observed in *V. rotundifolia* spongy mesophyll and palisade as a consequence of *P. viticola* infection [122].

Among phenolic compounds the synthesis of flavonoids besides by light intensity can be influenced by biotic elicitors [25]. Their accumulation in grapevine tissues was related to induced defence mechanisms as shown in different comparative studies on *Vitis* species. In downy mildew resistant *V. rotundifolia,* the rapid plant response to the infection and the inhibition of pathogen growth was associated with the occurrence of small tissue necrotic spots and the detection at 2 days post infection (dpi) of a high content of flavonoids in infected stomata and closer tissues. A similar accumulation of flavonoids was detected in *V. rupestris,* an intermediate resistant species to *P. viticola,* that at 8 dpi showed the presence of peroxidase activity and the occurrence of wide tissue necrosis, resveratrol accumulation and delayed synthesis of lignin (15 dpi). In *V. vinifera* cv Grenache any HR activity were observed after infection and delayed flavonoid accumulation, detected at 8 dpi, was not able to limit high pathogen sporulation. These data suggest a key role of flavonoids during downy mildew infection as their fast synthesis is able to limit pathogen growth [127].

Grapevine berries show a different resistance to *E. necator* during their growth, considering the development of berry ontogenetic resistance [60]. The presence of autofluorescent polyfenolic compounds induced by powdery mildew infection was monitored in *V. vinifera* during berry growth. The accumulation of phenols occurred in infected cells near fungal appressoria and in non infected contiguous cells with higher frequency in susceptible young berries compared to resistant older berries which showed the lowest rate of polyfenolic oxidization [63].

A different regulation of chalcone-flavonone isomerase, a key enzyme involved in the biosynthesis of flavones, a class of flavonoids, was also found in *V.vinifera* Nebbiolo variety as consequence of the Flavescence dorée disease, suggesting the possible involvement of polyphenols in plant response to phytoplasmas [128].

- Pathogenesis-Related Proteins

Pathogenesis-related (PR) proteins may be produced in host plants as response to biotic and abiotic stresses, chemical elicitors, tissue injured by the induction of specific PR genes [100, 129-131]. They are characterized by different structure and biological activity and include 17 families of proteins with low molecular mass, high resistance to proteolysis and soluble in acid buffers [132]. Different PR proteins families have been detected in grapevine: PR-2 proteins (β-1,3-glucanases) and PR-3 and 4 proteins (chitinases) are able to hydrolyse β-1,3-glucans and chitin respectively that are known to be components of cell wall of different higher fungi; PR-5 proteins (thaumatin-like proteins) which antifungal activity is associated with the permeabilization of fungal membrane or to chitinase activity [133]. Recently PR-10 proteins family was also described [134-135].

Some members of different PR families show antifungal activity strengthening their possible role in plant defence [129, 136]. Isoforms of grape berry chitinases proved to have high toxicity against *B.cinerea* as their *in vitro* reduction of fungal conidia germination and inhibition of hyphal growth [100, 137]. Also thaumatin-like protein derived from mature berries of *B. cinerea* resistant varieties inhibited hyphal growth of grape pathogen *Botrytis cinerea* [137].

Anyway, even though some classes of these PR proteins showed in vitro toxic activity against grape pathogens, their role in plant defence mechanisms must be elucidated. Several studies analyzed the synthesis of PR-like proteins in non infected grape berries during ripening. From veraison to harvest there is a significant increase in total content of berry proteins. During this period most induced soluble proteins are chitinase and a thaumatin-like proteins also considering the decrease of photosynthetic enzymes. The accumulation of antifungal proteins in berries during this period occurr in ripe berries as they acquire resistance to powdery and downy mildew. Experimental results show that the antifungal efficacy of PR-like proteins is enhanced by sugar concentrations, showing the possible role of berry hexoses in the preservation of protein structure [100, 137]. Transcriptional changes in pathogen susceptible and resistant grape varieties were observed after tissue infections and in several studies the largest proportion of common transcripts were related to disease resistance, including several encoding PR proteins such as chitinases and β-1,3-glucanases.

The variation of chitinase and of β-1,3-glucanase activities was analyzed during grape leaves infection with *B.cinerea*. Pathogen infection significantly elicited the biosynthesis of chitinases starting from 48 hpi. A similar trend was observed for glucanase activity which increased from 48 to 72 hpi. Both chitinases and β-1,3-glucanases presence was observed around leaf dead cells, were the accumulation of secondary metabolites, among which phenols, was detected [100]. High levels of chitinases and of β-1,3-glucanases, which showed a lytic activity against germinative tubes of *E. necator*, were detected in infected grape leaves and green berries [138]. Among defense-related proteins that accumulated in Cabernet Sauvignon infected leaves, two members of PR-10 familiy were identified at different times from inoculation as response to powdery mildew infection [139].

Some studies suggest that the different level of resistance to *P. viticola* between resistant and susceptible varieties is induced after infection and is not related to differences in basal gene expression. Transcriptional changes associated with *P. viticola* infection indicate that whereas in *V. riparia* the resistance is a post-infection condition related to the early activation of signal transduction and to the synthesis of defence metabolites, in susceptible *V. vinifera* only a weak and abortive defense response was shown after infection [140]. In downy mildew susceptible Pinot noir variety the induction of PR proteins occurred in the leaves at 48 hpi and the synthesis of most PR-10 defense related proteins increased significantly by 96 hpi, which was too late to produce an effective impact on the infection [141]. Anyway the increase of chitinase transcripts detected after *P.viticola* infection of susceptible young leaves of Pinot noir and the presence of a systemic induction of lytic enzyme activities were correlated with the expression of SAR [99].

The role of salicylic acid as molecular signal in the production of several chitinase isoforms in leaves and berries was showed [100] and recently in a comparative study between *V. riparia* and *V. vinifera* during the infection of the biotrophic pathogen *P. viticola* the significant increase of the basal level of jasmonic acid was detected only in resistant *V. riparia,* while in *V. vinifera* any difference between health and infected plants was observed [140]. A different regulation of thaumatin-like and osmotin-like proteins of the PR-5 family was also found in *V.vinifera* Nebbiolo variety as consequence of the phytoplasma disease Flavescence dorée [128].

It seems useful here to consider that the possibility to increase grapevine resistance to fungal pathogens by biotechnological techniques that can permit the overexpression of PR proteins could lead to the increase of the risks of wine turbidity.

3. Conclusion

In most suitable areas of grapevine cultivation a large number of hazaurdous pests and pathogens are able to compromise plant health and fruit quality. With the aim to protect vines from parasite attacks, viticulturists have developed agronomical strategies that include the use of chemical compounds, most of which have been successively found in mature grapes, causing the reduction of fruits and wine quality. The decrease of grape biodiversity and the present genetic homogeneity of most vineyards due to the wide cultivation of a restricted number of varieties, increase plant disease susceptibility and make difficult the implementation of protection strategies. The use of selective chemical compounds has significantly improved the control of some plant diseases, but different grape pathogens have developed resistant strains that reduced the effectiveness of plant chemical protection. At present the availability of disease resistant grape varieties or selected clones has became a key strategy in many viticultural areas. During last years the conservation of grapevine germplasm increased as the characterization of endangered genotypes can improve the study of grapevine natural defence mechanisms. Plants evolved different level of response against microbial attack and the studies on different disease mechanisms suggest that susceptible grapevine varieties show basal defences similarly to resistant genotypes, but in most cases delayed in time or weak for intensity. The study of morphological characteristics, genetic basis and chemical signals that regulated natural defence mechanisms in grapevine could allow us to develop significant advances in the exploitation of Vitis biological resources and in the use of marker assisted selection aimed to reduce the time to select resistant genotypes for fruit quality improvement and environmental costs reduction.

Author details

Massimo Muganu and Marco Paolocci

Department of Science and Technology for Agriculture, Forests, Nature and Energy, University of Tuscia, Viterbo, Italy

References

[1] McGovern P. The archeological and chemical hunt for the origin of viticulture in the Near East and Etruria. In: Ciacci A., Rendini P., Zifferero A. (eds.): proceedings of the International Symposium Archeologia della vite e del vino in Etruria, 9-10 Sept. 2005. pp. 108-122 Città del Vino, Siena, 2007.

[2] Brun J. P. Le tecniche di spremitura dell'uva: origini e sviluppo dell'uso del torchi nel Mediterraneo. In: Ciacci A., Rendini P., Zifferero A. (eds.): proceedings of the International Symposium Archeologia della vite e del vino in Etruria, 9-10 Sept. 2005. pp. 55-65 Città del Vino, Siena, 2007.

[3] Forni G. Quando e come sorse la viticoltra in Italia. In: Ciacci A., Rendini P., Zifferero A. (eds.): proceedings of the International Symposium Archeologia della vite e del vino in Etruria, 9-10 Sept. 2005. pp. 69-81 Città del Vino, Siena, 2007.

[4] De' Crescenzi P. De diversis speciebus vitium. In Liber ruralium commodorum 1305; IV; 4.

[5] H. Petrus, Basilea 1548.

[6] Maracchi G., Sirotenko O., Bindi M.. Impacts of present and future climate variability on agriculture and forestry in the temperate regions: Europe Climatic Change 2005; 70: 117-135.

[7] Matasci C. L., Gobbin D., Schärer H.-J., Tamm L., Gessler C. Selection for fungicide resistance throughout a growing season in populations of Plasmopara viticola. European Journal of Plant Pathology 2008; 120: 79-83.

[8] Boccacci P., Marinoni T.D., Gambino G., Schneider A. Genetic Characterization of Endangered Grape Cultivars of Reggio Emilia Province. Am. J. Enol. Vitic. 2005; 56:411-416.

[9] Muganu M., Dangl G., Aradhya M., Frediani M., Scossa A., Stover E. Ampelographic and DNA Characterization of local grapevine accessions of the Tuscia Area (Latium, Italy). Am. J. Enol. and Vitic. 2009; 60: 110-115.

[10] Lacombe T., Boursiquot J.M., Laucou V., Dechesne F., Varès D., This P. Relationships and Genetic Diversity within the Accessions Related to Malvasia Held in the Domaine de Vassal GrapeGermplasm Repository Am. J. Enol. Vitic. 2007; 58:124-131.

[11] Ministero Agricoltura, Industria e Commercio. Bullettino ampelografico, 1875-1887; Roma.

[12] Cinelli O. La cantina sperimentale di Viterbo. Società tipografica (Ed.), 1884; Bologna.

[13] Vannuccini L. I vitigni toscani. In "Annuario generale di Viticoltura ed Enologia, anno I, 1892.

[14] Gabler F.M., Smilanick J.L., Mansour M., Ramming D.W., B.E. Mackey. Correlations of morphological, anatomical and chemical features of grape berries with resistance to Botrytis cinerea. Phytopatology 2003; 93:1263-1273.

[15] Boso S., Martìnez M. C., Unger S., Kassemeyer H. H. Evaluation of foliar resistance to downy mildew in different cv. Albariño clones. Vitis 2006; 45, 23-27.

[16] Muganu M., Balestra G.M., Magro P., Pettinari G., Bignami C. Susceptibility of local grape cultivars to Plasmopara viticola and response to copper compounds with low cupric salts concentration in Latium (Central Italy). Acta Horticulturae 2007; 754: 373-378,

[17] Boso S., Kassemeyer H.H. Different susceptibility of European grapevine cultivars for downy mildew. Vitis 2008; 47: 39-49.

[18] Cadle-Davidson L., Chicoine D.R., Consolie N.H.,. Variation within and between Vitis species for foliar resistance to the powdery mildew pathogen Erysiphe necator. Plant Disease 2010; 95: 202-211.

[19] Gaforio L., Garcia-Munoz S., Cabello F., Munoz-Organero G. Evaluation of susceptibility to powdery mildew (Erysiphe necator) in Vitis vinifera varieties. Vitis 2011; 50 : 123-126.

[20] Boso S., Alonso-Villaverde V., Gago P., Santiago J.L., Martìnez M.C. Susceptibility of 44 grapevine (Vitis vinifera L.) varieties to downy mildew in the field. Australian Journal of Grape and Wine Research 2011; 17: 394-400.

[21] Wang Y., Liu Y., He P., Chen J., Lamikanra O., Lu J. Evaluation of foliar resistance to Uncinula necator in Chinese wild Vitis species. Vitis 1995; 34: 159-164.

[22] Staudt G., Kassemeyer H. H. Evaluation of downy mildew resistance in various accessions of wild Vitis species. Vitis 1995; 34: 225-228.

[23] Cadle-Davidson L.Variation Within and Between Vitis spp. for Foliar Resistance to the Downy Mildew Pathogen Plasmopara viticola. Plant Disese 2008; 92: 1577-1584.

[24] Feechan A., Kabbara S., Dry I.B. Mechanisms of powdery mildew resistance in the Vitaceae family. Molecular Plant Pathology 2011; 12: 263–274

[25] Prell H.H., Day P.R. Plant-fungal pathogen interaction – A classical and Molecular View. 2001. Springer Verlag, Germany.Treutter D. Significance of flavonoids in plant resistance: a review. Environmental Chemistry Letters 2006; 4: 147-157.

[26] OIV (Organisation Internationale de la Vigne et du Vin). Codes des caractères descriptifs des variétés et espèces de Vitis. Paris, 2009.

[27] Brewer C.A., Smith W.K.,Vogelmann T.C., 1991. Functional interaction between leaf trichomes, leaf wettability and optical properties of water droplets. Plant Cell and Environment 14; 995-962.

[28] Kortekamp A., Wind R., Zyprian E., 1999. The role of hairs on the wettability of grapevine (Vitis spp) leaves. Vitis 38; 101-105.

[29] Zaiter H.Z., Coyne D.P., Staedman J.R., Beaver J.S. Inheritance of abaxial leaf pubescence in beans. J. Amer. Soc. Horticult. Sci. 1990; 115: 1158-1160.

[30] Staedman J.R., Shaik M. Leaf pubescence confers apparent race-nonspecific rust resistance in bean (Phaseolus vulgaris). Phytopathology 1988; 78:1566.

[31] Levin D.A. The role of trichomes in plant defence. Quarterly Rewiev of Biology 1973; 48: 3-15.

[32] Staud G., Kassemeyer H. H. Evaluation of downy mildew resistance in various accessions of wild Vitis species. Vitis 1995; 34: 225-228.

[33] Kortekamp A., Zyprian E. Leaf hairs as a basic protective barrier against downy mildew of grape. J Phytopathology 1999; 147: 453-459.

[34] Boso S., Martínez M.C., Unger S. Kassemeyer H.H. Evaluation of foliar resistance to downy mildew in different cv. Albariño clones.Vitis 2006; 45, 23–27.

[35] Boso S., Alonso-Villaverde V., Santiago J.L., Gago P., Dürrenberger M., Düggelin M., Kassemeyer H.H., Martinez M.C. Macro and microscopic leaf characteristics of six grapevine genotypes (Vitis spp) with different susceptibilities to grapevine downy mildew. Vitis 2010; 49: 43-50.

[36] Rogiers S.Y., Hardie, Smith J.P. Stomatal density of grapevine leaves (Vitis vinifera L.) responds to soil temperature and atmospheric carbon dioxide. Australian Journal of Grape and Wine Research 2011; 17: 147–152

[37] Gómez-del-Campo M., Ruiz C., Baeza P., Lissarrague J.R. Drought adaptation strategies of four grapevine cultivars (Vitis vinifera L.): modification of the properties of the leaf area. Journal International des Sciences de la Vigne et du Vin 2003; 37: 131-143.

[38] Rogiers S.Y., Greer D.H., Hutton R.J., Landsberg, J.J. Does night time transpiration contribute to anisohydric behaviour in a Vitis vinifera cultivar? Journal of Experimental Botany 2009; 60: 3751-3763.

[39] Palliotti A., Cartechini A., Ferranti F. Morpho-anatomical and physiological characteristics of primary and lateral shoot leaves of Cabernet Franc and Trebbiano toscano grapevines under two irradiance regimes. Am. J. Enol. Vitic. 2000; 51: 122-130.

[40] Shaik M., Race-nonspecific resistance in bean cultivars to races of Uromyces appendiculatus var. appendiculatus and its correlation with leaf epidermal characteristic. Phytopathology 1985; 75: 478-481.

[41] Matta A. Basi generali della resistenza a patogeni. Petria 1996; 6: 29-38.

[42] Stenglein S. A., Arambarri A. M., Sevillano M. C. M., Balatti P. A. Leaf epidermal characters related with plant's passive resistence to pathogens vary among accessions of wild beans Phaseolus vulgaris var. aborigineus (Leguminosae-Phaseolae). Flora 2005; 200: 285-295.

[43] Gindro K., Pezet R., Viret O. Histological study of the responses of two Vitis vinifera cultivars (resistant and susceptible) to Plasmopara viticola infections. Plant Physiology and Biochemistry 2003; 41: 846-853.

[44] Kiefer B., Riemann M., Büche C., Kassemeyer H.H., Nick P. The host guides morphogenesis and stomatal targeting in the grapevine pathogen Plasmopara viticola. Planta 2002; 215: 387-393.

[45] Morris B.M., Nar G. Mechanism of electrotaxis of zoospores of phytopathogenetic fungi. Phytopathology 1993; 83: 877-882.

[46] Morris B.M., Reid B., Gow NAR. Tactic response of zoospores of the fungus Phytophthora palmivora to solutions of different pH in relation to plant infection. Microbiology 1995; 141; 1231-1237.

[47] Lalancette N., Ellis M.A., Madden L.V. Estimating infection efficiency of Plasmopara viticola on grape. Plant Disease 1987; 71 : 981-983.

[48] Thomas D.D, Peterson A.P. Chemotactic auto-aggregation in the water mold Achlya. J. Gen. Microbiol 1990; 136: 847-854.

[49] Kennelly M.M., Gadoury D.M., Wilcox W.F., Magarey P.A., Seem R.C. Seasonal development of ontogenetic resistance to downy mildew in grape berries and rachises. Phytopathology 2005; 95: 1445-1452.

[50] Bessis R. Etude de 1'évolution des stomates et des tissus péristomatiques du fruit de la vigne. Comptes Rendus de l'Académie des Sciences de Paris, Série D 1972; 274: 2158-2161.

[51] Rogiers S. Y., Whitelaw-Weckert M., Radovanonic-Tesic M., Greer L.A., White R.G., Steel C.C. Effects of spray adjuvants on grape (Vitis vinifera) berry microflora, epicuticular wax and susceptibility to infection by Botrytis cinerea. Australasian Plant Pathology 2005; 34: 221-228.

[52] Bernard A.C., Dallas J.P., Adheran F. Observations sur le nombre de stomates des baies de varietes de Vitis vinifera L. Relation avec leur comportement a l'egard de la pourriture grise (Botrytis cinerea Pers.). Le Progrès Agricole et Viticole 1981; 8: 230-232.

[53] Allègre M., Daire X., Héloir M.C., Trouvelot S., Mercier L., Adrian M., Pugin A. Stomatal deregulation in Plasmopara viticola-infected grapevine leaves. New Phytologist 2007; 173: 832-840.

[54] Percival D.C., Sullivan J.A., Fisner K.H. Effect of cluster exposure, berry contact and cultivar on cuticular membrane formation and occurrence of bunch rot (Botrvtis cinerea) with three Vitis vinifera L. cultivars. Vitis 1993; 32: 87-99.

[55] Riederer, M., Schreiber L.. Protecting against water loss: analysis of the barrier properties of plant cuticles. J. Exp. Bot. 2001, 52: 2023-2032.

[56] Rogiers S.Y, Hatfield J.M., Jaudzems V.G., White R.G., Keller M. Grape berry cv. Shiraz epicuticular wax and transpiration during ripening and preharvest weight loss. Am. J. Enol. Vitic. 2004; 55: 121-127

[57] Muganu M., Bellincontro A., Barnaba F.E., Paolocci M., Bignami C., Gambellini G., Mencarelli F. Influence of Bunch Position on Berry Epicuticular Wax During Ripening and on Weight Loss in Dehydration Process. Am. J. Enol. Vitic. 2011; 62, 91-98.

[58] Heintz C., Blaich., R., Structural characters of epidermal walls and resistance to powdery mildew of different grapevine cultivars. Vitis 1989, 28: 153-160.

[59] Heintz C., Blaich., R. Ultrastructural and histochemical studies on interactions between Vitis vinifera L. and Uncinula necator (Schw.) Burr. New Phytologist 1990; 115:107-117.

[60] Ficke A., Gadoury D. M., Seem R. C., Dry I. B. Effects of ontogenic resistance upon establishment and growth of Uncinula necator on grape berries. Phytopathology 2003; 93:556-563.

[61] Marois J.J., Bledsoe A.M., Gubler W.D. Effect of surfactants on epicutilcular wax and infection of grape berries by Botrytis cinerea. Phytopathology 1985; 75: 1329.

[62] Commenil P., Brunet L., Audran J.C. The development of the grape berry cuticle in relation to susceptibility to bunch rot disease. Journal of Experimental Botany 1997; 48: 1599-1607.

[63] Ficke A., Gadoury D. M., Seem R. C., Godfrey, D., Dry I. B. Host barriers and responses to Uncinula necator in developing grape berries. Phytopathology 2004; 94:438-445.

[64] Padgett M., Morrison J.C. Changes in grape berry exudates during fruit development and their effect on mycelial growth of Botrytis cinerea. Journal of the American Society for Horticultural Science 1990; 115: 269-73.

[65] 65] Commenil P., Belingheri L., Audran J.C., Collas A., Dehorter B. Mise en evidence d'une activite anti-Botrytis dans les cires epicuticulaires de jeunes baies de Vitis vinifera, variete Pinot noir. Journal International des Sciences de la Vigne et du Vin. 1996; 30: 7-13.

[66] Alonso-Villaverde V., Voinesco F., Viret O., Spring J.L., Gindro K. The effectiveness of stilbenes in resistant Vitaceae: Ultrastructural and biochemical events during Plasmopara viticola infection process. Plant Physiology and Biochemistry 2011, 49: 265-274.

[67] VanEtten H., Mansfield J. W., Bailey J. A., Farmer E. E. Two classes of plant antibiotics: phytoalexins versus "phytoanticipins". Plant Cell. 1994; 6:1191–1192.

[68] Muller, K.O., Borger, H. Experimentelle untersuchungen über die Phytophthora resistem der kartoffel. Arb. Biol. Reichsasnstalt. Landw. Forstw. Berlin 1940, 23: 189-231.

[69] Agati G., Zoran F., Cerovic G., Dalla Marta A., Di Stefano V., Pinelli P., Traversi M. L., Orlandini S. Optically-assessed preformed flavonoids and susceptibility of grapevine to Plasmopara viticola under different light regimes Functional Plant Biology 2008, 35: 77–84.

[70] Orsini M.C., Sansavini S. Determinazione delle component fenoliche associate alla resistenza alla ticchiolatura nel melo. Frutticoltura 2008; 2:51-59.

[71] Pezet R., Pont V. Mise en évidence de ptérostilbène dans les grappes de Vitis vinifera. Plant Physiol. Biochem. 1988; 26: 603-607.

[72] Goetz G., Fkyerat A., Métais N., Kunz M., Tabacchi R., Pezet R., Pont V. Resistance factors to grey mould in grape berries: identification of some phenolics inhibitors of Botrytis cinerea stilbene oxidase. Phytochemistry 1999; 52: 759-767.

[73] van Baarlen P., Legendre L., van Kann J.A.L. Plant defence compounds against Botrytis infection. In: Y.Elad et al. (eds.) Botrytis: Biology, Pathology and Control. Springer 2007; pp 143-161

[74] Pezet R., Pont V., Hoang-Van K. Evidence for oxidative detoxification of pterostilbene by a laccase-like stilbene oxidase produced Botrytis cinerea. Physiol. Mol. Plant Pathol. 1991; 39: 441- 450.

[75] Pezet R., Viret O., Perret C., Tabacchi R. Latency of Botrytis cinerea Pers.: Fr. and biochemical studies during growth and ripening of two grape berry cultivars, respectively susceptible and resistant to grey mould. Journal of Phytopathology 2003; 15: 208-214.

[76] Austin C.N., Wilcox W.F. Effects of fruit-zone leaf removal, training system, and variable irrigation on powdery mildew development on Vitis vinifera L. Chardonnay. Am. J. Enol. Vitic. 2011; 62: 193-198.

[77] Austin C.N., Mejers J., Grove J.J., Wilcox W.F.. Quantification of powdery mildew severity as a function of canopy variability and associated impacts on sunlight penetration and spray coverage within the fruit zone. Am. J. Enol. Vitic. 2011; 62: 23-31.

[78] Valdés-Gómez H., Gary C., Cartolaro P., Lolas-Caneo M., Calonnec A.. Powdery mildew development is positively influenced by grapevine vegetative growth induced by different soil management strategies. Crop Protection. 2011; 30 : 1168-1177.

[79] Zahavi T., Reuveni M. Effect of grapevine training systems on susceptibility of berries to infection by Erysiphe necator. European Journal of Plant Pathology 2012; 133: 511-515.

[80] Reddy G.V.P., Guerrero A. Interactions of insect pheromones and plant semiochemicals. Trends Plant Science 2004; 9: 253-261.

[81] Valterova I., Nehlin G., Borg-Karlsson A.K. Host plant chemistry and preferences in egg-laying Trioza apicalis (Homoptera, Psylloidea). Biochemical Systematics and Ecology 1997; 25: 448-491.

[82] Tasin M., Anfora G., Ioriatti C., Carlin S., De Cristofaro A., Schmidt S., Bengtsson M., Versini G., Witzgall P. Antennal and behavioral responses of grapeline moth Lobesia botrana females to volatiles from grapevine. J. Chem. Ecol. 2005; 31:77-87.

[83] Thiéry D., Moreau J. Relative performance of European grapevine moth (Lobesia botrana) on grapes and other hosts. Oecologia 2005; 143:548-557.

[84] Sharon R., Soroker V., Wesley S.D., Zahavi T., Harari A., Weintraub P.G. Vitex agnus-castus is a preferred host plant for Hyalesthes obsoletus. Journal of Chemical Ecology 2005; 31: 1051-1063.

[85] Bent A.F., Mackey D. Elicitors, effectors, and R genes: the new paradigm and a lifetime supply of questions. Annual Review of Phytopathology 2007; 45: 399-436.

[86] Dry I.B., Feechan A., Anderson C., Jermakow A.M., Bouquet A., Anne F., Adam Blondon A.F., Thomas M.R. Molecular strategies to enhance the genetic resistance of grapevine to powdery mildew. Australian Journal of Grape and Wine Research 2010; 16: 94-105.

[87] Ramming D.W., Gabler F., Smilanick J., Cadle-Davidson M., Barba P., Mahanil S., Cadle-Davidson L. 2011 A single dominant locus, Ren4, confers rapid non-race-specific resistance to grapevine powdery mildew. Phytopatology, 101: 502-508

[88] Feechan A., Kabbara S., Dry I.B. Mechanisms of powdery mildew resistance in the Vitaceae family. Molecular Plant Pathology 2011; 12: 263–274

[89] Pauquet J., Bouquet A., This P., Adam-Blondon A.F. Establishement of a local map of AFLP markers around the powdery mildew resistance gene Run1 in grapevine and assessment of their usefulness for marker assisted selection. Theoretical Applied Genetics 2001; 103: 1201-1210.

[90] Hoffmann S., Di Gaspero G.., Kovàcs L., Howard S., Kiss E., Galbàcs Z., Testolin R., Kozma P.. Resistance to Erysiphe necator in the grapevine "Kishmish vatkana" is controlled by a single locus through restriction of hyphal growth. Theoretical and applied genetics 2008; 116: 427-438.

[91] Cadle-Davidson L., Mahanil S., Gadoury D.M., Kozma P., Reisch B.I. Natural infection of Run1-positive vines by native genotypes of Erysiphe necator. Vitis 2011; 50: 173-175.

[92] Ryals J., Neuenschwander U., Willits M., Molina A., Steiner H. Y., Hunt M. Systemic acquired resistance. Plant Cell 1996; 8:1809-1819.

[93] Sticher L., Mauch-Mani B., Metraux J. P. Systemic acquired resistance. Annual Review of Phytopathology 1997; 35: 235-270.

[94] Durrant W.E., Dong X. Systemic acquired resistance. Annual Review of Phytopathology 2004; 42: 185-209.

[95] Beckers G.J.M., Spoel S.H. Fine -Tuning plant defence signalling: salicylate versus jasmonate Plant Biol. 2006; 8: 1-10.

[96] Bari R., Jones J.D. Role of plant hormones in plant defence responses. Plant. Mol. Biol. 2009; 69: 473-488.

[97] Raymond W.M.F., Gonzalo M., Fekete C., Kovacs L.G.., He Y., Marsh E., McIntyre L.M., Schachtman D.P., Qiu W. Powdery mildew induces defense-oriented reprogramming of the transcriptome in a susceptible but not in a resistant grapevine. Plant Physiol. 2008; 146: 236-249.

[98] Li X., Zheng X., Yan S., Li S. Effects of salicylic acid (SA), ultraviolet radiation (UV-B and UV-C) on trans-resveratrol inducement in the skin of harvested grape berries. Front. Agric. China 2008 2: 77-81.

[99] Busam G., Kassemeyer H.H., Matern U. Differential expression of chitinases in Vitis vinifera L. responding to systemic acquired resistance activators or fungal challenge. Plant Physiology 1997; 115: 1029-1038.

[100] Derckel J. P., Audran J.C., Haye B., Lambert B., Legendre L. Characterization, induction by wounding and salicylic acid and activity against Botrytis cinerea of chitinases and β-1,3-glucanases of ripening grape berries. Physiol. Plant. 1998; 104:56-64.

[101] Heintz C., Blaich R. Ultrastructural and histochemical studies on interactions between Vitis vinifera L. and Uncinula necator (Schw.) Burr. New Phytologist 1990; 115: 107-117.

[102] Kortekamp A., Wind R., Zyprian E. The role of callose deposits during infection of two downy mildew-tolerant and two susceptible Vitis cultivar. Vitis 1997; 36: 103-104.

[103] Gindro K., Spring J. L., Pezet R., Richter H., Viret O. Histological and biochemical criteria for objective and early selection of grapevine cultivars resistant to Plasmopara viticola. Vitis 2006; 45: 191-196.

[104] Hamiduzzaman M. M., Jakab G., Barnavon L., Neuhaus J.M., Mauch-Mani, B. β-amino butyric acid induced resistance against downy mildew in grapevine acts through

the potentiation of callose formation and JA signalling. Molecular Plant Microbe Interactions 2005; 18: 819-829.

[105] Bavaresco L. Role of viticultural factors on stilbene concentration of grape and wine. Drugs Under Experimental Clinical Research 2003; 29: 181-187.

[106] Mattivi F., Reniero F., Korhammer S. Isolation, characterization and evolution in red wine vinification of resveratrol monomers. Journal of Agricultural and Food Chemistry 1995; 43:1820-1823.

[107] Bavaresco L., Petegolli D., Cantù E., Fregoni M., Chiusa G., Trevisan M. Elicitation and accumulation of stilbene phytoalexins in grapevine berries infected by Botrytis cinerea. Vitis 1997; 36: 77-83

[108] Mattivi F., Vrhovsek U., Malacarne G., Masuero D., Zulini L., Stefanini M., Moser C., Velasco R., Guella G. Profiling of resveratrol oligomers, important stress metabolites, accumulating in the leaves of hybrid Vitis vinifera (Merzling x Terodelgo) genotypes infected with Plasmopara viticola. Journal of Agricultural and Food Chemistry 2011; 59: 5364-5375.

[109] Bavaresco L., Vezzulli S., Civardi S., Gatti M., Battimani P., Pietri A., Ferrari F. Effect of lime-induced leaf chlorosis on ochratoxin A, trans-resveratrol, and ε-viniferin production in grapevine (Vitis vinifera L.) berries infected by Aspergillus carbonarius. Journal of Agricultural and Food Chemistry 2008; 56: 2085-2089.

[110] Gatto P., Vrhovsek U., Muth J., Segala C., Romualdi C., Fontana P., Pruefer D., Stefanini M., Moser C., Mattivi F., Velasco R. Ripening and genotype control stilbene accumulation in healthy grapes. Journal of Agricultural and Food Chemistry 2008; 56: 11773-11785.

[111] Aggarwal B.B., Bhardwaj A., Aggarwal R.S., Seeram N.P., Shishodia S., Takada Y. Role of resveratrol in prevention and therapy of cancer: preclinical and clinical studies. Anticancer Research 2004; 24: 2783-2840.

[112] Aggarwal B., Shishodia S. Resveratrol in health and disease. CRC Press Taylor & Francis, Boca Raton FL, USA, 2006.

[113] Langcake P., Price R.J. Production of resveratrol by Vitis vinifera and other members of Vitaceae as a response to infection or injury. Physiological Plant Pathology 1976; 9: 77-86.

[114] Bavaresco L., Fregoni C., van Zeller de Macero Basto Gonçalves M.I., Pezzulli S. Physiology and molecular biology of grapevine stilbenes: an update, In: Roubelakis-Angelakis K.A. (ed.) Grapevine Molecular Physiology and Biotechonolgy, 2nd ed., Springer Science-Business Media B.V., pp. 341-364, 2009.

[115] Sarig P., Zutkhi Y., Monjauze A., Lisker N., Ben-Arie R. Phytoalexin elicitation in grape berries and their susceptibility to Rhizophus stolonifer. Physiological and molecular Plant Pathology 1997; 50: 337-347.

[116] Jeandet P., Douillet-Breuil A.C., Bessis R., Debord S., Sbaghi M., Adrian M. Phytoalexins from the Vitaceae: biosynthesis, phytoalexin gene expression in transgenic plants, antifungal activity, and metabolism. J. Agric. Food. Chem. 2002; 50: 2731-2741.

[117] Pezet R., Pont V.R. Mode of toxic action of Vitacee stilbenes on fungal cells. In Daniel M., Purkayastha (eds.) handbook of Phytoalexins Metabolism and action 1995 p317-331, M. dekker Inc New York, Basel, Hong-Kong.

[118] Adrian M., Jeandet P., Veneau J., Weston L.A., Bessis R. Biological activity of resveratrol, a stilbenic compound from grapevines, against Botrytis cinerea, the causal agent for gray mold. Journal of Chemical Ecology 1997; 23: 1689-1702.

[119] Sbaghi M., Jeandet P., Faivre B., Bessis R., Fournioux J.C. Development of methods using phytoalexin (resveratrol) assessment as a selection criterion to screen grapevine in vitro cultures for resistance to grey mould (Botrytis cinerea). Euphytica 1995; 86: 41-47

[120] Coutos-Thevenot P., Poinssot B., Bonomelli A., Yean H., Breda C., Buffard D., Esnault R., Hain R., Boulay M. In vitro tolerance to Botrytis cinerea of grapevine 41B rootstock in transgenic plants expressing the stilbene synthase Vst1 gene under the control of a pathogen-inducible PR 10 promoter. J. Exp. Bot. 2001; 52: 901-910.

[121] Pezet R., Gindro K., Viret O., Richter H. Effects of resveratrol, viniferins and pterostilbene on Plasmopara viticola zoospore mobility and disease development. Vitis 2004; 43: 145-148.

[122] Alonso-Villaverde V., Voinesco F., Viret O., Spring J.L., Gindro K. The effectiveness of stilbenes in resistant Vitaceae: Ultrastructural and biochemical events during Plasmopara viticola infection process. Plant Physiology and Biochemistry 2011; 49: 265–274.

[123] Schnee S., Spring J. L., Viret O., Dubuis P. H., Gindro K. Outils pour la sélection précoce de cépages résistants à l'oïdium. Revue suisse de Viticulture, Arboriculture et Horticulture 2009; 41: 87-93.

[124] Belhadj A., Saigne C., Telef N., Cluzet S., Bouscaut J., Corio-Costet M.F., Mérillon J.M. Methyl jasmonate induces defense reponses in grapevine and triggers protection against Erysiphe necator. Journal of Agricultural and Food Chemistry 2006; 54: 9119-9125.

[125] Schnee S.,Viret O.,Gindro K. Role of stilbenes in the resistance of grapevine to powdery mildew. Physiological and Molecular Plant Pathology 2008; 72: 128-133.

[126] Lattanzio V., Cardinali A., Linsalata V. Plant phenolics: a biochemical and physiological perspective. In: Cheynier V. (ed.) Recent advances in polyphenol research John Wiley and son, 2012 p 1-39

[127] Dai G.H., Andary C., Mondolot-Cosson L., Boubals D. Histochemical studies on the interaction between three species of grapevine, Vitis vinifera, V. rupestris and V. ro-

tundifolia and the downy mildew fungus, Plasmopara viticola. Physiological and Molecular Plant Pathology 1995; 4 :177-188.

[128] Margaria P, Palmano S. Response of the Vitis vinifera L. cv. 'Nebbiolo' proteome to Flavescence dorée phytoplasma infection. Proteomics 2011; 11: 212-24.

[129] L.C. Van Loon. Induced resistance in plants and the role of pathogenesis-related proteins. European Journal of Plant Pathology 1997; 103: 753–765.

[130] Kortekamp A. Expression analysis of defence-related genes in grapevine leaves after inoculation with a host and a non-host pathogen. Plant Physiology and Biochemistry 2006; 44: 58-67

[131] Aziz A., Gauthier A., Bézier A., Poinssot B., Joubert J.M., Pugin A., Heyraud H., Baillieul F. Elicitor and resistance-inducing activities of β-1,4 cellodextrins in grapevine, comparison with β -1,3 glucans and α-1,4 oligogalacturonides. Journal of Experimental Botany 2007; 58:1463–1472.

[132] van Loon L. C., Rep M., Pieterse C. M. J. Significance of inducible defense-related proteins in infected plants. Annu.Rev. Phytopathol. 2006, 44, 135-162.

[133] Ferreira R.B., Monteiro S.S., Piçarra-Pereira M.A., Teixeira A.R. Engineering grapevine for increased resistance to fungal pathogens without compromising wine stability. Trends in Biotechnology 2004; 22: 168-173.

[134] Liu J.J., Ekramoddoullah A.K.M. The family 10 of plant pathogenesis-related proteins: their structure, regulation, and function in response to biotic and abiotic stresses. Physiol. Mol. Plant. Pathol. 2006; 68: 3-13.

[135] Lebel S., Schellenbaum P., Walter B., Maillot P. Characterisation of the Vitis vinifera PR10 multigene family. BMC Plant Biol 2010; 10:184.

[136] Kasprzewska A. Plant chitinases - regulation and function. Cellular and Molecular Biology Letters 2003; 8: 809-824.

[137] Salzman R.A., Tikhonova I., Bordelon B.P., Hasegawa P.M., Bressan R.A. Coordinate Accumulation of Antifungal Proteins and Hexoses Constitutes a Developmentally Controlled Defense Response during Fruit Ripening in Grape. Plant Physiology 1998; 117:465-472.

[138] Giannakis C., Bucheli C.S., Skene K.G.M., Robinson S.P., Scott N.S., (1998). Chitinase and beta-1,3-glucanase in grapevine leaves: a possible defence against powdery mildew infection. Australian Journal of Grape and Wine Research 4, 14-22.

[139] Marsh E., Alvarez S., Hicks L.M., Barbazuk W.B., Qiu W., Kovacs L., Schachtman D. Changes in protein abundance during powdery mildew infection of leaf tissues of Cabernet Sauvignon grapevine (Vitis vinifera L.). Proteomics 2010; 10: 2057-64.

[140] Polesani M., Bortesi L., Ferrarini A., Zamboni A., Zadra C., Lovato A.,Pezzotti M., Delledonne M., Polverari A. General and species specific transcriptional responses to

downy mildew infection in a susceptible (Vitis vinifera) and resistant (V. riparia) grapevine species. BMC Genomics 2010; 11:117

[141] Milli A., Cecconi D., Bortesi L., Persi A., Rinalducci S., Zamboni A., Zoccatelli G., Lovato A., Zolla L., Polverari A. Proteomic analysis of the compatible interaction between Vitis vinifera and Plasmopara viticola. Journal of Proteomics 2012; 75: 1284-1302.

Production of Anthocyanins in Grape Cell Cultures: A Potential Source of Raw Material for Pharmaceutical, Food, and Cosmetic Industries

Anthony Ananga, Vasil Georgiev, Joel Ochieng,
Bobby Phills and Violeta Tsolova

Additional information is available at the end of the chapter

http://dx.doi.org/10.5772/54592

1. Introduction

Research continues to show that many artificial pigments are actually detrimental to our health. According to [1], there is an increasing consumer preference for healthy foods, which has invited considerable demand for the use of anthocyanins as natural colorants, because of their natural pedigree and healthful properties. Anthocyanins are the most widely distributed group of water-soluble plant pigments in nature. They are mainly responsible for the mauve, red, blue, and purple colors in flowers, fruits, leaves, seeds and other organs in most of the flowering plants. The other important class of water-soluble pigments are betalains, which are present only in plants belonging to 13 families of Caryophyllales order [2-5]. An interesting phenomenon is the existence of mutual exclusiveness of anthocyanins and betalains in plant kingdom [3, 5-9]. Recent research demonstrated that simultaneous production of anthocyanins and betalains is possible in cell cultures and seedlings of anthocyanin producing plants by introduction and expression of genes encoding dihydroxyphenylalanine (L-DOPA) dioxygenases in combination with substrate precursor feeding [10]. However, the co-occurrence of both pigments in the same plant species have never been found in nature and the plants which produce anthocyanins never produce betalains and vice versa [6]. The commercial production of anthocyanin pigments is one of the fastest growing segments of the food colorant industry [2, 11]. The only industrial sources for anthocyanin pigments are from whole plant extracts [1], with the most common source being grape skins from the wine industry. According to [1], the demand of natural colorants continues to rise by 5-15% every year and this translated to the sales of anthocyanins isolated from grape skins in 2002, which was estimated to be US$200 million worldwide. The increase in demand for processed foods and high health products has caused the manufacturers to look for alternative sources of colorants with antioxidant properties. One source is the production of anthocyanins through the use of plant cell cultures [2, 12, 13].

Anthocyanins are synthesized via the flavonoid pathway, and they are known to contribute red, blue and purple color to colored grapes, wines and other products [14-18]. Anthocyanins can be used not only as food and beverage additives to obtain attractive natural coloration [19], but also for generating pharmaceutical and cosmetic products. Most researchers are optimistic about utilizing them as bioactive compounds with the consideration that they have the potential to improve human health [1]. Anthocyanins have been implicated in lowering the risk of cardiovascular disease and certain cancers. Dietary anthocyanins can be obtained by humans through the ingestion of fresh colored fruits processed into food and beverages. For instance, the consumption of red grapes and wine is considered vital for bioavailable anthocyanins [20, 21]. To date, most anthocyanin colorants are extracted from grape skins, black carrots, red cabbage, and sweet potato [11]. However, researchers are also exploring the idea of cultivating plant cell cultures for the production of natural colorants. Therefore, there is an interest in improving the quantity and quality of anthocyanins produced in grape cells, and this means that commercially viable systems must be developed to produce anthocyanins in grape cell cultures.

Production of anthocyanins by plant cell cultures is a feasible technology being pursued by industrial and academic interests. Several strategies are being used to enhance anthocyanin biosynthesis in plant cells. This involves a proper selection of the cell strain and optimization of media as well as culture conditions. It is crucial to note that anthocyanins obtained directly from fresh plant materials has limitations such as low metabolite yield, variability, and seasonal availability of raw materials, fresh material losses, inconsistent product quality, and pigment degradation caused by storage and extraction process [22]. Therefore, it is prudent to use *in vitro* cell and tissue cultures for the production of anthocyanins as the potential alternative to synthetic coloring agents. In order to cultivate plant cell tissues using biological techniques, there needs to be two approaches; 1) cell cultures have to be studied, and 2) clonal propagation techniques have to be developed. According to [23], the study of cell cultures starts when the calli are initiated *in vitro*, for the purpose of finding the optimum media composition that best suits cultivation. It is important to note that during cultivation process, the calli can undergo somaclonal variation as they go through different steps of subculture. However, single lines needs to be screened when the genetic stability is reached, so that the productivity of each cell line can be evaluated with the purpose of using them in cell suspension cultures. In that regard, the production of anthocyanins can be increased in cell suspensions through different ways. The final step is the bioreactor cultivation and scale-up to commercial production of anthocyanins. The last one is a critical step since it is in direct correlation with the economical feasibility of the entire process. In our laboratory, we have more than 10 years experience with *in vitro* cultures of different Native American grape species. We have various types of cell suspensions, obtained from the super-epidermal cells of muscadine berry skins (Noble var.) at two phonological stages: veraison and physiological maturity. The long-term goal of our research is to use these cells for nutraceuticals, cosmeceuticals, and food additive studies.

2. Biology and chemistry of anthocyanins as pigments

As colored molecules, anthocyanins play a key role in survival and evolution of flowering plants by attracting pollinators, frugivores and seed dispersers on one hand, and by repelling herbivores and parasites on the other [24-26]. Moreover, anthocyanins execute several important physiological functions in plant cells, and their biosynthesis is strongly induced by biotic and abiotic stress factors. These factors include, light, UV radiation, high or low temperatures, wounding, osmotic stress, nutrient imbalance, ozone exposure, herbivores, microbial and viral attacks. In [24, 27] the major roles of anthocyanins in photoprotection of chloroplasts from photoinhibitory damage have been discussed in details. The authors have also clarified the involvement of anthocyanins in protection from UV-B radiation, as well as how anthocyanins decrease oxidative stress by scavenging free radicals and modulating reactive oxygen signaling cascades. These cascades are responsible for triggering the expression of stress-responsive genes as well as the regulation of plant growth and development [24, 27].

Structurally anthocyanins are substituted glycosides and acylglycosides of 2-phenylbenzopyrilium salts (anthocyanidins). The basic structure of anthocyanidins consist of a chromane ring (C-6 – ring A and C-3 – ring C) bearing a second aromatic ring (C-6 – ring B) in position 2 (Figure 1) [2, 5, 28-30]. The various anthocyanidins differs in number and position of the hydroxyl and /or methyl ether groups attached on 3, 5, 6, 7, 3′, 4′ and/or 5′ positions. Despite the fact that 31 different monomeric anthocyanidins have been identified (including 3-deoxyanthocyanidins, pyranoanthocyanidins and sphagnorubins), 90% of the naturally occurring anthocyanins are based on only six structures (30% on cyanidin **2**, 22% on delphinidin **3**, 18% on pelargonidin **1** and in summary 20% on peonidin **4**, malvidin **6** and petunidin **5**). Those six anthocyanidins are usually known as common anthocyanidins (Figure 1.) [29].

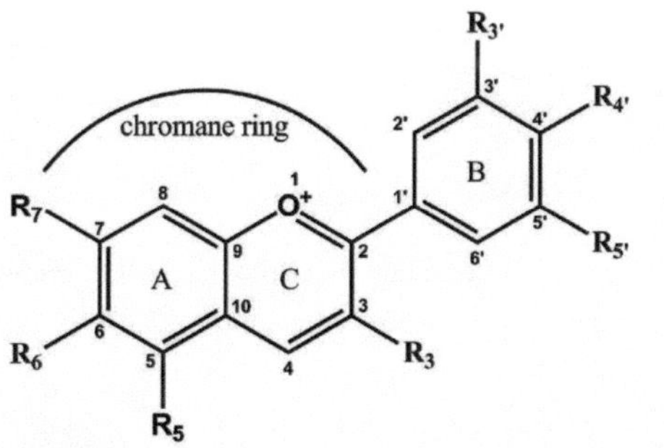

1	Pelargonidin (Pg)	**R_3**=OH; **R_5**=OH; **R_6**=H; **R_7**=OH; **$R_{3'}$**=H; **$R_{4'}$**=OH; **$R_{5'}$**=H
2	Cyanidin (Cy)	**R_3**=OH; **R_5**=OH; **R_6**=H; **R_7**=OH; **$R_{3'}$**=OH; **$R_{4'}$**=OH; **$R_{5'}$**=H
3	Delphinidin (Dp)	**R_3**=OH; **R_5**=OH; **R_6**=H; **R_7**=OH; **$R_{3'}$**=OH; **$R_{4'}$**=OH; **$R_{5'}$**=OH
4	Peonidin (Pn)	**R_3**=OH; **R_5**=OH; **R_6**=H; **R_7**=OH; **$R_{3'}$**=OMe; **$R_{4'}$**=OH; **$R_{5'}$**=H
5	Petunidin (Pt)	**R_3**=OH; **R_5**=OH; **R_6**=H; **R_7**=OH; **$R_{3'}$**=OMe; **$R_{4'}$**=OH; **$R_{5'}$**=OH
6	Malvidin (Mv)	**R_3**=OH; **R_5**=OH; **R_6**=H; **R_7**=OH; **$R_{3'}$**=OMe; **$R_{4'}$**=OH; **$R_{5'}$**=OMe

Figure 1. Structures of common anthocyanidins

The color of anthocyanidins differs with the number of hydroxyl groups, attached on their molecules (especially those substituted in ring B). With the increase of attached hydroxyl groups, the visible color of entire molecule shift from orange to violet (Figure 2) [2, 5, 29, 30]. Glycosylation of anthocyanidins results to additional reddening of obtained anthocyanins, whereas the presence of aliphatic or aromatic acyl moieties causes no color change or slight blue shift and has significant effect on their stability and solubility [5]. Changes in pH can also cause reversible structural transformations in anthocyanins molecules, which has a dramatic effect on their color (Figure 3) [30-34].

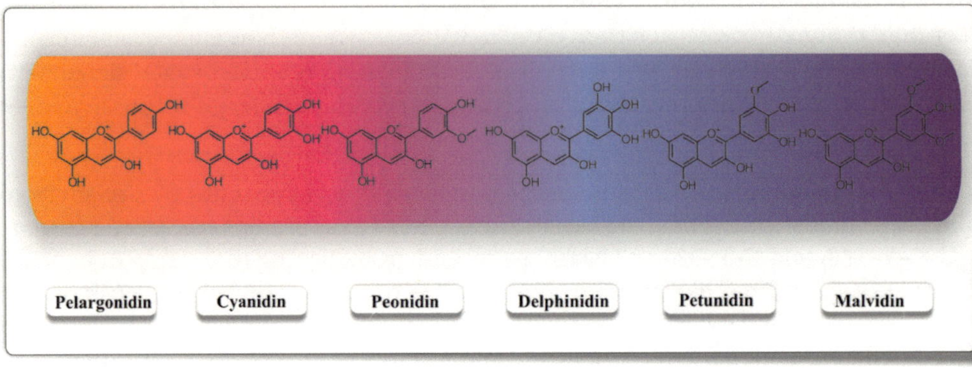

Figure 2. Visible color range of common anthocyanidins

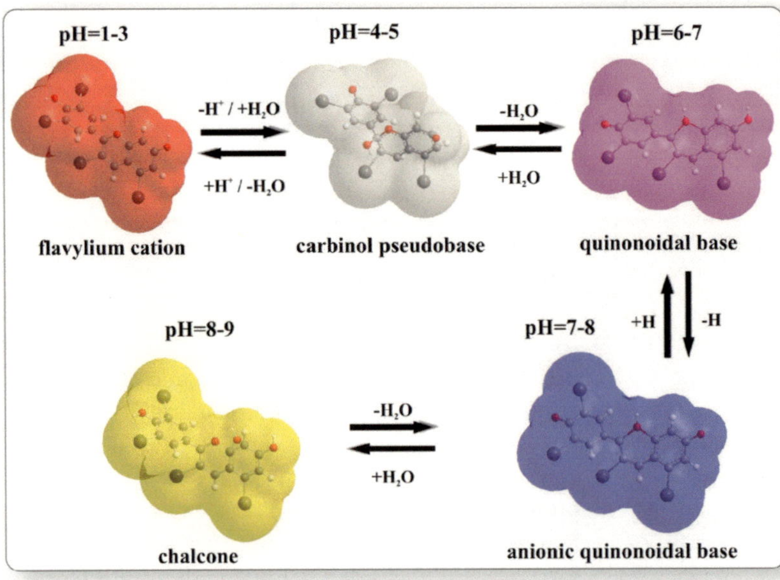

Figure 3. Structural transformations of anthocyanins in aqueous medium with different pH.

Most of the anthocyanins are *O*-glycosylated at 3 (except those based on 3-deoxyanthocyanidins and sphagnorubinns), 5 or 7 positions and in some cases at 3′, 4′ and 5′ positions [24, 35]. However, 8-*C*-glycosylanthocyanins have been found only in *Tricyrtis formosana* Baker [36, 37]. Anthocyanins contain two, one or tree monosaccharide units in their molecules. The usual monosaccharide residues are glucose, galactose, arabinose, ramnose, xylose and glucuronic acid. However, anthocyanins containing disaccharides and trisaccharides were also found in nature but no tetrasaccharides have been discovered yet [24]. Different anthocyanins based on cyanidin **2** aglycone, found in nature are presented on Figure 4

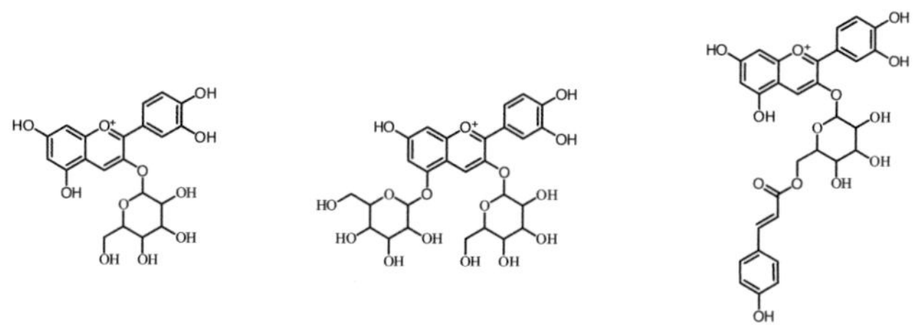

7 Cyanidin-3-*O*-monoglucoside **8** Cyanidin-3,5-*O*-diglucoside **9** Cyanidin-3-*O*-*trans*-*p*-coumaryglucoside

Figure 4. Anthocyanins based on cyanidin aglycone.

2.1. Anthocyanin biosynthesis in grapes

As a major flavonoid group, anthocyanins are products of phenylpropanoid metabolism of plant cells [28, 29, 38]. Anthocyanins in grapes are synthesized via flavonoid pathway. The biosynthetic pathway can be divided into two sections, the basic flavonoid upstream pathway, which includes early biosynthetic genes (EBGs), and the specific anthocyanin downstream branch, which includes late biosynthetic genes (LBGs) (Figure 5). Studies have shown that the basic flavonoid upstream pathway is restricted in many plants [39, 40, 41, 42] and that large gene families encodes the enzymes that act early in the flavonoid pathway, while the enzymes acting late in the pathway are encoded by single active gene [43]. The flavonoid pathway starts with phenylalanine, produced via shikimate pathway and transformed to 4-coumaroyl-CoA. The key enzyme, chalcone synthase (CHS) produce a naringenin chalcone by condense of one molecule 4-coumaroyl-CoA and three malonyl-CoA molecules (derived from citrate produced by The Krebs cycle) (Figure 5) [44]. In this case, the rings A and C are derived from the acetate pathway, whereas the ring B is derived from shikimate pathway [45]. Currently, there are three genes encoding CHS in grapes: *Chs1*(AB015872), *Chs 2* (AB066275), and *Chs 3* (AB066274), which are transcribed under different controls [46, 47]. The three genes act to synthesize naringenin chalcone, which is used in the formation of anthocyanins, proanthocyanidins, and other phenolic compounds. According to [47], the three different CHSs may act in three different pathways to produce different secondary metabolites. In the next step, chalcone isomerase (CHI) converts stereospecifically the naringenin chalcone to its isomer naringenin. Ring B of the naringenin undergoes further hydroxylation by the enzymes flavonoid 3′-hydroxylase (F3′H), flavonoid 3′5′-hydroxylase (F3′5′H) or flavanon 3ß-hydroxylase (F3H) [48]. Then, the obtained dihydroflavonols are reduced by the enzyme dihydroflavonol 4-reductase (DFR) to the corresponding leucoanthocyanidins.

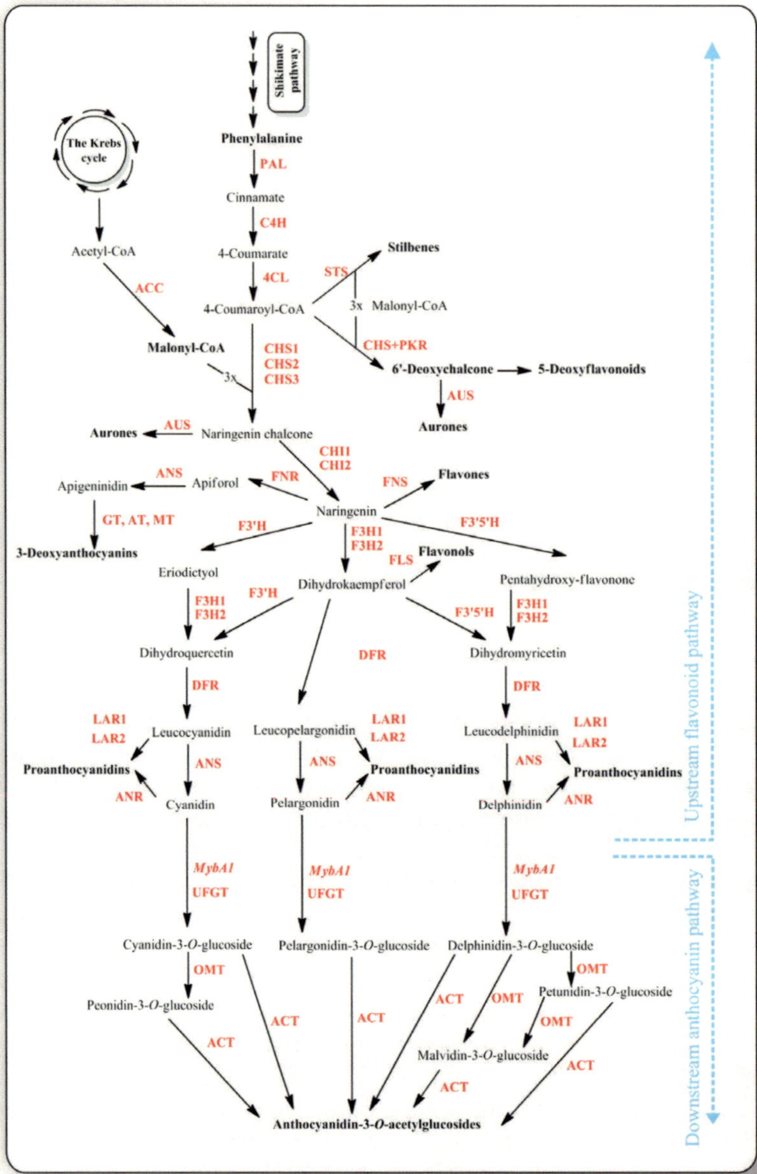

Figure 5. Flavonoids biosynthetic pathways and biosynthetic pathway of anthocyanins in grape: PAL - phenylalanine ammonia-lyase; C4H – cinnamate 4-hydroxylase; 4CL – 4-coumarate:CoA ligase; ACC – acetyl-CoA carboxylase; STS - stilbene synthase; CHS1, CHS2, and CHS3 - chalcone synthase 1, 2, and 3, respectively; PKR – polyketide reductase; AUS – aureusidin synthase; CHI1 and CHI2 - chalcone isomerase 1 and 2, respectively; FNS – flavone synthaes; FNR – flavonone 4-reductase; ANS – anthocyanidin synthase; GT – glucosyltransferases; AT – acyltransferases; MT – methyltransferases; F3'H – flavonoid 3'-hydroxylase; F3'5'H - flavonoid 3'5'-hydroxylase; F3H1 and F3H2 - flavanon 3β-hydroxylase 1 and 2, respectively; FLS – flavonol synthase; DFR – dihydroflavonol 4-reductase; LAR1 and LAR2 - leucoanthocyanidin reductase 1 and 2, respectively; ANR - anthocyanidin reductase; *MybA1* - MYB transcription factor gene [49]; UFGT - UDP-glucose: anthocianidin: flavonoid glucosyltransferase; OMT – *O*-methyltransferase; ACT – anthocyanin acyltransferase.

After this reduction, anthocyanidin synthase (ANS) oxidize leucoanthocyanidins to their corresponding anthocyanidins. Anthocyanidins are inherently unstable under physiological conditions and were immediately glycosylated to anthocyanins by UDP-glucose: Anthocianidin: Flavonoid glucosyltransferase (UFGT) [48]. Anthocyanins, containing methylated anthocyanidins (peonidin **4**, petunidin **5** and malvidin **6**) as aglycone can be obtained by methylation of hydroxyl groups on the ring B of the cyanidin-3-*O*-glucoside **7**, delphinidin-**3**-*O*-glucoside and petunidin-3-*O*-glucoside by the enzyme *O*-methyltransferase (OMT). Future acylation of produced anthocyanins is possible by the action of different anthocyanin acyltransferases (ACT).

2.2. Anthocyanins storage in grape cells

Once anthocyanins have been produced, they are transported and stored into the cell vacuole. Inside of vacuole, anthocyanins could be connected to specific proteins forming nomembrane intravacuolar bodies, known as anthocyanic vacuolar inclusions (AVI) (Figure 6) [50]. It has been confirmed that AVI plays a critical role in formation of color in flowering plants [50]. Recently, AVI from grape cell suspension were isolated and analyzed [51]. In contradiction with other plants, it was demonstrated that in grape cell suspension AVI consist of complex mix of tannins, anthocyanins (predominantly acylated derivates), proteins and other organic compounds, encased by lipid membrane [32, 51]. It was observed that a strong correlation between the prevalence of AVI structures in grapevine cell suspensions and the increase of their anthocyanin accumulation exist [51]. However, the enhancement of AVI prevalence does affect neither the number of available pigmented cells nor the overall growth rate of suspension cultures. Since AVI plays an important function in the storage and concentration of anthocyanins in cell vacuoles, their perspective role as enhancers of anthocyanin accumulation in grape cell suspensions have been proposed [51].

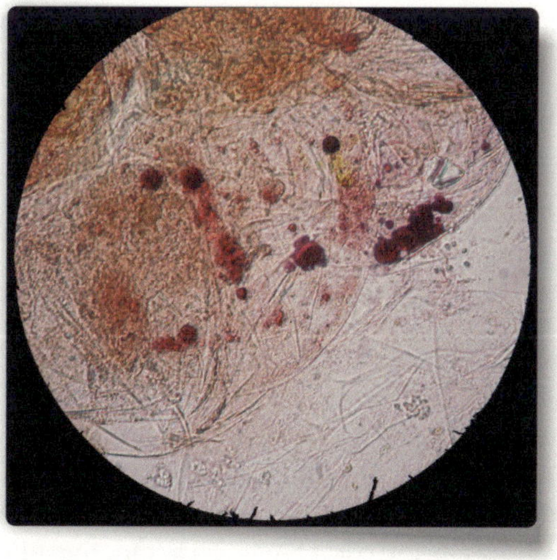

Figure 6. Anthocyanic vacuolar inclusions (AVI) in cell vacuoles of muscadine callus culture.

2.3. Characterization of anthocyanins in grapes

Grapes are rich sources of anthocyanins and bioavailable flavonoids. Grapevines are one of the world's most grown economically important fruit crops. Currently, there are more than 10000 grape cultivars deposited in germplasm collection [52, 53]. Among them, the cultivars of *Vitis vinifera* L. and *Muscadinia rotundifolia* (Michx.) Small., are the most important grapes in Europe and United States, respectively [48]. As a result of crop domestication, numerous changes in grapevine genome have occurred, leading to great variation in berry color [54]. Great differences exist between the vine varieties determined by the presence or absence of anthocyanins in their berries, as well as due to the different anthocyanin compositions of colored berries. It was found that retrotransposon-induced mutations in *Myb*-related *VvmybA1* gene are responsible for appearance of white-fruited *Vitis* varieties [54-56]. The color of grape berries are unique based on the anthocyanin accumulation, and it can be used as "fingerprint" for variety recognition [57]. Thus, *Vitis vinifera* L. varieties produce only anthocyanidin-3-*O*-monoglucosides, anthocyanidin-3-*O*-acetylglucosides and anthocyanidin-3-*O*-*p*-coumarylglucosides, whereas the other *Vitis* species and hybrids produce also anthocyanidin-3,5-*O*-diglucosides [48, 58-61]. Moreover, some *V. vinifera* L. cultivars such as Pinot Noir, red "Chardonnay" and pink "Sultana" produced only nonacylated anthocyanidin-3-*O*-monoglucosides and *M. rotundifolia* (Michx.) Small. accumulates only nonacylated anthocyanidin-3,5-*O*-diglucosides [48, 58, 62-64]. It was demonstrated that the lack of anthocyanidin-3,5-*O*-diglucosides in *V. vinifera* L. varieties is due to occurrence of double mutation in their anthocyanin 5-*O*-glucosyltransferase gene [65]. However, hybrid varieties always produce mixtures of anthocyanidin-3-*O*-monoglucosides and anthocyanidin-3,5-*O*-diglucosides which is used to monitor their usage in winemaking [48, 57]. The aglycones of common anhocyanins found in grapes are malvidin **6**, delphinidin **3**, petunidin **5**, cyanidin **2**, peonidin **4** and pelargonidin **1**. Malvidin **6** is the predominant anthocyanidin in most of the *V. vinifera* L. varieties (with exception of some "Muscat" cultivars in which cyanidin **2** is the major aglycone), whereas pelargonidin-**3**-*O*-glucoside was detected only in trace amounts [48, 66]. Recently trace amounts of pelargonidin-**3**-*O*-glucoside and pelargonidin-3,5-*O*-diglucoside were found to present also in *Vitis labrusca* L. and *Vitis amurensis* Rupr., respectively [57, 67]. The absence of pelargonidin 1 derivates in detectable concentrations in grape could be explained with the higher activities of F3'H and F3'5'H in *Vitis* species, which redirects the metabolite flux to production of cyanidin **2** and delphinidin **3** instead of pelargonidin **1** [48].

2.4. Anthocyanins relevant to berry quality

During their growth, grape berries follow a double sigmoid curve [68]. Veraison is the unique stage of berry development, representing the transition from growth stage to ripening. Once the grape berries enter to veraison, many physiological and biochemical changes occur. The grape cells completely redirect their metabolism to production of secondary metabolites, necessary to prepare berries for reaching the stage of physiological maturity. During this stage, the chlorophyll in berries has been completely lost and the biosynthesis of flavonoids, including anthocyanins is promoted [68]. Microarray analysis showed remarkable overexpression of genes involved in flavonoid biosynthesis and particularly in anthocyanin production during veraison and ripening in *vinifera, rotundifolia* and *aestivalis* species [47, 69, 70]. With the progress of anthocyanins accumulation the color of grape berries changes from green through red and then to purple due to the subsequent methylation of produced anthoyanins. This results in the increment of the relative shear of more metabolically evolved anthocyanins in the mature berries [68, 71]. Anthocyanin composition of grape variety is of great importance for estimation of technological properties of berries in winemaking. Thus the varieties, which accumulates predominantly anthocyanins build on anthocyanidins having ortho-positioned hydroxyl groups (cyanidin **2**, delphinidin **3** and petunidin **5**) are sources of more unstable color, compared to varieties, which accumulates anthocyanins based on malvidin **6**, peonidin **4** and pelargonidin **1** [48, 63]. Moreover, the increased amount of acylated anthocyanins significantly contributes to the stabilities of their color [63, 72, 73]. The quantity and quality of anthocyanins plays a crucial role in evaluation of berry qualities and determination of the right time for their collection. To monitor accumulation of pigments in situ, a rapid and non-invasive method based on application of fluorescence sensor has been developed [74, 75]. The method allows anthocyanin content in grape bunches to be monitored non-destructively on the vine in the field and was found to be effective in detecting the earlier ripening processes [74, 75].

3. Transcriptional regulation of anthocyanin biosynthesis in grapes

Anthocyanin biosynthesis in grapes commences only when ripening of the berry begins (termed véraison) and normally continues throughout the ripening phase of growth. Anthocyanin biosynthesis pathway in grapes has been greatly investigated, including intracellular transportation and accumulation [18, 76-78]. Most of the structural genes have been isolated, cloned and characterized, and there is valuable information available on the mechanisms that regulate their expression within the plant cell [29, 44, 79]. Multiple regulatory genes under the complex regulation are responsible for the synthesis of anthocyanin at the transcriptional level [17, 18, 80]. The early biosynthetic genes (EBGs), which are upstream of the anthocyanin biosynthetic pathway, are regulated by several different families of genes called the Myb transcriptional factors, Myc transcriptional factors (encoding basic helix–loop-helix proteins, bHLH) and WD40-like proteins [40-42, 81]. On the other hand, late biosynthetic genes (LBGs), which are downstream leading to anthocyanin formation through glycosylation and subsequent modification (methylation and acylation) are under the specific control of several regulatory factors. Specific regulatory genes have been identified and characterized in *A. thaliana* and they include *PAP1 & PAP2* (Myb family), *GL3 & EGL3* (Myc family) and *TTG1* (WD40 family). These genes are known to regulate the expression of the structural genes involved in the anthocyanin biosynthesis [82-84]. According to [85], some genes belonging to the Myb and Myc factors (e.g MYBL2, MYB4 and BHLH32) also act negatively to regulate biosynthetic pathway of anthocyanins in *A. thaliana*. However, in grapes, R2R3-Myb transcriptional factors have been implicated to control different branches of the phenylpropanoid pathway including anthocyanins, flavonols, and proanthocyanidins. Deluc *et al.* [86, 87], reported that transcriptional factors, *Vvmyb5a* and *Vvmyb5b* belongs to this group. However, still little is known about the transcriptional regulation of the structural genes involved in anthocyanin biosynthesis throughout berry development.

Recent studies in grapevine indicates that VvMYBA1 and VvMYBA2 transcription factors regulate UFGT gene, which plays a crucial role in the synthesis and accumulation of anthocyanins [56, 88-90], also identified another key R2R3-MYB protein that regulates proanthocyanidins (PA) synthesis in berry skin and seeds. But it is important to note that regulatory genes that control the expression of genes that encodes enzymes located upstream of UFGT have not been identified. It has been suggested that there is a contribution of at least two distinct regulatory complexes involved in the early and late steps of berry development [61]. Another study [86] also revealed that a MYB gene named *VvMYB5a* is associated with the regulation of the flavonoid pathway during the early phase of berry development. The study of transcription factors involved in the later steps of berry ripening is crucial, so that the coordinate regulatory mechanisms of the biosynthetic pathway throughout the berry development can be understood. R2R3-Myb related transcriptional factors such as *VlmybA1-1*, *VlmybA1-2* and *VlmybA2* have been identified as specifically regulating anthocyanin accumulation. Kobayashi et al. [56] and This et al. [54] revealed that, a retrotransposon, *Gret1*, is inserted in the 5'-flanking region of its related non-functional *VvmybA1* gene, and this contributes to the function lose of its transcriptional factor in white *V. vinifera* L. grapes. However, their research indicated that, the *Gret1* insertion is missing from the *VvmybA1* genes in red skinned spots of white cultivars, leaving behind a solo long terminal repeat (LTR), which becomes the functional regulatory gene. This means that red and white spots seen in the skins of grape berries are the result of deletion of the inception region of the retrotransposon, *Gret1* [91]. *VvmybA1* factor is regarded to be the major gene determining the synthesis of anthocyanin in the grape skin, thus the red and white color of the berries [55, 92]. In addition, *VvmybA1* and *VvmybA2* genes have also been reported to belong to the *VvmybA* family regulator genes in grapes, and they are also responsible for the color accumulation of the grape berries [91].

4. Muscadine grape cells as model system

Muscadines (*Muscadinia rotundifolia* (Michx.) Small.) are native grapes of North America and they are considered the most important cultivated grapevine species in the Southeastern part of the United States. Muscadine grapes are well known for their characteristic flavor and popularity in making juices, jellies, jams, and wine. It is important to note that muscadines are the only grapes that contain ellagic acid, which is known for its anticarcinogenic activity. In recent years, their production has increased significantly because of health promoting effects. They are known to have thick, tough skins that result in relatively low yield in juice, and therefore, 40% to 50% of the berry skins, pulp and seeds have not been traditionally used. The utilization of muscadine cell lines could have an important economic influence on the muscadine industry by expanding the uses of muscadine beyond the traditional jellies, jams, juices, and wines. Our laboratory (Viticulture lab at Florida A&M University) has started to investigate the feasibility of using *in vitro* red cell cultures [93] to improve human health. These cells were established from super-epidermal cells of muscadinia 'Noble' var [93]. They have been cultivated from a hard calli to fine cells both in suspension and in solid media. Current functional genomic studies suggest that there is an elevated expression of anthocyanin biosynthetic pathway genes in the *in vitro* red cells of muscadine [69, 94]. Other studies are also underway to evaluate their phenolic compounds and transcriptomics. However, it is important to note that gene expression studies of flavonoid biosynthetic genes are crucial for anthocyanin biosynthesis in the red cells of muscadine.

4.1. Genetic modification: A promising strategy to up-regulate or down-regulate the production of anthocyanins in muscadine grape cells

Genetic engineering of plants has lots of benefit in the agricultural field [95-97]. It contributes to an efficient and cost-effective way to produce a wide array of novel, value-added plant and food products in an environmentally friendly manner. Most scientists including Butelli et al. [98] have highlighted the interest of producing crop plants and their products enriched with health-promoting natural compounds. These compounds include anthocyanins and flavonoids, which have become the targets for improving the nutritional value of foods. This requires an in depth knowledge of the molecular mechanisms underlying the biosynthetic pathways of secondary metabolites in plants as demonstrated by [98]. Even though there are some plants that contain high levels of anthocyanins such as blueberries, there are some species where the accumulation of secondary metabolites is not enough. This is why genetic engineering has been used as a strategy to modify flavonoid biosynthesis in order to enhance flower pigmentation in ornamentals and fruit plants [99-103]. Genetic engineering has become increasingly important worldwide because it provides significant improvements in the quantity, quality, and acceptability of the world's food supply and may be the best source for food security [5, 104]. Currently there is an increased production of plant-based products with an enhanced antioxidant capacity, which is facilitated by this technology [103]. An example of this approach is in tomato, which is also an important vegetable crop worldwide [98, 105, 106]. Several transgenic approaches have been used to enhance the accumulation of flavonoid levels in grape berries and tomato fruit by overexpressing either the structural or regulatory genes involved in the biosynthetic pathway [87, 107]. Most of these studies have been carried out *in vivo*, however, none of these have been exploited in the cell models. Although a significant increase in the final content of some flavonoids (flavonols in particular) has been achieved [108-111], the production of anthocyanins in cell cultures still needs to be exploited. Commercial application of grape cell cultures for production of anthocyanin-based colorants is being delayed due to the following main problems: 1.) biosynthetic instabilities of grape cell suspensions; 2.) yield and productivity are too low to justify commercial production economically; 2.) the capacity of cultured cells for anthocyanin accumulation is limited; 3.) anthocyanins are exclusively intracellular and the development of two-phase process based on their continuous secretion and recovery from the medium is difficult; 4.) anthocyanins with desirable application properties may not be achieved using conventional cell line selection and process manipulation. Majority of these problems can be overcome using genetic engineering strategies by focusing on the functional genomics, transcriptomics, and proteomic studies. This will give us the ability to select a fast growing cell lines with the ability to produce high yield of target anthocyanin types. Some of the most effective genetic engineering approaches for manipulation of flavonoid biosynthesis pathway in grape cell suspension are discussed bellow by presenting our experience with Native American muscadine grape cells.

4.2. Modification of the anthocyanin pathway using regulatory genes

The final assembly of secondary metabolites in plants is determined by the coordinate transcriptional control of structural biosynthetic genes. Based on the information provided in [112], modulation of the rate of initiation of mRNA synthesis depends on the specific transcription factors, which interact with promoter regions of targeted genes. The regulatory genes that control the pattern and intensity of anthocyanin pigmentations through regulating the expression of several flavonoid-anthocyanin structural genes have been identified in many plants [16, 113-115]. There are two families of transcription factors i.e *MYB* and *MYC* that are involved in tissue-specific regulation of the structural genes found in the anthocyanin biosynthesis [14, 116-118]. It is notable that these transcription factors have a sequence homology from different plant species that shows that they have a common ancestor. In our previous study, an ectopic expression of the *MYB* gene in embryonic cell lines of muscadine grapes confirmed that these regulatory genes can be used for genetic modification of cell cultures in order to increase the final anthocyanin accumulation in cells [119]. However, it is vital to note that the quantity and class of anthocyanin produced might depend on several factors including 1.) How the transcription factor binds to the specific promoter site of the targeted structural gene, 2.) The ability of the endogenous transcription factors to cooperate with the introduced regulators, and 3.) How functional the endogenous transcription factors are [8, 116]. It is important to note that the enhancement of anthocyanin production using genetic engineering was achieved more than a decade a go in model plants such as *Arabidopsis*, tobacco, maize, and tomato [109, 120]. In *Arabidopsis* and tobacco, this was done by introducing the maize regulatory genes R and C1 [121]. The over-expression of the regulatory gene R by itself was able to trigger anthocyanin accumulation in the tissues that originally did not produce anthocyanin in *Arabidopsis* and tobacco [121, 122]. But, when the C1 gene was expressed by itself, there was no pigmentation. In another study carried out by Lloyd et al. [121], the over-expression of CI and LC gene in *Arabidopsis* triggered the accumulation of anthocyanins in tissues that normally do not contain them. The same applies to transgenic cherry tomatoes that showed the accumulation of anthocyanins in the leaves, stems, sepals, and veins, when LC gene was over-expressed [122]. Recently, Deluc et al. [87], isolated *VvMYB5b*, a transcription factor that encodes a protein that belongs to the R2R3-MYB family of transcription factors in *V. vinifera* L. This protein displays significant similarities with *VvMYB5a*, another MYB factor that has been shown to regulate flavonoid synthesis in grapevine [86]. Transient expression of the cDNAs for VvMYB5a and VvMYB5b in grape cells confirmed that they can activate the grapevine promoters of several structural genes of the flavonoid pathway [87]. Deluc et al. [87], also determined that the over-expression of VvMYB5b gene in tobacco lead to an up-regulation of genes encoding enzymes of the flavonoid pathway and also triggered the accumulation of anthocyanin and proanthocyanidin compounds.

In addition to the over-expression of the transcription factors, the suppression or negative regulators of flavonoid biosynthesis have been described [123]. For instance, high pigmented phenotypes (hp-2) in tomato were revealed when Bino et al. [123], mutated the DE-ETIOLATED1 gene (DET1). The fruits produced from these mutants are dark, and it is because of the elevated levels of flavonoids and carotenoids. The suppression of the regulatory gene *Det-1* resulted in increased levels of secondary metabolite groups. Further study of the mutants by Davuluri et al. [124], indicated that flavonoid levels were increased up to 3.5 fold, lycopene content was two-fold higher and *b*-carotene levels accumulated up to ten-fold compared to wild type fruits. This is an indication that we can use both over-expression and suppression (RNAi) techniques to improve anthocyanin production in grape cells.

4.3. Modification of the anthocyanin pathway using structural genes

The regulatory gene families MYB and MYC control the structural genes within the grape anthocyanin biosynthetic pathway [87, 107]. But it is important to note that, the way in which the structural genes are regulated in grape berry skins appears to be different from the patterns observed in snapdragon, petunia, and maize [125-127]. There are two ways in which the pattern of gene expression in grape berry skins could be explained in relation to regulatory genes; 1.) early biosynthetic genes, which induces the expression of all of the structural genes except UFGT, and 2.) late biosynthetic genes that results in the induction of expression of all structural genes [1]. Alternatively, two types of regulatory genes may be present, one that controls expression of PAL, CHS, CHI, F3H, DFR and LDOX and another that induces UFGT gene expression [43]. This means that the regulatory gene that controls expression of PAL, CHS, CHI, F3H, DFR, and LDOX is expressed early in berry development. But it is crucial to note that many studies have identified UFGT, as the major control point to anthocyanin biosynthesis in grape berry skins, and this control is later in the pathway than has been observed in the studies of maize, petunia, and snapdragon anthocyanin biosynthesis.

Irrespective of the function of the regulatory genes, ectopic expression of the structural genes can also enhance the accumulation of anthocyanins. A study carried out by Muir et al. [108] determined that an ectopic expression of the Petunia *CHI* gene in tomato fruits increased total flavonoids up to 70-fold in tomato fruit peel. Particular flavonoids increased consisted mainly of the flavonols rutin (quercetin 3-rutinoside) and isoquercetin (quercetin-3-glucoside), and to a smaller but still substantial extent of kaempferol glycosides. In another separate study, Colliver et al. [128] increased the amount of flavonoids in the flesh of tomato fruit by introducing a four-gene construct that lead to concomitant ectopic expression of structural genes, *CHS, CHI, F3H* and *FLS* in tomato fruit. This resulted in increased levels of flavonols in both peel (primarily quercetin glycosides) and flesh (primarily kaempferol glycosides) [128]. When expressed separately, none of these four genes was sufficient to lead to flavonol production in fruit flesh. This kind of approach can be replicated in grape cell cultures, by concomitantly over-expressing structural genes such as *DFR* and *LDOX* to enhance the accumulation of anthocyanins. These studies indicate that transgenic approaches can be taken to increase anthocyanin production levels in muscadine grape cell cultures by overexpressing either the structural or regulatory genes involved in the biosynthetic pathway.

4.4. Blocking specific steps in the anthocyanin biosynthetic pathway using RNAi: control of the flavonoid flux

Silencing the structural or regulatory genes on the anthocyanin pathway in muscadine requires the following steps: 1) isolation of the endogenous structural or regulatory gene; 2) construction of the transformation cassettes using structural or regulatory-gene fragments as transgenes; 3) transformation of transgenic red cells via *Agrobacterium tumefaciens* and regeneration; 4) molecular analyses to identify transgenic muscadine cell lines; 5) Protein analyses of transgenic cells to confirm the suppression of flavonoid-related proteins.

The flow of genetic information dictates that "DNA is transcribed into RNA that is translated into a protein" (Fig 7A). Flavonoid biosynthetic proteins are produced using this concept. To shift the metabolic flux in muscadine grape cells, one can consider either 1) over-expressing the genes on the flavonoid pathway or 2) to knock out the production of the flavonoid proteins. Blocking the production of flavonoid proteins can be done by interfering with the flow of the genetic information. For example, to eliminate the production of the muscadine DFR protein (Fig 7B), we could interfere either at the mRNA transcription level (transcriptional gene silencing [TGS]) or at the post-transcriptional level (PTGS). The advancements made in genetic engineering have led to the possibility of knocking out the production of specific proteins in organisms by downregulating and/or silencing the genes encoding these proteins. Strategies developed to downregulate genes in plants include mutation-based reverse genetics [129], gene targeting [130], antisense RNA [131], cosuppression [95, 132], and RNA interference (RNAi) [133]. Genetic and biochemical evidences suggest that antisense-mediated gene silencing, co-suppression, and RNAi are all inputs into a common RNA silencing pathway triggered by the formation of a double-stranded RNA (dsRNA). This pathway, called PTGS, is characterized by accumulation of 21 to 25 nucleotides, small-interfering RNAs (siRNAs), sequence-specific degradation of target mRNA, and methylation of target gene sequences [134]. A typical example is demonstrated by Muir et al. [108], where they used RNAi to blocked specific metabolic conversions in the endogenous tomato flavonoid biosynthesis pathway by down-regulating the expression of specific structural flavonoid genes. In another study, Schijlen [135] also used RNAi technique to inhibit tomato *CHS1* gene and this resulted in a strong reduction of total flavonoid levels (naringenin chalcone and quercetin rutinoside). Based on these data, they decided to use an RNAi-mediated gene construct to block the flavonoid pathway leading to flavonols at CHS, F3H and FLS [103]. In all these studies, a clear reduction of flavonols was obtained by introducing an RNAi construct. But it is important to note that when the FLS RNAi-construct was introduced, there was high accumulation of anthocyanins in the vegetative tissues such as stems, leaves, and flower buds as a result of the decreased activity of FLS. Therefore, based on these studies, it is possible that dihydroflavonols as the natural substrates for both FLS and DFR were efficiently converted into anthocyanins [103]. This is because the decrease of FLS activity may have caused less competition between the flavonol and anthocyanin branches in vegetative tissue of FLS RNAi-tomatoes, thereby improving the metabolic flux towards anthocyanin end products. This is a demonstration that RNAi technique can be used to re-direct the metabolic flax in cell cultures to produce more anthocyanins or flavonoids based on the consumer's interest.

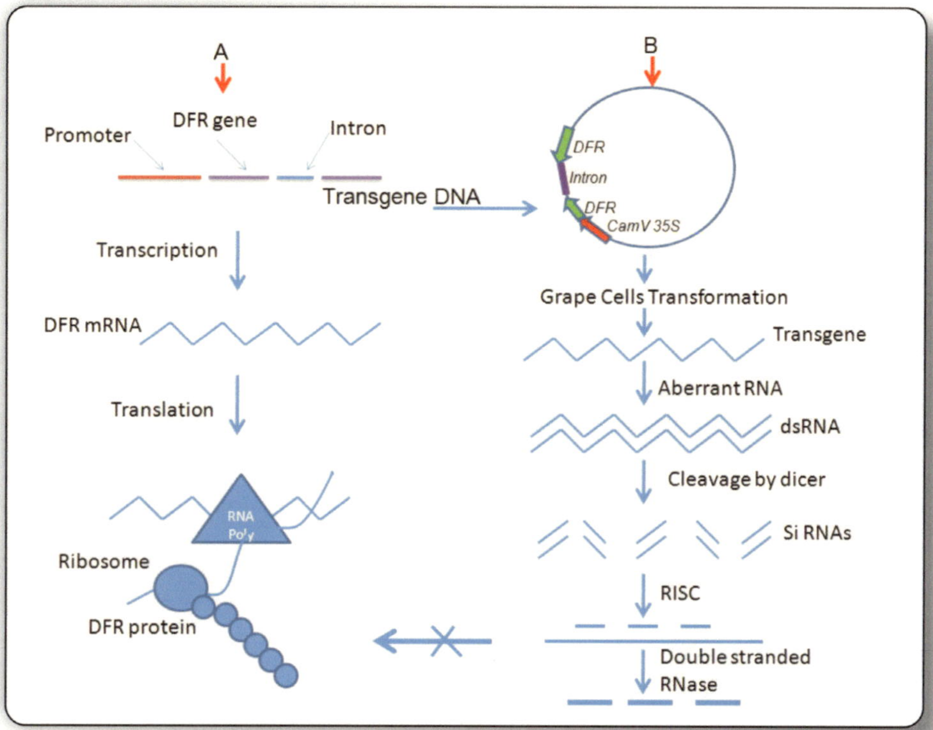

Figure 7. Schematic representation describing the flow of genetic information. **A,** The Central Dogma of Molecular Biology: *dfr* gene is transcribed into a messenger RNA, which is translated into a DFR protein. **B,** Schematic representation highlighting key steps in silencing structural or regulatory genes based on the RNAi model using DFR as an example. dsRNA—double-stranded RNA; RNAi—RNA inteference; RISC—RNA-interfering silencing complex; RNase—ribonuclease. Post-transcriptional gene silencing is initiated by dsRNA molecules that mediate the degradation of homologous transcripts. **B** gives a schematic representation of the molecular process that would be involved in the degradation of mRNA transcripts from muscadine *DFR* genes, based on the PTGS model. A fragment of the *DFR* gene is amplified by polymerase chain reaction (PCR) and cloned into a transformation vector. This vector is a DNA vehicle, which transfers the *DFR* gene fragment (called transgene) into muscadine grape cells following *Agrobacterium*-mediated transformation. The transgene integrates into the endogenous DNA of muscadine cells for expression. Transcripts from the transgene initiate long dsRNA molecules, which would be processed into small 21 to 25 nucleotides (siRNAs) by an endogenous RNAIII enzyme called "Dicer" [136]. These siRNAs would guide the RNA-interfering silencing complex (RISC), which contains the proteins necessary for unwinding the double-stranded siRNAs, and cleave the endogeneous *DFR* mRNAs at the site where the antisense RNAs are bound [137]. Sources of dsRNA formation from *DFR* transgene transcripts include: 1) pairing of transcripts transcribed from an inverted repeat (IR) transgene; 2) pairing of the normal sense RNAs and antisense RNAs (asRNA) arising from aberrant transcription of the transgene; 3) pairing of complementary regions of RNA degradation products; and 4) pairing of transcripts with antisense RNA (asRNA) produced by RNAdirected RNA polymerase (RdRP) [138, 139].

5. Key concepts for enhancing anthocyanin biosynthesis in grape cell lines

Grape cell suspension cultures have been extensively studied as an model for elucidation of anthocyanin biosynthesis pathway, for performing functional genetic studies, somatic embryo development and most importantly as an alternative source of natural colorants [1, 69, 140-142]. Cultivation of plant cells in controlled conditions offers advantages of continuous supply of high quality anthocyanin pigments. However, scientists have tried for over 40 years to produce anthocyanins in different cell systems, but until now, no commercially feasible anthocyanin producing system has been developed. Switching the production of natural pigments from the traditional approach (involving implementation of numerous agricultural activities) to modern industrial biotech factories is not an easy task. But it is obvious that for the successful realization of such biotechnological advances, research on both empirical and rational levels have to be performed [143]. To succeed in such challenges, we propose to follow a simple integrated approach based on consecutive conduction of various multidisciplinary experiments, optimization and monitoring procedures (Figure 8). As we already discussed some of the rational approaches for manipulation of anthocyanin biosynthesis on genetic level, in the next few pages, the basic principles of empirical studies are highlighted and the current progress on them has been reviewed.

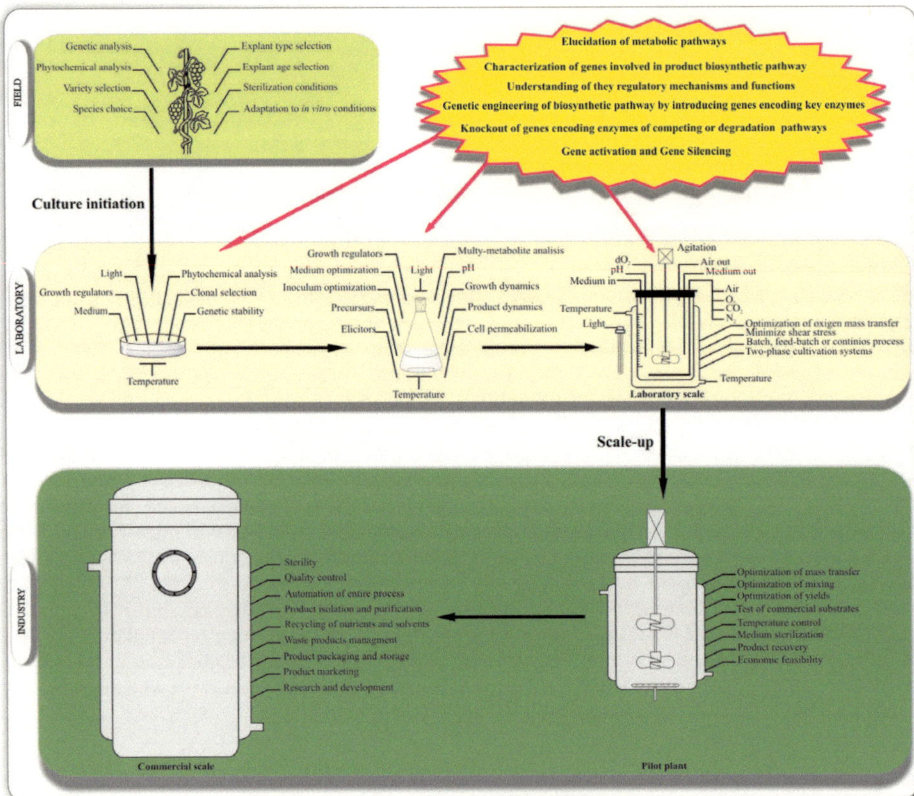

Figure 8. Basic concepts for the development of biotechnological system for anthocyanins production based on grape cell suspension culture. The key steps in bioprocess engineering and optimization of anthocyanins production, involving application of both empirical and rational approaches are presented on each technological stage.

5.1. Cell line development and improvement

The first step for creation of biotechnology process for anthocyanins production is the development and selection of high producing cell line. To facilitate this process several important preliminary questions needs to be answered. The most important is the right choice of plant species (respectively, the appropriate cultivar) having the necessary anthocyanins profile in both quantity and quality aspects. Recently Lazar and Petolescu [144], generated cell suspension cultures of six grapevine varieties (Burgund Mare, Cabernet Sauvignon, Merlot, Oporto, Negru Tinctorial and Pinot Noir). They cultivated them in a laboratory bioreactor and demonstrated that the growth rates and biosynthetic potential for anthocyanins production were in strong dependence of cultivar used for culture initiation [144]. Currently, most of the research on grape cell suspensions have been performed with cell lines derived from two sources – *V. vinifera* L. *cv.* Gamay Fréaux and *Vitis* hybrid Bailey Alicant A (*V. lincocumii* x *V. labrusca* x *V. vinifera*) x (*V. vinifera* x *V. vinifera*) [1, 48]. Both of them are teinturier cultivars which usually produce malvidin-derivatives in berry skin and peonidin-derivatives in the pulp [145]. Cell suspension cultures of *V. vinifera* L. *cv.* Cabernet Sauvignon (CS4 and CS6) were also reported [146]. In our laboratory we have patented anthocyanin producing cell suspension cultures of *Muscadinia* spp. and *V. aestivalis* var. Cynthiana, which produces mainly anthocyanidin-3,5-*O*-diglucosides [93].

After the choice of targeted variety, the critical step is the selection of appropriate explants, which are used for callus initiation. The type and the age of explants, as well as the environmental conditions at which they are collected are critical factors for successful initiation of *in vitro* cultures [147]. Sterilization procedure, applied to the explants should be as gentle as possible and the necrotic ends of the sterilized cuttings must be removed to avoid the secretion of phenolic compounds into the induction medium. In some cases, the addition of activated charcoal or antioxidant mixture into the culture medium is necessary [147]. Two-phase cultivation systems with adsorbent resin like Amberlite XAD4, which have been recently applied in shoot regeneration and hairy root induction, could be also a prospective alternative for reducing explants browning [148, 149]. One of the most typical characteristic of grapevine callus cultures is their high heterogeneity concerning coloration. The obtained calli usually exist as a mixture of colorless, yellow and red clusters [1, 142, 150]. Therefore, the repeated selection and subcultivation of different-colored clusters can be applied for screening of new high producing lines [142, 146, 150, 151]. As a result of such selection procedure, significant changes in anthocyanin profiles of isolated lines could be observed [142, 146]. Usually those changes lead to the elevation of the 3′-methylated and acylated anthocyanins in grape cell suspension cultures and therefore greatly improve the quality of overall pigments with regards to their color stability [72, 142, 152]. As an example, in cell suspension of *V. vinifera* L. *cv.* Gamay Fréaux, the major anthocyanins were found to be cyanidin-3-*O*-glucoside **7**, peonidin-3-*O*-glucoside and cyanidin-3-*O*-*p*-coumarylglucoside **9** [153]. By repeated selection of colored aggregates from the same culture, Krisa and colleagues obtained a line, which produce remarkably high level of malvidin-3-*O*-glucoside (63% of total anthocyanin content) [146]. However, in analog with callus culture, the grape cell suspensions usually consist of a mixture of uncolored cells and red colored anthocyanin producing cells [51, 142, 150, 151]. Moreover, the colorless cell fraction often has a high growth rate compared to the red fraction and this is the main factor that contributes to instability of the supensions as regards anthocyanin production [150, 154]. With time, anthocyanin producing grape cell suspensions could significantly decrease their ability to produce anthocyanins and often the yields, obtained by the cultivation at equal conditions in the same cultivation vessels are unpredictable [142]. The other complication is due to the existence of strong correlation between cell differentiation, anthocyanin accumulation and cell growth [1]. Anthocyanins are produced only in cells that undergo some level of differentiation (growing as small aggregates), whereas the fast growing, undifferentiated fraction of cells (growing as single cells) does not produce the pigments [1, 155]. It was found that the fraction of aggregates larger than 0.6 mm accumulates high amounts of anthocyanins, whereas with the reduction of aggregate size to 0.2 mm the concentration of produced anthocyanins decreases with 50 % [155]. To control the ratios between non-pigmented and pigmented cells, as well as between aggregated and single cells, the period of sub-culturing and the size and age of used inoculums should be precisely adjusted [1, 150]. Optimization of the parameters for the inoculums has often been underestimated by researchers, working with plant cells. However, the establishment of right parameters for the inoculum has been found to have significant impact on growth and secondary metabolite production by different plant cell suspensions and hairy root cultures [3, 7, 156, 157]. The size of inoculum could significantly shorten the lag phase of grape cell suspension, as well as to affect the culture response to different changes in nutrient composition [158]. In our laboratory, we investigated the effect of different inoculum sizes on growth of anthocyanin producing *M. rotundifolia* var. "Noble" cell suspension cultures (Figure 9). The experiment was performed in submerged flasks by using 9, 15, 20 and 30 % (v/v) inoculum of 14 days old cell suspension. The maximum amount of accumulated fresh biomass (78.7±1.7 g FW/L) was achieved when 20 % of the inoculum was used. At these conditions, the amount of accumulated fresh biomass was 3.18-fold higher, compared to the experiment at which 9% of the inoculum was used, without observing any decrease in anthocyanin production (unpublished data).

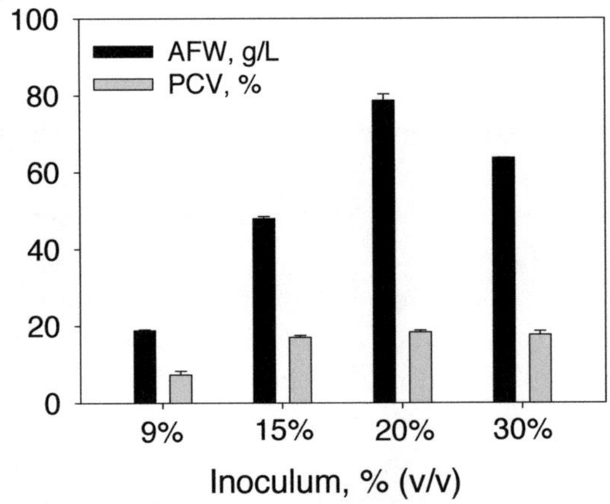

Figure 9. Effects of the inoculum sizes (14 days old culture) on accumulated fresh weight (AFW) and packed cell volume (PCV) by *M. rotundifolia* var. "Noble" cell suspension culture cultivated in shaking flasks (250 ml/50 mL) for 18 days. The presented values are means with standard deviations of two independent experiments repeated twice (n=4).

However, the non-homogenous growth of grape cell suspensions could be a serious issue for large-scale cultivation of these in *vitro* cultures because of the complicated mass transfer in bioreactor systems. It has been observed that grape callus culture can undergo dramatic changes in their anthocyanin profiles, because of their high heterogeneity and somaclonal variability. To prevent those negative effects, a periodical implementation of deep phytochemical, genetic and metabolomic analyses are requires. This will retain acceptably high biosynthetic potential in selected lines during their maintenance.

5.2. Medium optimization

The optimal balance of nutrients in cultivation medium has been found to be an essential factor, determining the success on *in vitro* cultivation of plant cells and tissues. Plant cell suspensions are exceptionally sensitive to concentration of macronutrients, microelements, growth regulators, nitrogen and carbon sources. Even insignificant changes in composition of cultivation medium could promote the appearance of significant changes in cell morphology, growth and secondary metabolite profiles. For example, transferring of anthocyanin producing cell suspension of *M. rotundifolia* var. "Noble" from the original B5 medium into LS medium completely change the color pattern from red to yellow and the cell growth pattern from small colored aggregates to single colorless cells (Figure 10) (unpublished data).

Figure 10. Cell suspension of *M. rotundifolia* var. "Noble" cultivated in shaking flasks on: A - B5 medium; B – LS medium.

As the grape cell suspensions exist as mixture of colored and colorless cells, the optimal cultivation medium should be developed by the way to provide a right balance between the growth rates of both cell populations [1, 142]. Finding the right nutrients balance is often a complicated task mainly because of the observation that the colorless population usually has better growth characteristics and the colored cells showed slow growth [1, 142, 150]. Dedaldechamp and Uhel [154] isolated a cell line from colorless cells of *V. vinifera* L. *cv.* Gamay Fréaux and demonstrated that under reduction of cell division by phosphate deficiency, the anthocyanin production was reactivated in colorless cells [154]. Removing of phosphate ions from the medium leads to significant decrease of cell growth but remarkably increase anthocyanin production due to the increasing transcript levels of *UFGT* and *VvmybA1* genes [159]. Reduction of nitrate ions in the cultivation medium to a critical level was also found to enhance the anthocyanin production in cell suspension of *Vitis* hybrid Bailey Alicant A [160]. Increased osmotic pressure in medium, caused either by the increased sugar concentrations or by the addition of osmotic-active compounds as D-manitol, sorbitol, poliol or carboxymethyl cellulose has been found to play a critical role for anthocyanin production by different grape cell suspension cultures [155, 160-163]. Because of the significant differences in nutrient media compositions required to provide optimal cell growth and that to activate anthocyanin production, most of the authors used two-stage cultivation involving usage of "maintenance" medium, following by next transfer on "production" medium [1, 142, 146, 150, 160]. The maintenance medium provides rapid growth of the grape cell culture but the lines often lost their colors. When transferred on production medium, the growth was usually almost completely inhibited, but the anthocyanin accumulation was significantly enhanced. The production medium usually differs from the maintenance medium by the increased sugar concentration, decreased phosphate and nitrate concentrations, supplement of osmotic-active compounds, growth regulator compositions etc. [1, 142, 150, 146, 160]. However, it is obviously that the commercial realization of such two-stage based cultivation process is accompanied with numerous complications from technological point of view. In fact, this is one of the main reasons for the lack of industrial process for anthocyanins production by grape cell suspensions.

5.3. Effect of the pH

Since the pH is an important factor for anthocyanin stability and activities of enzyme systems in plant cells, its value in the cultivation medium is critically important for the regulation of both pigments and biomass yields. Recently, the effect of pH in culture medium was investigated on callus cultures of three grapevine varieties (Coarnă neagră, Fetească neagră and Cadarcă) [164]. The authors observed that the largest amount of accumulated anthocyanins (13.5 mg/g FW) were registered in callus culture of Fetească neagră, cultivated on the medium with the pH=4.5. When cultivated on medium with pH=9.0, the anthocyanin production by the same culture was significantly decreased (up to 3.2 mg/g FW) [164]. Suzuki and colleagues [162], investigated the growth and production of anthocyanin in grape cell suspension from *Vitis* hybrid Bailey Alicant A cultivated on media with different pH (4.5, 5.5, 7.0 and 7.5) [162]. The authors found that the best growth and pigment production were recorded when cultivation was done on medium with the pH=4.5. In contrary to that, when cultivated on medium with pH=7.0 the cell growth and the anthocyanin production were almost completely inhibited. Moreover, the percentage of pigmented cells were significantly decreased with the increase of pH (from 50% at pH=4.5 to 4% at pH=7.5) [162]. Because the pH value of culture medium during plant cell cultivation processes correlates with the metabolism of nitrogen sources, it is important to render an account of this factor during creation of experimental matrices for optimization procedures of nutrient medium composition (especially ammonia/nitrate ratio).

5.4. Effect of temperature

Temperature has a strong effect on anthocyanin biosynthetic pathway, since some cold regulation genes are involved on it [165, 166]. Anthocyanin accumulation in berry skins of "Aki Queen" (*V. labrusca* x *V. vinifera*) after temperature treatment during ripening was significantly higher at 20 °C compared to that at 30 °C [167]. Similar effect was reported for berries of *V. vinifera* L. *cv.* Cabernet Sauvignon, where the anthocyanins concentration was reduced more than 50 % with the increase of cultivation temperature from 25 °C to 35 °C [168]. The authors suggest that the observed decrease in anthocyanin concentration under high temperature was as a result of both anthocyanin degradation and inhibition of mRNA transcription of the anthocyanin biosynthetic genes [168]. Recently, a qRT-PCR analysis have been carried out to compare the expression levels of MYB-related transcription factor genes in the berry skins of "Pione" (*V. vinifera* x *V. labruscana*) during temperature and light treatments [169]. The authors demonstrated that both low temperature and light irradiation were needed to induce the expression of *V1mybA1-3* (gene, encoding MYB-related transcription factors that regulate anthocyanin biosynthesis pathway genes), whereas the expression of *Myb4* (the repressor of UFGT) was up-regulated only by high temperature, independently of the light levels [169]. However, in optimization procedures for anthocyanin production by grape cell suspension cultures, the influence of temperature was underestimated. The new data, discovering the importance of those critical parameters in regulation of anthocyanin biosynthetic pathways outline the necessity for conduction of more experiments concerning optimization of temperature during cultivation of grape cell suspensions.

5.5. Light

Light is an important controlling agent in anthocyanin biosynthesis [166]. Light has been found to induce the expression of genes, responsible for activation of the promoters of the flavonoid pathway genes (*MybpA1*), but it has no effect on expression of repressor of UFGT (*Myb4*) in grape berries of vine 'Pione' (*V. vinifera* x *V. labruscana*) [169]. Recently it was demonstrated that UV-A light significantly stimulate the expression of structural genes, encoding the entry enzymes of the shikimate pathway, whereas only the UV-B and UV-C irradiation triggers the production of various anthocyanins in grape berries of *V. vinifera* L. *cv.* Cabernet Sauvignon [170]. Illumination of grape callus and cell suspension cultures is of great importance for expression of their anthocyanin producing potential. In our laboratory, we investigated the influence of light on anthocyanin production in callus culture of *M. rotundifolia* var. "Noble" (Figure 11). The red colored line was selected by 3 months repeated selection of dark red colored clusters, grown under illumination (16 h light : 8 dark). When the red colored line was transferred for cultivation in darkness, the culture completely lost its color for one sub-cultivation cycle (31 days). The obtained colorless culture was supported on darkness for 3 more sub-cultivation cycles and then transferred back under illumination (16 h light: 8 dark). Fourteen days after the light treatments began, the culture started to form red colored clusters and the total anthocyanin concentration reached the original levels, detected in initial red colored lines after 31 days of light exposure (unpublished data).

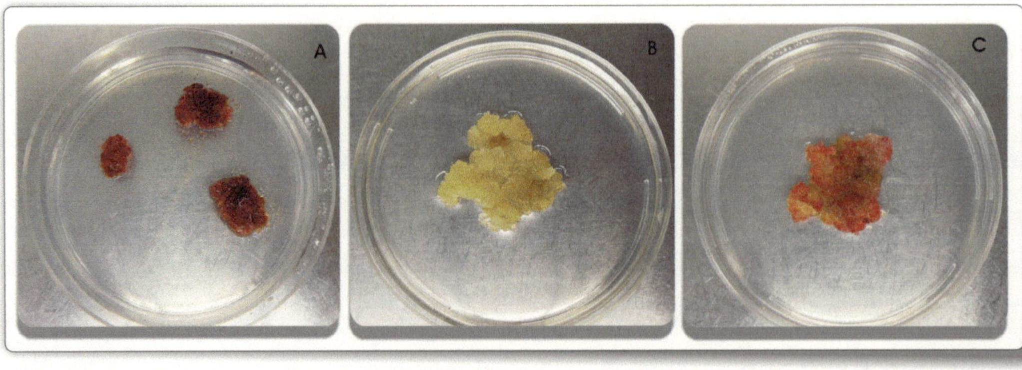

Figure 11. Anthocyanin accumulation in callus culture of *M. rotundifolia* var. "Noble": A - under illumination (16 h light : 8 dark); B – on darkness; C – roll-back under illumination (16 h light : 8 dark) of the culture, cultivated on darkness. The presented pictures are on the 14 days old cultures.

Lazar and colleagues investigated the effect of light on anthocyanin accumulation by callus cultures from six grapevine varieties (Burgund Mare, Cabernet Sauvignon, Merlot, Oporto, Negru Tinctorial and Pinot Noir) [171]. They found that the light has a stimulating effect on anthocyanin production in all calli studied, but the amount of accumulated pigments was in strong correlation with the genotype of variety used for callus initiation [171]. In grape cell suspension of *V. vinifera* L. *cv.* Gamay Fréaux, treatment with light leads to additional increase of peonidin-3-*O*-glucoside with 0.6 mg/g DW [172]. The same group reported that the continuous light irradiation (8000–8300 LUX) can contribute up to 4.8-fold increase in anthocyanin accumulation compared to no illuminated control [173]. Combined treatment with light and elicitor (jasmonic acid) additionally increased anthocyanins production (13.9-fold compared to non-treated control) [173]. It is obvious that the light requirements of grape cell suspension cultures can cause a serious complication during the scale up of cultivation process. Cormier and colleagues reported the isolation of red colored clusters growing on darkness, but the line has been very unstable and easily changed to variegated culture [142]. However the authors succeeded to obtain stable anthocyanin producing cell line in suspension culture but the production of pigments was relatively low (1.32 mg/g FW) [142]. Therefore, much effort is needed to initiate light-independent high anthocyanin producing cell lines, which will greatly facilitate the transfer of the process to large-scale bioreactors.

5.6. Effect of growth regulators

Availability of growth regulators (auxins and cytokinins) in cultivation medium are essential for ensuring the growth and to determine the levels of produced secondary metabolite by *in vitro* cultivated plant cells. In the case of anthocyanin biosynthesis in grape cell suspension, the composition of growth regulators in the medium can be of great importance for the production pigments [1]. Hirasuna and colleagues [160], tested the effects of different auxins (IAA, NAA, IBA and 2,4-D) and cytokinins (Kinetin and BAP) on anthocyanin biosynthesis by *Vitis* hybrid Bailey Alicant A cell suspension, cultivated on specially composed production medium [160]. The authors found that the synthetic auxin 2,4-D has stimulating effect on anthocyanin biosynthesis over a wide range of concentrations, whereas the addition of cytokinins even in low concentrations completely inhibit the pigments production. However, the effects of growth regulators are unpredictable and should be evaluated experimentally for every individual cell line. As an example, Krisa and colleagues [146], reported that for cell suspension of *V. vinifera* L. *cv.* Gamay Fréaux, the addition of NAA as auxin leads to better anthocyanin accumulation compared to when 2,4-D was used [146]. Moreover, their suspension culture requires the addition of cytokinin, Kinetin for ensuring the culture growth and production of pigments. The change of Kinetin with BAP results to inhibition of anthocyanin production [146]. Recently, Gagne and colleagues demonstrated that abscisic acid (ABA) promotes anthocyanin production in grape cell culture of *V. vinifera* L. *cv.* Cabernet Sauvignon (CS6) by the expression of some genes in the upstream part of the flavonoid biosynthesis pathway [174].

5.7. Effect of elicitors and precursors

Application of different (biotic or abiotic) elicitors has been proved to be an effective strategy for enhancement of the production of secondary metabolites related to the plant defense system [175]. Anthocyanin biosynthetic pathway as a part of phenylpropanoid metabolism of plant cells could be significantly manipulated by application of different elicitors or feeding with specific precursors. Treatment of cell suspension culture of *V. vinifera* L. *cv.* Gamay Fréaux with a combination of phenylalanine as a precursor and methyl jasmonate as an elicitor resulted in a 3.4-fold increase in anthocyanin yield compared to the control [151]. Cell suspension culture of *V. vinifera* L. *cv.* Gamay Fréaux was recently used as a model system for evaluating the effects of different chemical (streptomycin, activated charcoal, ethephon, indanoyl-isoleucine and N-linolenoyl-l-glutamine), biotic (insect saliva, chitosan, pectin, alginate, yeast extract and gum arabic) and physical (hydrostatic pressure and pulsed electric field) elicitors on anthocyanin biosynthesis [176-180]. It was found that the combined treatment of samples with pulsed electric field and ethephon, lead to a 2.5-fold increase in anthocyanin content, whereas the combination of hydrostatic pressure and ethephon does not alter anthocyanin production, but increases the other flavonoids [176, 177, 180]. The application of indanoyl-isoleucine enhanced the anthocyanin production with 2.6-fold, whereas the insect saliva stimulated the production of phenolic acids [178]. In contrast to insect saliva, the other investigated biotic elicitors (chitosan, pectin, and alginate) had a significant effect on anthocyanin production (resulted on 2.5-fold, 2.5-fold, and 2.6-fold increase compared to control, respectively) [179]. The most widely used elicitors are methyl jasmonate and jasmonic acids, which seems to have better beneficial effects on accumulation of phenolic acids and stilbenes than the accumulation of anthocyanins [181, 182]. However, jasmonic acid was found to increase preferably the level of peonidin-3-*O*-glucoside (from 0.3 to 1.7 mg/g DW) and the relative share of acylated anthocyanins (from 32% to 45%) in cell suspension of of *V. vinifera* L. *cv.* Gamay Fréaux, whereas the concentrations of the other major anthocyanins were insignificantly increased [172]. The effect of jasmonic acid was significantly increased when combined treatment with light irradiation was applied [172]. Magnesium treatment of cell suspension of *V. vinifera* L. *cv.* Gamay Red was found to increase the anthocyanin concentration by inhibiting the degradation of pigments but not by promoting induction of biosynthetic-related genes [183].

6. Commercialization and applications of grape cell products

The increased demand of natural colorants and nutraceuticals determines the needs for development of alternative technologies for supply of such additives. The anthocyanins, produced by grape cell suspensions, represent a very attractive class of natural compounds, which could find application in food industry (as colorants), pharmacy (as nutraceuticals and therapeutic compounds) and in cosmetics (as UV protectors, antioxidant and anti cancer compounds). Biotechnological production of grape anthocyanins presents significant economical benefits. Cormier and colleagues calculated that the cost of 1 kg anthocyanins, produced by two-stage cultivation process of grape cell suspension in bioreactor with working volume of 155,000 L can cost almost the half of the price of such amount of anthocyanins, produced by the extraction of grape skins ($ 931 per 1 kg of anthocyanins from grape cell suspension, compared to $ 2,083 per 1 kg of anthocyanins, produced by grape skin extraction) [142]. However, the specific requirements of the available grape cell suspensions significantly complicate the scale up of the cultivation process, which is the serious restriction for realization of such biotechnological process.

6.1. Pharmaceutical applications

Anthocyanins have great potential for application in pharmaceutical products both as nutraceuticals and as therapeutic compounds. Frequent ingestion of anthocyanins could provide various health benefits including reduced risk of coronary heart diseases, anti-carcinogenic activity, antioxidant activity, reduced risk of stroke, anti-inflammatory effects etc. [13, 34, 184-186]. Biological activities of anthocyanin pigments have been already discussed in several excellent reviews [184, 187, 188]. Their pharmaceutical value has been additionally increased due to their high bioavailability. However, the administration and metabolism of anthocyanins *in vivo* have been investigated in details mostly in rats, whereas the detailed studies on humans still are scantly presented in scientific literature [60, 189]. For better understanding and investigation of anthocyanins absorption and *in vivo* metabolism in human and animal bodies, grape cell suspension culture of *V. vinifera* L. *cv.* Gamay Fréaux, was adapted to produce ^{13}C-labeled anthocyanins (delphinidin-3-*O*-glucoside, cyanidin-3-*O*-glucoside **7**, petunidin-3-*O*-glucoside, peonidin-3-*O*-glucoside and malvidin-3-*O*-glucoside) [190]. Development of reliable sources of isotopically labelled anthocyanins could have remarkable impact on advancement in diagnostic of metabolomic assimilation studies of these compounds *in vivo*.

6.2. Food industry

The world market of natural food colorants expands with the annual growth rate of 4-6% [142]. In USA 4 of the 26 colorants approved by the food administration, that are exempt from certification, are based on anthocyanin pigments [34]. In European Union, all anthocyanin-containing colorants are classified as natural colorants under the classification E163 [191]. Currently most of the worldwide anthocyanins supply comes from processing of grape pomace, which is a waste product from winemaking. But in European Union other plant sources such as red cabbage, elderberry, black currant, purple carrot, sweet potato, and red radish are also allowed [192]. Anthocyanins, produced by grape cell suspensions can be a promising alternative supply of natural colorants. It has already been demonstrated that the produced pigments by the grape cell suspensions undergo significant structural modifications. Grape cell suspensions accumulates higher levels of metabolically more evolved structures (methylated and acylated anthocyanins). Acylated anthocyanins are suitable for application in food products, mainly because of the improved color stability compared to non-acylated structures [72]. Moreover, the grape cell suspensions can also produce elevated levels of beneficial phenolic compounds such as flavonoids, stilbenes, phenolics, etc., which are capable of increasing the added value of the final additive. The overall metabolite profile of grape cells in combination with the lack of microbial and toxic contaminations will give the potential for development of new types of food additives if the entire cell suspension biomass are utilized.

6.3. Cosmetic industry

The commercial interest of cosmetic companies to apply plant additives, derived by biotechnological cultivation of plant cells to their products has increased remarkably in the last few years [193]. The addition of plant cell derived extracts in cosmetic products has been considered as a powerful approach used to increase their health benefits. Several plant extracts have been added to various cosmetic products as moisturizers, antioxidants, whitening agents, colorants, sunscreens, preservatives etc. [193]. With the advancement of plant cell biotechnology, more and more cosmetic companies have been attracted for application of additives, based on plant cell suspensions. Recently the application of so-called plant "steam" cells attracts industry's attention [193]. In the last few years, the French company "Sederma" launched the product "Resistem™" based on application of *in vitro* cultivated plant cells (www.sederma.fr). The other company, "Mibelle Biochemistry", situated in Switzerland, developed a "PhytoCellTec" product, based on grape cell suspension of *V. vinifera* L. *cv.* Gamay Fréaux, which was processed by high-pressure homogenizer to produce liposomes for application in cream products (www.mibellebiochemistry.com). According to the company, the grape cell derived liposomes contained higher amounts of anthocyanins and when applied on skins serve as strong UV protectors and fight photoaging. The presented examples clearly demonstrate the commercial interest to application of grape cell suspension derived products. However, it is a matter of time for the scientists to develop the biotechnological approach of producing anthocyanins by grape cell suspensions from the frame of experimental scale to commercially applicable products.

7. Conclusion and future prospects

The approaches described in this chapter can be effective in improving novel anthocyanin-derived metabolites in grape cell suspensions. Continuous study and exploitation of the knowledge of grape cell lines and their control mechanisms will open up new possibilities for metabolic engineering of the anthocyanin biosynthesis pathway. In parallel, the recent achievements in bioengineering with plant cell suspensions and the improvements of the existed bioreactor designs discovers new prospectives for commercial realization of anthocyanin producing technology based on cultivation of grape cells. This is a research area that is growing and gaining interest in the analysis of plant-based health-related compounds. Therefore, the full impact of metabolomics on muscadine research is yet to be experienced. But this chapter serves as a starting point for scientists who are interested in cell cultures from muscadine grapes.

Acknowledgements

The authors are grateful to the Florida A&M University College of Agriculture and Food Science. The research work has been done with the financial support of USDA/NIFA/AFRI Plant Biochemistry Program Grant # 2009-03127.

Author details

Anthony Ananga[1], Vasil Georgiev[1], Joel Ochieng[2], Bobby Phills[1] and Violeta Tsolova[1]

1 School of Agriculture and Food Sciences, Center for Vituculture and Small Fruit Research, Florida A&M University, Tallahassee, FL, USA

2 Faculties of Agriculture and Veterinary Medicine, University of Nairobi, Nairobi, Kenya

References

[1] Deroles S. Anthocyanin biosynthesis in plant cell cultures: A potential source of natural colourants. In: C Winefield, K Davies and K Gould (Eds.) The Anthocyanins. Springer New York. 2009. pp108-167.

[2] Delgado-Vargas F., Jimenez A. R. and Paredes-Lopez O. Natural pigments: carotenoids, anthocyanins, and betalains - characteristics, biosynthesis, processing, and stability. *Crit Rev Food Sci Nutr.* 2000; 40(3) 173-289

[3] Georgiev V., Ilieva M., Bley T. and Pavlov A. Betalain production in plant *in vitro* systems. *Acta Physiologiae Plantarum.* 2008; 30(5) 581-593

[4] Pavokovic′ D. and Krsnik-Rasol M. Complex biochemistry and biotechnological production of betalains. *Food Technology and Biotechnology.* 2011; 49(2) 145-155

[5] Tanaka Y., Sasaki N. and Ohmiya A. Biosynthesis of plant pigments: anthocyanins, betalains and carotenoids. *The Plant Journal.* 2008; 54(4) 733-749

[6] Brockington S. F., Walker R. H., Glover B. J., Soltis P. S. and Soltis D. E. Complex pigment evolution in the Caryophyllales. *New Phytologist.* 2011; 190(4) 854-864

[7] Georgiev V. G., Bley T. and Pavlov A. I. Bioreactors for the cultivation of red beet hairy roots. In: B Neelwarne (Ed.) Red Beet Biotechnology. Springer US. 2012. pp251-281.

[8] Grotewold E. The genetics and biochemistry of floral pigments. *Annu Rev Plant Biol.* 2006; 57 761-780

[9] Mabry T. J. Selected topics from forty years of natural products research: Betalains to flavonoids, antiviral proteins, and neurotoxic nonprotein amino acids. *Journal of Natural Products.* 2001; 64(12) 1596-1604

[10] Harris N. N., Javellana J., Davies K. M., Lewis D. H., Jameson P. E., Deroles S. C., Calcott K. E., Gould K. S. and Schwinn K. E. Betalain production is possible in anthocyanin-producing plant species given the presence of DOPA-dioxygenase and L-DOPA. *BMC Plant Biol.* 2012; 12 34

[11] Downham A. and Collins P. Colouring our foods in the last and next millennium. *International Journal of Food Science & Technology.* 2000; 35(1) 5-22

[12] Ramachandra Rao S. and Ravishankar G. A. Plant cell cultures: Chemical factories of secondary metabolites. *Biotechnology Advances.* 2002; 20(2) 101-153

[13] Lila M. A. Anthocyanins and human health: An *in vitro* investigative approach. *Journal of Biomedicine and Biotechnology.* 2004; 2004(5) 306-313

[14] Koes R. E., Quattrocchio F. and Mol J. N. M. The flavonoid biosynthetic pathway in plants: Function and evolution. *BioEssays.* 1994; 16(2) 123-132

[15] Mazza G. and Francis F. J. Anthocyanins in grapes and grape products. *Critical Reviews in Food Science and Nutrition.* 1995; 35(4) 341-371

[16] Holton T. A. and Cornish E. C. Genetics and biochemistry of anthocyanin biosynthesis. *Plant Cell.* 1995; 7(7) 1071-1083

[17] Winkel-Shirley B. Flavonoid biosynthesis. A colorful model for genetics, biochemistry, cell biology, and biotechnology. *Plant Physiol.* 2001; 126(2) 485-493

[18] Koes R., Verweij W. and Quattrocchio F. Flavonoids: a colorful model for the regulation and evolution of biochemical pathways. *Trends in Plant Science.* 2005; 10(5) 236-242

[19] Jackman R. L., Yada R. Y., Tung M. A. and Speers R. A. Anthocyanins as food colorants - review. *J. Food Biochem.* . 1987; 11(3) 201-247

[20] Bitsch R., Netzel M., Frank T., Strass G. and Bitsch I. Bioavailability and biokinetics of anthocyanins from red grape juice and red wine. *Journal of Biomedicine and Biotechnology.* 2004; 2004(5) 293-298

[21] Castellarin S. D., Pfeiffer A., Sivilotti P., Degan M., Peterlunger E. and Di Gaspero G. Transcriptional regulation of anthocyanin biosynthesis in ripening fruits of grapevine under seasonal water deficit. *Plant, Cell & Environment.* 2007; 30(11) 1381-1399

[22] Zhang W. and Furusaki S. Production of anthocyanins by plant cell cultures. *Biotechnology and Bioprocess Engineering.* 1999; 4(4) 231-252

[23] Yesil-Celiktas O., Gurel A. and Vardar-Sukan F. Large scale cultivation of plant cell and tissue culture in bioreactors. *Transworld Research Network* 2010; 1-54

[24] Gould K. S. and Lister C. Flavonoid functions in plants. In: Ø M Andersen and K R Markham (Eds.) Flavonoids: chemistry, biochemistry and applications. Boca Raton: CRC Press LLC. 2006. pp397-441.

[25] Lev-Yadun S. and Gould K. S. Role of anthocyanins in plant defence. In: C Winefield, K Davies and K Gould (Eds.) The Anthocyanins. Springer New York. 2009. pp22-28.

[26] Steyn W. J. Prevalence and functions of anthocyanins in fruits. In: C Winefield, K Davies and K Gould (Eds.) The Anthocyanins. Springer New York. 2009. pp86-105.

[27] Hatier J.-H. B. and Gould K. S. Anthocyanin function in vegetative organs. In: C Winefield, K Davies and K Gould (Eds.) The Anthocyanins. Springer New York. 2009. pp1-19.

[28] Andersen Ø. M. Recent advances in the field of anthocyanins – main focus on structures. In: F Daayf and V Lattanzio (Eds.) Recent advances in polyphenol research. Wiley-Blackwell. 2009. pp167-201.

[29] Andersen Ø. M. and Jordheim M. The Anthocyanins. In: Ø M Andersen and K R Markham (Eds.) Flavonoids: chemistry, biochemistry and applications. Boca Raton: CRC Press LLC. 2006. pp471-553.

[30] Delgado-Vargas F. and Parades-López O (Eds.) Natural colorants for food and nutraceutical uses. Washington, D.C.: CRC Press Inc. ; 2003.

[31] Brat P., Tourniaire F. and Amiot-Carlin M. J. Stability and analysis of phenolic pigments. In: C Socaciu (Ed.) Food colorants, chemical and functional properties. New York: CRC Press. 2008. pp71-87.

[32] Conn S., Zhang W. and Franco C. Anthocyanic vacuolar inclusions (AVIs) selectively bind acylated anthocyanins in *Vitis vinifera* L. (grapevine) suspension culture. *Biotechnol Lett.* 2003; 25(11) 835-839

[33] Mercadante A. Z. and Bobbio F. O. Anthocyanins in foods: Occurrence and physicochemical properties. In: C Socaciu (Ed.) Food colorants, chemical and functional properties. New York: CRC Press. 2008. pp241-277.

[34] Wrolstad R. E. Anthocyanin pigments - bioactivity and coloring properties. *Journal of Food Science.* 2004; 69(5) C419-C425

[35] Pereira D., Valentão P., Pereira J. and Andrade P. Phenolics: From chemistry to biology. *Molecules.* 2009; 14(6) 2202-2211

[36] Saito N., Tatsuzawa F., Miyoshi K., Shigihara A. and Honda T. The first isolation of C-glycosylanthocyanin from the flowers of *Tricyrtis formosana*. *Tetrahedron Letters.* 2003; 44(36) 6821-6823

[37] Tatsuzawa F., Saito N., Miyoshi K., Shinoda K., Shigihara A. and Honda T. Diacylated 8-C-Glucosylcyanidin 3-Glucoside from the Flowers of *Tricyrtis formosana*. *Chemical and Pharmaceutical Bulletin.* 2004; 52(5) 631-633

[38] Vogt T. Phenylpropanoid biosynthesis. *Molecular Plant.* 2010; 3(1) 2-20

[39] Dixon R. A. and Steele C. L. Flavonoids and isoflavonoids – a gold mine for metabolic engineering. *Trends in Plant Science.* 1999; 4(10) 394-400

[40] Dixon R. A., Xie D.-Y. and Sharma S. B. Proanthocyanidins – a final frontier in flavonoid research? *New Phytologist.* 2005; 165(1) 9-28

[41] Xie D.-Y. and Dixon R. A. Proanthocyanidin biosynthesis – still more questions than answers? *Phytochemistry.* 2005; 66(18) 2127-2144

[42] Tian L., Kong W. F., Pan Q. H., Zhan J. C., Wen P. F., Chen J. Y., Wan S. B. and Huang W. D. Expression of the chalcone synthase gene from grape and preparation of an anti-CHS antibody. *Protein Expression and Purification.* 2006; 50(2) 223-228

[43] Sparvoli F., Martin C., Scienza A., Gavazzi G. and Tonelli C. Cloning and molecular analysis of structural genes involved in flavonoid and stilbene biosynthesis in grape (*Vitis vinifera* L.). *Plant Molecular Biology.* 1994; 24(5) 743-755

[44] Davies K. M. and Schwinn K. E. Molecular biology and biotechnology of flavonoid biosynthesis. In: Ø M Andersen and K R Markham (Eds.) Flavonoids: chemistry, biochemistry and applications. Boca Raton: CRC Press LLC. 2006. pp143-219.

[45] Veitch N. C. and Grayer R. J. Chalcones, dihydrochalcones, and aurones. In: Ø M Andersen and K R Markham (Eds.) Flavonoids: chemistry, biochemistry and applications. Boca Raton: CRC Press LLC. 2006. pp1003-1101.

[46] Goto-Yamamoto N., Wan G. H., Masaki K. and Kobayashi S. Structure and transcription of three chalcone synthase genes of grapevine (*Vitis vinifera*). *Plant Science.* 2002; 162(6) 867-872

[47] Ageorges A., Fernandez L., Vialet S., Merdinoglu D., Terrier N. and Romieu C. Four specific isogenes of the anthocyanin metabolic pathway are systematically co-expressed with the red colour of grape berries. *Plant Science.* 2006; 170(2) 372-383

[48] He F., Mu L., Yan G.-L., Liang N.-N., Pan Q.-H., Wang J., Reeves M. J. and Duan C.-Q. Biosynthesis of anthocyanins and their regulation in colored grapes. *Molecules.* 2010; 15(12) 9057-9091

[49] Gutha L. R., Casassa L. F., Harbertson J. F. and Naidu R. A. Modulation of flavonoid biosynthetic pathway genes and anthocyanins due to virus infection in grapevine (*Vitis vinifera* L.) leaves. *BMC Plant Biol.* 2010; 10 187

[50] Markham K. R., Gould K. S., Winefield C. S., Mitchell K. A., Bloor S. J. and Boase M. R. Anthocyanic vacuolar inclusions — their nature and significance in flower colouration. *Phytochemistry.* 2000; 55(4) 327-336

[51] Conn S., Franco C. and Zhang W. Characterization of anthocyanic vacuolar inclusions in *Vitis vinifera* L. cell suspension cultures. *Planta.* 2010; 231(6) 1343-1360

[52] Leão P. C. d. S., Cruz C. D. and Motoike S. Y. Genetic diversity of table grape based on morphoagronomic traits. *Scientia Agricola.* 2011; 68 42-49

[53] Leão P. C. d. S. and Motoike S. Y. Genetic diversity in table grapes based on RAPD and microsatellite markers. *Pesquisa Agropecuária Brasileira.* 2011; 46 1035-1044

[54] This P., Lacombe T., Cadle-Davidson M. and Owens C. Wine grape (*Vitis vinifera* L.) color associates with allelic variation in the domestication gene *VvmybA1*. *TAG Theoretical and Applied Genetics.* 2007; 114(4) 723-730

[55] Azuma A., Kobayashi S., Yakushiji H., Yamada M., Mitani N. and Sato A. *VvmybA1* genotype determines grape sin color. *Vitis.* 2007; 46(3) 154-155

[56] Kobayashi S., Goto-Yamamoto N. and Hirochika H. Retrotransposon-induced mutations in grape skin color. *Science.* 2004; 304(5673) 982

[57] Zhao Q., Duan C.-Q. and Wang J. Anthocyanins profile of grape berries of *Vitis amurensis*, Its hybrids and their wines. *International Journal of Molecular Sciences.* 2010; 11(5) 2212-2228

[58] Boss P. K. and Davies C. Molecular biology of sugar and anthocyanin accumulation in grape berries. In: K A Roubelakis-Angelakis (Ed.) Molecular biology & biotechnology of the grapevine. London: Kluwer Academic Publishers. 2001. pp1-34.

[59] Boss P. K., Davies C. and Robinson S. P. Anthocyanin composition and anthocyanin pathway gene expression in grapevine sports differing in berry skin colour. *Australian Journal of Grape and Wine Research.* 1996; 2(3) 163-170

[60] He J. and Giusti M. M. Anthocyanins: natural colorants with health-promoting properties. *Annu Rev Food Sci Technol.* 2010; 1 163-187

[61] Boss P. K., Davies C. and Robinson S. P. Analysis of the expression of anthocyanin pathway genes in developing *Vitis vinifera* L. *cv* Shiraz grape berries and the implications for pathway regulation. *Plant Physiol.* 1996; 111(4) 1059-1066

[62] Ballinger W. E., Maness E. P., Nesbitt W. B. and Carroll D. E. Anthocyanins of black grapes of 10 clones of *Vitis rotundifolia*, Michx. *Journal of Food Science.* 1973; 38(5) 909-910

[63] Huang Z., Wang B., Williams P. and Pace R. D. Identification of anthocyanins in muscadine grapes with HPLC-ESI-MS. *LWT - Food Science and Technology.* 2009; 42(4) 819-824

[64] Sandhu A. K., Gray D. J., Lu J. and Gu L. Effects of exogenous abscisic acid on antioxidant capacities, anthocyanins, and flavonol contents of muscadine grape (*Vitis rotundifolia*) skins. *Food Chemistry.* 2011; 126(3) 982-988

[65] Jánváry L. s., Hoffmann T., Pfeiffer J., Hausmann L., Töpfer R., Fischer T. C. and Schwab W. A double mutation in the anthocyanin 5-*O*-glucosyltransferase gene disrupts enzymatic activity in *Vitis vinifera* L. *Journal of Agricultural and Food Chemistry.* 2009; 57(9) 3512-3518

[66] Cravero M. C., Guidoni S., Schneider A. and De Stefano R. Caractérisation variétale de cépages musqués à raisin coloré au moyen de paramètres ampélographiques descriptifs et biochimiques. *Vitis.* 1994; 33(2) 75-80

[67] Wu X. and Prior R. L. Systematic identification and characterization of anthocyanins by HPLC-ESI-MS/MS in common foods in the United States: fruits and berries. *Journal of Agricultural and Food Chemistry.* 2005; 53(7) 2589-2599

[68] Lücker J., Martens S. and Lund S. T. Characterization of a *Vitis vinifera* cv. Cabernet Sauvignon 3′,5′-O-methyltransferase showing strong preference for anthocyanins and glycosylated flavonols. *Phytochemistry.* 2010; 71(13) 1474-1484

[69] Davis G., Ananga A., Krastanova S., Sutton S., Ochieng J. W., Leong S. and Tsolova V. Elevated gene expression in chalcone synthase enzyme suggests an increased production of flavonoids in skin and synchronized red cell cultures of North American native grape berries. *DNA Cell Biol.* 2012; 31(6) 939-945

[70] Samuelian S. K., Camps C., Kappel C., Simova E. P., Delrot S. and Colova V. M. Differential screening of overexpressed genes involved in flavonoid biosynthesis in North American native grapes: 'Noble' *muscadinia* var. and 'Cynthiana' *aestivalis* var. *Plant Science.* 2009; 177(3) 211-221

[71] Fournier-Level A., Hugueney P., Verries C., This P. and Ageorges A. Genetic mechanisms underlying the methylation level of anthocyanins in grape (*Vitis vinifera* L.). *BMC Plant Biology.* 2011; 11(1) 179

[72] Bakowska-Barczak A. Acylated anthocyanins as stable, natural food colorants - a review. *Polish Journal of Food and Nutrition Sciences.* 2005; 14(2) 107-116

[73] Flora L. F. Influence of heat, cultivar and maturity on the anthocyanidin-3,5-diglucosides of *Muscadine* grapes. *Journal of Food Science.* 1978; 43(6) 1819-1821

[74] Baluja J., Diago M., Goovaerts P. and Tardaguila J. Assessment of the spatial variability of anthocyanins in grapes using a fluorescence sensor: relationships with vine vigour and yield. *Precision Agriculture.* 2012; 13(4) 457-472

[75] Tuccio L., Remorini D., Pinelli P., Fierini E., Tonutti P., Scalabrelli G. and Agati G. Rapid and non-destructive method to assess in the vineyard grape berry anthocyanins under different seasonal and water conditions. *Australian Journal of Grape and Wine Research.* 2011; 17(2) 181-189

[76] Broun P. Transcriptional control of flavonoid biosynthesis: a complex network of conserved regulators involved in multiple aspects of differentiation in *Arabidopsis*. *Current Opinion in Plant Biology*. 2005; 8(3) 272-279

[77] Ramsay N. A. and Glover B. J. MYB–bHLH–WD40 protein complex and the evolution of cellular diversity. *Trends in Plant Science*. 2005; 10(2) 63-70

[78] Petroni K. and Tonelli C. Recent advances on the regulation of anthocyanin synthesis in reproductive organs. *Plant Science*. 2011; 181(3) 219-229

[79] Schwinn K. E. and Davies K. M. Flavonoids. In: K Davies (Ed.) Plant Pigments and their Manipulation. Oxford: Blackwell Publishing Ltd. 2004. pp92–149.

[80] Springob K., Nakajima J.-i., Yamazaki M. and Saito K. Recent advances in the biosynthesis and accumulation of anthocyanins. *Natural Product Reports*. 2003; 20(3) 288-303

[81] He F., Pan Q.-H., Shi Y. and Duan C.-Q. Biosynthesis and genetic regulation of proanthocyanidins in plants. *Molecules*. 2008; 13(10) 2674-2703

[82] Abrahams S., Tanner G. J., Larkin P. J. and Ashton A. R. Identification and biochemical characterization of mutants in the proanthocyanidin pathway in *Arabidopsis*. *Plant Physiol*. 2002; 130(2) 561-576

[83] Ramsay N. A., Walker A. R., Mooney M. and Gray J. C. Two basic-helix-loop-helix genes (*MYC-146* and *GL3*) from *Arabidopsis* can activate anthocyanin biosynthesis in a white-flowered *Matthiola incana* mutant. *Plant Molecular Biology*. 2003; 52(3) 679-688

[84] Gonzalez A., Zhao M., Leavitt J. M. and Lloyd A. M. Regulation of the anthocyanin biosynthetic pathway by the TTG1/bHLH/Myb transcriptional complex in *Arabidopsis* seedlings. *The Plant Journal*. 2008; 53(5) 814-827

[85] Dubos C., Le Gourrierec J., Baudry A., Huep G., Lanet E., Debeaujon I., Routaboul J.-M., Alboresi A., Weisshaar B. and Lepiniec L. *MYBL2* is a new regulator of flavonoid biosynthesis in *Arabidopsis thaliana*. *The Plant Journal*. 2008; 55(6) 940-953

[86] Deluc L., Barrieu F., Marchive C., Lauvergeat V., Decendit A., Richard T., Carde J. P., Merillon J. M. and Hamdi S. Characterization of a grapevine R2R3-MYB transcription factor that regulates the phenylpropanoid pathway. *Plant Physiol*. 2006; 140(2) 499-511

[87] Deluc L., Bogs J., Walker A. R., Ferrier T., Decendit A., Merillon J. M., Robinson S. P. and Barrieu F. The transcription factor *VvMYB5b* contributes to the regulation of anthocyanin and proanthocyanidin biosynthesis in developing grape berries. *Plant Physiol*. 2008; 147(4) 2041-2053

[88] Kobayashi S. K., Ishimaru M. I., Hiraoka K. H. and Honda C. H. *Myb*-related genes of the Kyoho grape (*Vitis labruscana*) regulate anthocyanin biosynthesis. *Planta*. 2002; 215(6) 924-933

[89] Walker A. R., Lee E., Bogs J., McDavid D. A. J., Thomas M. R. and Robinson S. P. White grapes arose through the mutation of two similar and adjacent regulatory genes. *The Plant Journal*. 2007; 49(5) 772-785

[90] Bogs J., Downey M. O., Harvey J. S., Ashton A. R., Tanner G. J. and Robinson S. P. Proanthocyanidin synthesis and expression of genes encoding leucoanthocyanidin reductase and anthocyanidin reductase in developing grape berries and grapevine leaves. *Plant Physiol.* 2005; 139(2) 652-663

[91] Kobayashi S., Goto-Yamamoto N. and Hirochika H. Association of *VvmybAl* gene expression with anthocyanin production in grape (*Vitis vinifera*) skin-color mutants. *Journal of the Japanese Society for Horticultural Science* 2005; 74(3) 196–203

[92] Azuma A., Kobayashi S., Mitani N., Shiraishi M., Yamada M., Ueno T., Kono A., Yakushiji H. and Koshita Y. Genomic and genetic analysis of *Myb*-related genes that regulate anthocyanin biosynthesis in grape berry skin. *TAG Theoretical and Applied Genetics.* 2008; 117(6) 1009-1019

[93] Colova V. Synchronized strains of subepidermal cells of muscadine (muscadine sp.) grapevine pericarp for use as a sourse of flavonoids (nutraceuticals). United States. 2011 20110054195.

[94] Ananga A., Krastanova K., Sutton S. and Colova V. Molecular assessments of synchronized *in vitro* red cell cultures of American native grapes. *Proc. Fla. State Hort. Soc.* 2011; 124 7-12

[95] van der Krol A. R., Mur L. A., Beld M., Mol J. N. and Stuitje A. R. Flavonoid genes in *Petunia*: addition of a limited number of gene copies may lead to a suppression of gene expression. *Plant Cell.* 1990; 2(4) 291-299

[96] van der Krol A. R., Lenting P. E., Veenstra J., van der Meer I. M., Koes R. E., Gerats A. G. M., Mol J. N. M. and Stuitje A. R. An anti-sense chalcone synthase gene in transgenic plants inhibits flower pigmentation. *Nature.* 1988; 333(6176) 866-869

[97] Courtney-Gutterson N., Napoli C., Lemieux C., Morgan A., Firoozabady E. and Robinson K. E. P. Modification of flower color in florist's chrysanthemum: Production of a white-flowering variety through molecular genetics. *Nat Biotech.* 1994; 12(3) 268-271

[98] Butelli E., Titta L., Giorgio M., Mock H.-P., Matros A., Peterek S., Schijlen E. G. W. M., Hall R. D., Bovy A. G., Luo J. and Martin C. Enrichment of tomato fruit with health-promoting anthocyanins by expression of select transcription factors. *Nat Biotech.* 2008; 26(11) 1301-1308

[99] Mol J., Cornish E., Mason J. and Koes R. Novel coloured flowers. *Current Opinion in Biotechnology.* 1999; 10(2) 198-201

[100] Suzuki H., Nakayama T., Yonekura-Sakakibara K., Fukui Y., Nakamura N., Yamaguchi M. A., Tanaka Y., Kusumi T. and Nishino T. cDNA cloning, heterologous expressions, and functional characterization of malonyl-coenzyme a:anthocyanidin 3-o-glucoside-6"-o-malonyltransferase from dahlia flowers. *Plant Physiol.* 2002; 130(4) 2142-2151

[101] Zuker A., Tzfira T., Ben-Meir H., Ovadis M., Shklarman E., Itzhaki H., Forkmann G., Martens S., Neta-Sharir I., Weiss D. and Vainstein A. Modification of flower color and

fragrance by antisense suppression of the flavanone 3-hydroxylase gene. *Molecular Breeding.* 2002; 9(1) 33-41

[102] Fukui Y., Tanaka Y., Kusumi T., Iwashita T. and Nomoto K. A rationale for the shift in colour towards blue in transgenic carnation flowers expressing the flavonoid 3′,5′-hydroxylase gene. *Phytochemistry.* 2003; 63(1) 15-23

[103] Bovy A., Schijlen E. and Hall R. Metabolic engineering of flavonoids in tomato (*Solanum lycopersicum*): the potential for metabolomics. *Metabolomics.* 2007; 3(3) 399-412

[104] Schijlen E. G. W. M., Ric de Vos C. H., van Tunen A. J. and Bovy A. G. Modification of flavonoid biosynthesis in crop plants. *Phytochemistry.* 2004; 65(19) 2631-2648

[105] Khachik F., Carvalho L., Bernstein P. S., Muir G. J., Zhao D.-Y. and Katz N. B. Chemistry, distribution, and metabolism of tomato carotenoids and their impact on human health. *Experimental Biology and Medicine.* 2002; 227(10) 845-851

[106] Mes P. J., Boches P., Myers J. R. and Durst R. Characterization of tomatoes expressing anthocyanin in the fruit. *Journal of the American Society for Horticultural Science.* 2008; 133(2) 262-269

[107] Gonzali S., Mazzucato A. and Perata P. Purple as a tomato: towards high anthocyanin tomatoes. *Trends in Plant Science.* 2009; 14(5) 237-241

[108] Muir S. R., Collins G. J., Robinson S., Hughes S., Bovy A., Ric De Vos C. H., van Tunen A. J. and Verhoeyen M. E. Overexpression of petunia chalcone isomerase in tomato results in fruit containing increased levels of flavonols. *Nat Biotech.* 2001; 19(5) 470-474

[109] Bovy A., de Vos R., Kemper M., Schijlen E., Almenar Pertejo M., Muir S., Collins G., Robinson S., Verhoeyen M., Hughes S., Santos-Buelga C. and van Tunen A. High-flavonol tomatoes resulting from the heterologous expression of the maize transcription factor genes LC and C1. *Plant Cell.* 2002; 14(10) 2509-2526

[110] Verhoeyen M. E., Bovy A., Collins G., Muir S., Robinson S., de Vos C. H. R. and Colliver S. Increasing antioxidant levels in tomatoes through modification of the flavonoid biosynthetic pathway. *Journal of Experimental Botany.* 2002; 53(377) 2099-2106

[111] Luo J., Butelli E., Hill L., Parr A., Niggeweg R., Bailey P., Weisshaar B. and Martin C. *AtMYB12* regulates caffeoyl quinic acid and flavonol synthesis in tomato: expression in fruit results in very high levels of both types of polyphenol. *The Plant Journal.* 2008; 56(2) 316-326

[112] Ranish J. A. and Hahn S. Transcription: basal factors and activation. *Current opinion in genetics & development.* 1996; 6(2) 151-158

[113] Hernandez J. M., Heine G. F., Irani N. G., Feller A., Kim M.-G., Matulnik T., Chandler V. L. and Grotewold E. Different mechanisms participate in the R-dependent activity of the R2R3 MYB transcription factor C1. *Journal of Biological Chemistry.* 2004; 279(46) 48205-48213

[114] Schwinn K., Venail J., Shang Y., Mackay S., Alm V., Butelli E., Oyama R., Bailey P., Davies K. and Martin C. A small family of *MYB*-regulatory genes controls floral pigmentation intensity and patterning in the genus *Antirrhinum*. *The Plant Cell Online*. 2006; 18(4) 831-851

[115] Morita Y., Saitoh M., Hoshino A., Nitasaka E. and Iida S. Isolation of cDNAs for R2R3-MYB, bHLH and WDR transcriptional regulators and identification of c and ca mutations conferring white flowers in the Japanese morning glory. *Plant and Cell Physiology*. 2006; 47(4) 457-470

[116] Quattrocchio F., Wing J. F., Leppen H., Mol J. and Koes R. E. Regulatory genes controlling anthocyanin pigmentation are functionally conserved among plant species and have distinct sets of target genes. *The Plant Cell Online*. 1993; 5(11) 1497-1512

[117] Aharoni A., De Vos C. H. R., Wein M., Sun Z., Greco R., Kroon A., Mol J. N. M. and O'Connell A. P. The strawberry FaMYB1 transcription factor suppresses anthocyanin and flavonol accumulation in transgenic tobacco. *The Plant Journal*. 2001; 28(3) 319-332

[118] Mathews H., Clendennen S. K., Caldwell C. G., Liu X. L., Connors K., Matheis N., Schuster D. K., Menasco D. J., Wagoner W., Lightner J. and Wagner D. R. Activation tagging in tomato identifies a transcriptional regulator of anthocyanin biosynthesis, modification, and transport. *The Plant Cell Online*. 2003; 15(8) 1689-1703

[119] Sutton S., Krastanova S., Ananga A., Leong S. and Tsolova V. Genetic transformation for overexpression of flavonoid compounds in *Muscadinia* grape cell cultures. *Proc. Fla. State Hort. Soc.* 2011; 124 13-17

[120] Le Gall G., DuPont M. S., Mellon F. A., Davis A. L., Collins G. J., Verhoeyen M. E. and Colquhoun I. J. Characterization and content of flavonoid glycosides in genetically modified tomato (*Lycopersicon esculentum*) fruits. *Journal of Agricultural and Food Chemistry*. 2003; 51(9) 2438-2446

[121] Lloyd A. M., Walbot V. and Davis R. W. *Arabidopsis* and *Nicotiana* anthocyanin production activated by maize regulators R and C1. *Science (New York, N.Y.)*. 1992; 258(5089) 1773-1775

[122] Goldsbrough A. P., Tong Y. and Yoder J. I. Lc as a non-destructive visual reporter and transposition excision marker gone for tomato. *The Plant Journal*. 1996; 9(6) 927-933

[123] Bino R. J., De Vos C. H. R., Lieberman M., Hall R. D., Bovy A., Jonker H. H., Tikunov Y., Lommen A., Moco S. and Levin I. The light-hyperresponsive high pigment-2dg mutation of tomato: alterations in the fruit metabolome. *New Phytologist*. 2005; 166(2) 427-438

[124] Davuluri G. R., van Tuinen A., Fraser P. D., Manfredonia A., Newman R., Burgess D., Brummell D. A., King S. R., Palys J., Uhlig J., Bramley P. M., Pennings H. M. J. and Bowler C. Fruit-specific RNAi-mediated suppression of *Det-1* enhances carotenoid and flavonoid content in tomatoes. *Nat Biotech*. 2005; 23(7) 890-895

[125] Beld M., Martin C., Huits H., Stuitje A. R. and Gerats A. G. M. Flavonoid synthesis in *Petunia hybrida*: partial characterization of dihydroflavonol-4-reductase genes. *Plant Molecular Biology.* 1989; 13(5) 491-502

[126] Cone K. C., Burr F. A. and Burr B. Molecular analysis of the maize anthocyanin regulatory locus C1. *Proceedings of the National Academy of Sciences.* 1986; 83(24) 9631-9635

[127] Coombe B. and Bishop G. Development of the grape berry. II. Changes in diameter and deformability during veraison. *Australian Journal of Agricultural Research.* 1980; 31(3) 499-509

[128] Colliver S., Bovy A., Collins G., Muir S., Robinson S., de Vos C. H. R. and Verhoeyen M. E. Improving the nutritional content of tomatoes through reprogramming their flavonoid biosynthetic pathway. *Phytochemistry Reviews.* 2002; 1(1) 113-123

[129] Miyao A., Tanaka K., Murata K., Sawaki H., Takeda S., Abe K., Shinozuka Y., Onosato K. and Hirochika H. Target site specificity of the Tos17 retrotransposon shows a preference for insertion within genes and against insertion in retrotransposon-rich regions of the genome. *Plant Cell.* 2003; 15(8) 1771-1780

[130] Terada R., Urawa H., Inagaki Y., Tsugane K. and Iida S. Efficient gene targeting by homologous recombination in rice. *Nat Biotechnol.* 2002; 20(10) 1030-1034

[131] Zamecnik P. C. and Stephenson M. L. Inhibition of Rous sarcoma virus replication and cell transformation by a specific oligodeoxynucleotide. *Proceedings of the National Academy of Sciences.* 1978; 75(1) 280-284

[132] Napoli C., Lemieux C. and Jorgensen R. Introduction of a chimeric chalcone synthase gene into *Petunia* results in reversible co-suppression of homologous genes in trans. *Plant Cell.* 1990; 2(4) 279-289

[133] Fire A., Xu S., Montgomery M. K., Kostas S. A., Driver S. E. and Mello C. C. Potent and specific genetic interference by double-stranded RNA in *Caenorhabditis elegans*. *Nature.* 1998; 391(6669) 806-811

[134] Waterhouse P. M., Wang M.-B. and Finnegan E. J. Role of short RNAs in gene silencing. *Trends in Plant Science.* 2001; 6(7) 297-301

[135] Schijlen E. G. W. M. Genetic engineering of flavonoid biosynthesis in tomato. PhD Thesis. Amsterdam, Universiteit van Amsterdam. 2007

[136] Bernstein E., Caudy A. A., Hammond S. M. and Hannon G. J. Role for a bidentate ribonuclease in the initiation step of RNA interference. *Nature.* 2001; 409(6818) 363-366

[137] Sharp P. A. RNA interference. *Genes Dev.* 2001; 15(5) 485-490

[138] Meins F. RNA degradation and models for post-transcriptional gene silencing. *Plant Molecular Biology.* 2000; 43(2) 261-273

[139] Hammond S. M., Bernstein E., Beach D. and Hannon G. J. An RNA-directed nuclease mediates post-transcriptional gene silencing in *Drosophila* cells. *Nature.* 2000; 404(6775) 293-296

[140] Colova (Tsolova) V. M., Bordallo P. N., Phills B. R. and Bausher M. Synchronized somatic embryo development in embryogenic suspensions of grapevine *Muscadinia rotundifolia* (Michx.) Small. *Vitis.* 2007; 46(1) 15-18

[141] Conn S., Curtin C., Bézier A., Franco C. and Zhang W. Purification, molecular cloning, and characterization of glutathione S-transferases (GSTs) from pigmented *Vitis vinifera* L. cell suspension cultures as putative anthocyanin transport proteins. *Journal of Experimental Botany.* 2008; 59(13) 3621-3634

[142] Cormier F., Brion F., Do C. B. and Moresoli C. Development of process strategies for anthocyanin-based food colorant production using *Vitis vinifera* cell cultures. In: F DiCosmo and M Misawa (Eds.) Plant cell culture secondary metabolism toward industrial application. CRC Press LLC. 1996. pp167-186.

[143] Zhang W., Franco C., Curtin C. and Conn S. To stretch the boundary of secondary metabolite production in plant cell-based bioprocessing: anthocyanin as a case study. *Journal of biomedicine & biotechnology.* 2004; 2004(5) 264-271

[144] Lazar A. and Petolescu C. Experimental results concerning the synthesized anthocyanin amount in the *Vitis vinifera* L. suspension cell culture in the laboratory bioreactor. *Journal of Horticulture, Forestry and Biotechnology.* 2009; 13 443-446

[145] Guan L., Li J.-H., Fan P.-G., Chen S., Fang J.-B., Li S.-H. and Wu B.-H. Anthocyanin accumulation in various organs of a teinturier grape cultivar (*V. vinifera* L.) during the growing season. *American Journal of Enology and Viticulture.* 2012; DOI: 10.5344/ajev.2011.11063

[146] Krisa S., Vitrac X., Decendit A., Larronde F., Deffieux G. and Mérillon J.-M. Obtaining *Vitis vinifera* cell cultures producing higher amounts of malvidin-3-*O*-β-glucoside. *Biotechnology Letters.* 1999; 21(6) 497-500

[147] Torregrosa L., Bouquet A. and Goussard P. G. *In vitro* culture and propagation of grapevine. In: K A Roubelakis-Angelakis (Ed.) Molecular biology & biotechnology of the grapevine. London: Kluwer Academic Publishers. 2001. pp281-326.

[148] Marchev A., Georgiev V., Ivanov I., Badjakov I. and Pavlov A. Two-phase temporary immersion system for *Agrobacterium rhizogenes* genetic transformation of sage (*Salvia tomentosa* Mill.). *Biotechnology Letters.* 2011; 33(9) 1873-1878

[149] Velcheva M., Faltin Z., Vardi A., Hanania U., Eshdat Y., Dgani O., Sahar N. and Perl A. *Aloe vera* transformation: the role of Amberlite XAD-4 resin and antioxidants during selection and regeneration. *In Vitro Cellular & Developmental Biology - Plant.* 2010; 46(6) 477-484

[150] Qu J., Zhang W., Yu X. and Jin M. Instability of anthocyanin accumulation in *Vitis vinifera* L. var. Gamay Fréaux suspension cultures. *Biotechnology and Bioprocess Engineering*. 2005; 10(2) 155-161

[151] Qu J., Zhang W. and Yu X. A combination of elicitation and precursor feeding leads to increased anthocyanin synthesis in cell suspension cultures of *Vitis vinifera*. *Plant Cell, Tissue and Organ Culture*. 2011; 107(2) 261-269

[152] Baublis A., Spomer A. R. T. and Berber-JimÉNez M. D. Anthocyanin pigments: comparison of extract stability. *Journal of Food Science*. 1994; 59(6) 1219-1221

[153] Mewis I., Smetanska I., Müller C. and Ulrichs C. Specific poly-phenolic compounds in cell culture of *Vitis vinifera* L. cv. Gamay Fréaux. *Applied Biochemistry and Biotechnology*. 2011; 164(2) 148-161

[154] Dédaldéchamp F. and Uhel C. Induction of anthocyanin synthesis in nonpigmented grape cell suspensions by acting on DFR substrate availability or precursors level. *Enzyme and Microbial Technology*. 1999; 25(3–5) 316-321

[155] Nagamori E., Hiraoka K., Honda H. and Kobayashi T. Enhancement of anthocyanin production from grape (*Vitis vinifera*) callus in a viscous additive-supplemented medium. *Biochemical Engineering Journal*. 2001; 9(1) 59-65

[156] Pavlov A., Georgiev V. and Kovatcheva P. Relationship between type and age of the inoculum cultures and betalains biosynthesis by *Beta vulgaris* hairy root culture. *Biotechnology Letters*. 2003; 25(4) 307-309

[157] Nagella P., Chung I.-M. and Murthy H. N. *In vitro* production of gymnemic acid from cell suspension cultures of *Gymnema sylvestre* R. Br. *Engineering in Life Sciences*. 2011; 11(5) 537-540

[158] Decendit A. and Mérillon J. M. Condensed tannin and anthocyanin production in *Vitis vinifera* cell suspension cultures. *Plant Cell Reports*. 1996; 15(10) 762-765

[159] Yin Y., Borges G., Sakuta M., Crozier A. and Ashihara H. Effect of phosphate deficiency on the content and biosynthesis of anthocyanins and the expression of related genes in suspension-cultured grape (*Vitis* sp.) cells. *Plant Physiology and Biochemistry*. 2012; 55(0) 77-84

[160] Hirasuna T. J., Shuler M. L., Lackney V. K. and Spanswick R. M. Enhanced anthocyanin production in grape cell cultures. *Plant Science*. 1991; 78(1) 107-120

[161] Larronde F., Krisa S., Decendit A., Chèze C., Deffieux G. and Mérillon J. M. Regulation of polyphenol production in *Vitis vinifera* cell suspension cultures by sugars. *Plant Cell Reports*. 1998; 17(12) 946-950

[162] Suzuki M. Enhancement of anthocyanin accumulation by high osmotic stress and low pH in grape cells (*Vitis* hybrids). *Journal of Plant Physiology*. 1995; 147(1) 152-155

[163] Hiroyuki H., Kousuke H., Eiji N., Mariko O., Yoshihito K., Setsuro H. and Takeshi K. Enhanced anthocyanin production from grape callus in an air-lift type bioreactor using

a viscous additive-supplemented medium. *Journal of Bioscience and Bioengineering.* 2002; 94(2) 135-139

[164] Iercan C. and Nedelea G. Experimental results concerning the effect of culture medium pH on the synthesized anthocyanin amount in the callus culture of *Vitis vinifera* L. *Journal of Horticulture, Forestry and Biotechnology.* 2012; 16(2) 71-73

[165] Christie P. J., Alfenito M. R. and Walbot V. Impact of low-temperature stress on general phenylpropanoid and anthocyanin pathways: Enhancement of transcript abundance and anthocyanin pigmentation in maize seedlings. *Planta.* 1994; 194(4) 541-549

[166] Chalker-Scott L. Environmental significance of anthocyanins in plant stress responses. *Photochemistry and Photobiology.* 1999; 70(1) 1-9

[167] Yamane T., Jeong S. T., Goto-Yamamoto N., Koshita Y. and Kobayashi S. Effects of temperature on anthocyanin biosynthesis in grape berry skins. *American Journal of Enology and Viticulture.* 2006; 57(1) 54-59

[168] Mori K., Goto-Yamamoto N., Kitayama M. and Hashizume K. Loss of anthocyanins in red-wine grape under high temperature. *Journal of Experimental Botany.* 2007; 58(8) 1935-1945

[169] Azuma A., Yakushiji H., Koshita Y. and Kobayashi S. Flavonoid biosynthesis-related genes in grape skin are differentially regulated by temperature and light conditions. *Planta.* 2012; 236(4) 1067-1080

[170] Zhang Z.-Z., Li X.-X., Chu Y.-N., Zhang M.-X., Wen Y.-Q., Duan C.-Q. and Pan Q.-H. Three types of ultraviolet irradiation differentially promote expression of shikimate pathway genes and production of anthocyanins in grape berries. *Plant Physiology and Biochemistry.* 2012; 57(0) 74-83

[171] Lazar A., Petolescu C. and Popescu S. Experimental results concerning the effect of photoperiod and callus culture duration on anthocyanin amount. *Journal of Horticulture, Forestry and Biotechnology.* 2010; 14(2) 153-157

[172] Curtin C., Zhang W. and Franco C. Manipulating anthocyanin composition in *Vitis vinifera* suspension cultures by elicitation with jasmonic acid and light irradiation. *Biotechnol Lett.* 2003; 25(14) 1131-1135

[173] Zhang W., Curtin C., Kikuchi M. and Franco C. Integration of jasmonic acid and light irradiation for enhancement of anthocyanin biosynthesis in *Vitis vinifera* suspension cultures. *Plant Science.* 2002; 162(3) 459-468

[174] Gagné S., Cluzet S., Mérillon J.-M. and Gény L. ABA initiates anthocyanin production in grape cell cultures. *Journal of Plant Growth Regulation.* 2011; 30(1) 1-10

[175] Zhao J., Davis L. C. and Verpoorte R. Elicitor signal transduction leading to production of plant secondary metabolites. *Biotechnol Adv.* 2005; 23(4) 283-333

[176] Cai Z., Riedel H., Thaw Saw N., Kütük O., Mewis I., Jäger H., Knorr D. and Smetanska I. Effects of pulsed electric field on secondary metabolism of *Vitis vinifera* L. cv. Gamay

Fréaux suspension culture and exudates. *Applied Biochemistry and Biotechnology.* 2011; 164(4) 443-453

[177] Cai Z., Riedel H., Saw N. M. M. T., Mewis I., Reineke K., Knorr D. and Smetanska I. Effects of elicitors and high hydrostatic pressure on secondary metabolism of *Vitis vinifera* suspension culture. *Process Biochemistry.* 2011; 46(7) 1411-1416

[178] Cai Z., Knorr D. and Smetanska I. Enhanced anthocyanins and resveratrol production in *Vitis vinifera* cell suspension culture by indanoyl-isoleucine, N-linolenoyl-l-glutamine and insect saliva. *Enzyme and Microbial Technology.* 2012; 50(1) 29-34

[179] Cai Z., Kastell A., Mewis I., Knorr D. and Smetanska I. Polysaccharide elicitors enhance anthocyanin and phenolic acid accumulation in cell suspension cultures of *Vitis vinifera. Plant Cell, Tissue and Organ Culture.* 2012; 108(3) 401-409

[180] Saw N., Riedel H., Cai Z., Kütük O. and Smetanska I. Stimulation of anthocyanin synthesis in grape (*Vitis vinifera*) cell cultures by pulsed electric fields and ethephon. *Plant Cell, Tissue and Organ Culture.* 2012; 108(1) 47-54

[181] Santamaria A. R., Mulinacci N., Valletta A., Innocenti M. and Pasqua G. Effects of elicitors on the production of resveratrol and viniferins in cell cultures of *Vitis vinifera* L. cv Italia. *Journal of Agricultural and Food Chemistry.* 2011; 59(17) 9094-9101

[182] Riedel H., Akumo D. N., Saw N. M. M. T., Kütük O., Neubauer P. and Smetanska I. Elicitation and precursor feeding influence phenolic acids composition in *Vitis vinifera* suspension culture. *African Journal of Biotechnology.* 2012; 11(12) 3000-3008

[183] Sinilal B., Ovadia R., Nissim-Levi A., Perl A., Carmeli-Weissberg M. and Oren-Shamir M. Increased accumulation and decreased catabolism of anthocyanins in red grape cell suspension culture following magnesium treatment. *Planta.* 2011; 234(1) 61-71

[184] Lila M. A. Plant pigments and human health. In: K Davies (Ed.) Plant Pigments and their Manipulation. Oxford: Blackwell Publishing Ltd. 2004. pp248–274.

[185] Wallace T. C. Anthocyanins in cardiovascular disease. *Adv Nutr.* 2011; 2(1) 1-7

[186] Mazza G. and Kay C. D. Bioactivity, absorption, and metabolism of anthocyanins. In: F Daayf and V Lattanzio (Eds.) Recent Advances in Polyphenol Research. Wiley-Blackwell. 2009. pp228-262.

[187] Stintzing F. C. and Carle R. Functional properties of anthocyanins and betalains in plants, food, and in human nutrition. *Trends in Food Science & Technology.* 2004; 15(1) 19-38

[188] Amiot-Carlin M. J., Babot-Laurent C. and Tourniaire F. Plant pigments as bioactive substances. In: C Socaciu (Ed.) Food colorants, chemical and functional properties. New York: CRC Press. 2008. pp127-147.

[189] Yue X., Zhang W. and Deng M. Hyper-production of 13C-labeled trans-resveratrol in *Vitis vinifera* suspension cell culture by elicitation and in situ adsorption. *Biochemical Engineering Journal.* 2011; 53(3) 292-296

[190] Aumont V., Larronde F., Richard T., Budzinski H., Decendit A., Deffieux G., Krisa S. and Mérillon J.-M. Production of highly 13C-labeled polyphenols in *Vitis vinifera* cell bioreactor cultures. *Journal of Biotechnology*. 2004; 109(3) 287-294

[191] Socaciu C. Natural pigments as food colorants. In: C Socaciu (Ed.) Food colorants, chemical and functional properties. New York: CRC Press. 2008. pp583-603.

[192] Mortensen A. Carotenoids and other pigments as natural colorants. *Pure and Applied Chemistry* 2006; 78(8) 15

[193] Schürch C., Blum P. and Zülli F. Potential of plant cells in culture for cosmetic application. *Phytochemistry Reviews*. 2008; 7(3) 599-605

Chapter 12

From the Olive Flower to the Drupe: Flower Types, Pollination, Self and Inter-Compatibility and Fruit Set

Catherine Breton and André Bervillé

Additional information is available at the end of the chapter

http://dx.doi.org/10.5772/55312

1. Introduction

1.1. Frame of the chapter

Although the tree development directs the olive tree yield we focus here to describe the main phases, stages and key steps from blossoming to harvest. The olive tree produces much more flowers than all other trees, however, fruit set and final harvest are limited by several parameters.

Most articles in literature deal with the physiological aspects of the transformation of the olive flower into the drupe. Little attention has been given to genetic involvements that underlay the physiology and the biochemistry. A comprehensive survey of the literature as much as we can is given with genetic commentaries based upon recent genetic progresses made in the field.

The olive tree derived from the oleaster and the oleaster belongs to an intertropical (Asian, African) genus and a major difference between northern species (*Fraxinus, Syringa, Ligustrum*) is the bare axillary buds at the basis of leaves. This explains why the olive is more susceptible to freeze than other Oleaceae as *Fraxinus*. Olive flower buds appear on one-year-old twig in terminal position or at the axillary of the leaves.

2. Blossoming induction

2.1. Wood bud versus flower bud

There are many disputes in literature to determine when the induction takes place and when differentiation between vegetative and reproductive bud occurs. The first cytological traces of differentiation between the two types of buds occur in February –March [1] in the north hemisphere that infers the physiological signals had triggered the mechanism early, and physiologists agree the induction takes place about the preceding July [2]. Guayneychya [3] from the P. Villemur' team in Montpellier has delimited the stage by sequential removal of fruits in inflorescence, and he showed that it is really induction since without enough cold in fall (November-December) the buds will not turn to flower, but stay vegetative. Conversely, if the load is excessive most buds will not further turn to blossom next year, which is responsible of alternance. All buds will not give a twig or an inflorescence. Some are quiescent without apparent reason. It appears that this depend on the age of the twig and as old it is between 1 and 5 years and the older it is the mower the chance to give a new twig. The main buds will not maintain up to 3 years, but the supernumerary buds survive and may develop. This explains why the external branches of a tree are pending and when it is not pruned old twigs will not give new buds either for a new twig or for an inflorescence. Thus, the specific shape of an old olive tree is due to this mechanism.

2.2. Genetic basis: Gene for architecture, gene for ramification

The natural development of an olive tree from a seedling or a one year cutting leads to shape the tree depending on the genotype. Moutier et al. [4] have recorded different shapes and the tree architecture has consequence on the appearance of the first flower buds. The repartition of the flower buds on a twig the second year is also depending on the genotype: *Lucques* has the lower number of flower buds whereas *Olivière* has the highest number.

The question is therefore whether pruning trees may change the density of flower buds. Pruning has effect to regulate alternance bearing, but not to change the fate of the buds. Deep trimming on an old tree sill suppress old twigs and enables the development of supernumerary buds.

2.3. Environmental factors: Temperature, cold, latitude, diseases

Surprisingly, the main factor that enables the development of inflorescences is temperature. Cold 7.2°C are required to induce inflorescence appearance in April-May as experimented by [1]. A long cold exposure is required to obtain the optimal level of flower buds. There also the effect is genotype dependent. Latitude has not effect on the induction of flower buds. Drought stresses do not affect the proportion of flower buds, but may affect severely fruit set by increasing the proportion of staminate flowers.

3. Types of flowers

3.1. Inflorescence structure

Either short, compact and long is the classic way to describe olive inflorescences. The type is a raceme panicle that as the leaves are at each node both opposite and decussate Obviously, long inflorescence is apparently better for fruit production, but we will see that it is not so obvious. The inflorescence shape is genotype dependent. The number of flowers per inflorescence is between 10 to 30 flowers depending upon the genotype and the position of the inflorescence and the twig that bears it. The olive tree is characterized by bearing three types of flowers: Hermaphrodite, staminate, pistillate. Pistillate flowers – some lack stamens and most of them are hermaphrodite -- and staminate flowers that have no pistil.

Pistillate flowers are completed flowers with 4 sepals and 4 petals, two stamens and two fuse ovaries.

Some flowers may have a pistil without stamens, but they are seldom and probably due to accident in the development. In contrast, staminate flowers that do not have pistil or traces of pistil, are numerous and may be prevalent. They are results of abortion of the pistil, or the pistil was there in the enlage of bud and it degenerate rapidly [5]. They concluded from their study "It is proposed that the main advantage provided by production of staminate flowers in olive is to enhance male fitness by increasing pollen output at the whole plant level, although a relict function of attracting pollinators cannot be completely discarded.».

Varieties have stamens that may abort also more or less rapidly. They may display black anthers as *Lucques*, which is fully male sterile, *Olivière* that displays empty brown anthers and *Tanche* that contains pollen only in one of the stamen bag. Other variety may display more or less yellow pollen (a fine yellow powder escapes from the stamen) [6].

3.2. Function

Literature displays many hypotheses to explain the reasons for two types of flowers. To summarize most hypotheses are based on the limiting nutrient (nutrient economy) to produce more pollen without the load of producing ovules that could not be fed. Wind dispersion of pollen in all species requires a high pollen production and it is logic to think that the olive tree manages high pollen production and it saves hermaphrodite flowers. It is probably not so simple due to the diversity of flower bearing in the different Oleaceae.

Since fruits can be produced only by pistillate flowers the final load of the tree will depend on those flowers. The proportion of hermaphrodite flowers is depending upon several factors that can be organized into a hierarchy as the genotype, and the environment factors such as the position of the branch, the position of the inflorescences on the twig, and the position of the twig, the temperature at the meiosis stage [5].

Hartmann [1] has reported to see olive orchards with only staminate flowers and of course they can't produce any fruits. Thus, it is clear that the environment may determine a part of the trait. However, we noticed that selection of such variety was probably done due to quality of the fruits. However, this example seems exceptional. The amplitude of variation for the trait "ratio staminate/ hermaphrodite flowers" in one tree seems is not so high when choosing a branch with enough inflorescences. As for fruits (1kg fruits are required for oil analysis) the number of flowers to be examined has to be thousands unless the determination will be uncertain. Usually, the terminal flower(s) of one inflorescence is(are) hermaphrodite and the lateral ones may more or less staminate. Furthermore the determination has to be done at different stages following white flower bud stages due to many flowers may fall quickly leaving the hermaphrodite flowers prevalent [7].

3.3. Genetic basis of the proportion between complete and staminate flowers

Only few diversity studies lead to determine the proportion of staminate and pistillate flowers. In the Villemur team, [8] and [9],[10] have determined the proportion of each type of flowers to estimate fruit set based on hermaphrodite flowers only. From their observations the percentage of staminate flowers may vary from 95% as in *Lucques* to 5% as in *Salonenque* with little year variation effect. However, the pollen released from an olive tree is not well correlated with the proportion of staminate flowers, whereas it is better correlated with the size of the inflorescence and the numbers of flowers [11]. [11; 12;13]. Ferrara et al, [14] shows the variation in pollen production and pollen viability between varieties. Damialis et al, [15] show that *O europaea* does not produce more pollen than other tree species. Zafra et al, [16] have studied enzymes that eliminate free oxygen toxic for cells (ROS and NO) in developing anther tissue showing the activity is correlated with receptivity stages.

Consequently, cultivars are ranked based on the proportion of hermaphrodite flowers and in the whole the proportion of inflorescence leading to a fruit is about 1 and therefore hermaphrodite flowers are enough to ensure enough harvest.

Flowering period, amplitude and variation year a year

In a given place the date of full blossoming varies year a year in a range of 6 weeks. Early blossoming is reported to ensure better yield that late blossoming in the people belief. Other factors given below may modulate the harvest yield. Orlandi et al, [17] have studied several factors to model spatial pollen spread in orchards that may help olive growers to locate pollinisers in orchards.

3.4. Expectation from climate changes

As for grape the average date of blossoming is earlier than thirty years back in a given place. But the risk of climatic incident increases with the earliness of the blossoming and statistics are more or less neutral on yields, thus the effects of climate changes affect surely pathogens and pests of the olive. Late blossoming increases the risk of heat shock on flower that may prevent flowers to open, and if they open pollen may be damaged (Villemur P. Pers comm.).

3.5. Cases of related genera in Oleaceae

Wild olive or oleaster

The olive is said andromonoiecious and on wild in nature trees the proportion of complete flowers to staminate flowers may vary in a wide range: 95% staminate flowers, even in some cases, complete absence of hermaphrodite flowers, whereas on some trees hermaphrodite flowers are exclusive [18]. Consequently, from this key data one can infer that selection of varieties along centuries was relatively neutral as for the proportion of staminate flowers, and thus was based upon fruits and oil characteristics

3.6. *Olea europaea* subsp. *cuspidata*

Is a related subspecies to the olive belonging to the same genus. It looks like the wild olive, and twigs of the year before bears different proportion of hermaphrodite and male flowers [19].

In *Olea* and *Fraxinus* genera both anemophilous, flowers appeared before the leaf and after the leaf, respectively, and they both carry hermaphrodite flowers without calyx as corolla and with calyx and corolla, respectively. Besides hermaphrodite flowers made of 1 pistil and 2 stamens. They both develop panicles (inflorescence) with 50 – 400 flowers from lateral one-year-old twigs The olive tree individuals display different proportion of hermaphrodite (95 to 5%) and flowers lacking pistil (staminate flowers) (95-5%).

Other genera such as Phillyrea, Jasminum, and Fraxinus display different interesting and intriguing situations for the proportion of staminate and hermaphrodite flowers. *Chionanthus virginicus* is dioecious, having male (staminate) and female (pistillate) flowers on separate plants. Phillyrea angustifolia displays high proportion of individuals with only staminate flowers whereas the counterpart is hermaphrodite, but among hermaphrodite there are two classes called G1 and G2 that can mate each other and with the male individuals, whereas individuals belong to G1 cannot mate each other as individuals belong to G2 [20.

Fraxinus includes several species. *Fraxinus* individuals are said polygamous because they may even be different flower types within an inflorescence *and* the sex expression of individual trees shows a continuum of gender, i.e. they have either staminate flowers, pistillate flowers, both unisexual flower types, only hermaphroditic flowers, or a combination of uni- and bisexual flowers on different branches. The male flowers for both the olive and *Fraxinus* are always on lateral twigs. The difference between *Olea* and *Fraxinus* is that all the olive trees carry always at least some hermaphrodite flowers.

The present knowledge on flower development leads to believe that staminate flowers are due to pistil abortion determined by genetic regulation plans about one month before blossom [1]. For the olive several authors sustain that nutritional resources are insufficient to make all flowers complete, and that pollen production is prioritized because there is enough hermaphrodite flowers to ensure fruit production [5].

4. The reproduction system of the olive tree

4.1. Pollination

4.1.1. Pollen structure and organization

As for other Oleaceae the pollen grain is slightly ovoid about 20-30 μm in diameter [21]. It is not possible to identify the species which produces pollen grains in archaeological remains. Thus, Archaeologists have frequently noticed 'Oleaceae'. In most cases it is olive pollen. An adult olive tree produces about $2\ 10^9$ pollen grains that give yellow colour to every surface around an olive grove.

Pollen transport is made by winds onto long distances – over 1000 km – but loss in pollen viability is not well analyzed and there are many divergent assertions on time and distance for loss viability. Probably, as estimated by [22] loss in pollen viability occurs after about one hour due to exposition to rays in the atmosphere [23].

Pollen wind role is very important to not only disperse and diffuse pollen grains, but also to transport them fast to reach stigma before they loose their viability. Thus, the pollen physiology is not well understood in the olive tree, and most environmental factors – hot temperature stress, rain, remain to be studied in details to explain variation in fruit set when pollination has appeared deficient [24].

4.1.2. Pollen sterility types are numerous in the olive cultivars

The main type of male sterility is determined by the mitochondria of the CCK cytoplasm [25]. The anthers are empty and if some pollen seems present it is fully sterile, but it may exist some restorer alleles (Rf) encountered in the wild olive trees in North of Africa [26]. In some environmental conditions the olive trees carrying this cytoplasm, which gives the cytoplasmic male sterility (CMS) may produce a few of yellow powder, but the pollen is never functional. *Olivière* a variety from France (Languedoc-Roussillon) displays such CMS. Varieties carrying this CMS are vigorous as Chemlal de Kabylie due to probably to the economy they make in pollen production.

Lucques, a table variety from France (Languedoc-Roussillon) displays empty anthers and is characterized by dark anthers. *Lucques* display a high proportion of staminate flowers and thus the pollination has to be very effective on the few hermaphrodite flowers to ensure fruit sets.

Tanche a table variety from France (Rhone - Alpes Region) displays a partial sterility (about one bag of the anther is full of pollen per flower) that infers olive growers have to help pollination by adding pollinisers varieties that are included in the appellation "Olives de Nyons".

4.2. Pollen and archaeology

The Oleaceae pollen in Egyptian tombs, in sediment of lakes have been widely used to trace the colonisation of the Mediterranean basin by the oleaster [27] and to trace the diffusion of the olive trees by Phoenicians and Romans. However, due to the long displacement by the wind some reserves are made on these studies.

4.3. Pollen and allergy

The most common effect of the olive pollen is its allergenic effect on about 30% of the population, but most people ignore that is due to the olive pollen. About ten allergens have been characterized [28] with a huge diversity in the responsible proteins [29,30]. Many researchers forecast that plantation of the olive trees for decorative purposes in the towns is not neutral to the populations and should be probably limited to reduce the allergenic effects in the future. Many researches dealt with the allergens to understand from an evolution point of view the reason why the Oleaceae have developed so many allergens in the pollen grains (Ash tree, Lilac, Privet, Jasmines, Phillyrea, and so on). Is there interference between allergens and S-alleles in varieties? Although the role of allergens in the pollen is not well understood, their variation described at protein level seems huge and not corresponding to clear cut classes as S-alleles. However, because the physiological mechanism involved in pollen-stigma reaction is ignored yet, it is too early to respond to this question.

4.4. Pollen to forecast harvests

Pollen grains from many plants and in peculiar from the olive tree are trapped in filters from the atmosphere to forecast the epidemic peak in the populations [31]. Moreover, researchers have developed models to forecast olive harvest based on pollen grain density in the atmosphere [14]. The method is running in different countries (Italy, Portugal, Spain and Tunisia) [32].

Indeed, the pollen should land in most of stigma from the olive tree and because there is no attractive effect but only chance that a pollen grain lands on an olive stigma, one may think that is the reason to explain the abundant pollen production by the olive tree. For other fruit trees (Peach, Apple, Cherry) bees and wind can transport the pollen and due to bees visit only flowers from the same species during its journey, the exchanges of pollen are targeted onto the same species. It is not the case for the olive.

5. Style organization

5.1. Stigma structure and organization

The size of the stigma (a cone, see pictures from [33 34] even small for us, is enough feathered once opened to capture about 20,000 pollen grains [14, 35]. The Italian team has made deep studies in pollination between trees in an orchard and their data are very important to understand what may occur at theses stages.

Stigma receptivity is probably a stage difficult to surround because fruit set in the olive tree is very peculiar. In any case, most hermaphrodite flowers will not give fruits whatever the proportion of hermaphrodite flowers. Recently, gene expression studies have throw some light on this key stage showing the role of oxygen bound enzymes that reveal whether the stigma is receptive or not. From literature data, by hand pollination studies the receptivity of the stigma appears to be a week, but some varieties as *Lucques* displays a shorter while by about 3-4 days, that may cause poor fruit set if pollination does not coincide with pollen receptivity.

Pollen stigma interaction revealed strong metabolism intensity. Several Key enzymes have been studied during different stages of pollen stigma interaction [36]. Peroxidase, esterase, and acid phosphatase activities are considered as signs of stigma receptivity. However, data on RNAse activities in the pollen tube growing, although its absence is considered as evidence for pollen compatibility in the frame of the GSI model (see paragraph below), are not convincing since in other species harbouring a GSI type as in Solanaceae, the RNAse activity is omnipresent, even in inter-compatible pollen.

Many questions remain without response on the quality of the pollen that has landed on the stigma. The quality means two thinks: 1) whether the pollen is able to germinate by itself if all other conditions enable it to do so. 2) whether the pollen and the style reaction is compatible because of the self-incompatibility mechanism that exists in the olive. If the pollen - style reaction is not compatible – it is said cross-incompatible- then the pollen tube will be blocked and further destroyed before it reached the embryonic bag.

6. Pollen style interaction

6.1. Physiology in the style

The first role played by the olive style is to hydrate pollen grains to enable them to germinate. Thus, there is probably no regulation by the stigma for pollen germination at this step. However, little is know on which pollen grain may germinate and several observations suggest that several can germinate, but only one reaches the embryo bag region [8, 10, 34]. Temperature has been shown to influence interaction [37]. The olive carries a strong self-incompatibility system, which also exists under different forms in other Oleaceae, and the hermaphrodite and male flower proportion is widely variable both in the crop and the wild [8, 18].

The self-incompatibility locus in other plants directs several genes that enable pollen stigma recognition and then pollen tube growth until the ovule. We would just stress the main features of these systems to lead readers to question about the mechanism in the olive.

6.2. Genetic bases for self-incompatibility and inter-incompatibility

Either one pollen grain harbours on the exine and the pollen tube one of the two determinant of the S-locus (each S-allele specify one determinant: a trans-membrane protein with more or less glycosyl- radicals) it is called the gametophytic type (abbreviated GSI) or the two determinants; it is called the sporophytic type (abbreviated SSI).

The GSI type is present in Solanaceae, Plantaginaceae, Rosaceae and Papilionaceae where the incompatible pollen tube undergone a programmed cell death (PCD) or apoptosis and disappeared in few hours [38, 39, 40]. Researchers working in the olive have a priori (probably due to other fruit trees belonging to Rosaceae harbour such GSI type) considered the olive tree belongs to this type, but in our opinion there is no convincing evidence to sustain this model for the olive tree [41].

The GSI type also exists in some other plant families without RNAse activity as in Papaver [42], and the IIC reaction occurs on the style in one second due to a strong flux of calcium. In Gramineae [43], Papilionaceae and Chenopodiaceae with the GSI types the mechanisms does not involved an S-RNAse, and little is known in these species. Many S-alleles have been sequenced in *Prunus* species (Rosaceae) [44] and if the size of the S-allele varies between 100 to 350 Kb, for the allele expressed in the pollen as to the one expressed in the style side, little is understand on the exact reaction leading to ICO and IIC between the pollen and the style.

The GSI model has genetic involvements that have to be exposed:

1. a *SvSw* plant cannot self-pollinate
2. it cannot be pollinated by pollen grains carrying either Sv or Sw.
3. the mutation of *Sv* or *Sw* to *S0* may lead to a self-compatible allele. S0 is the conventional name given to a self-compatible allele.

The SSI type exists in Brassicaceae and Asteraceae. The pollen grain harbours the two determinants, thus leaving dominance relationships between the two S-alleles. We would stress that even if the pollen grain genome carries only one of the two alleles (*Sx* or *Sy*) on the surface of the exine and the pollen tube both S-allele products (Px and Py) are present on the pollen coat leading to the self-incompatibility reaction if pollen and style carry the same alelle. In Brassicaceae, the self-incompatible pollen grain triggers a cascade of events leading to PCD in a few hours. This cascade involves a protein kinase with a SRK-motive (SRK-protein kinase) that add a phosphorus to a template protein whereas if self-compatible there is no reaction [45].

The dominance relationship between two S-alleles makes (as example *Sx*>*Sy* means *Sx* dominant over *Sy*) that the recessive determinant of the pollen is hinted and consequently the same determinant on the female side on the the stigma or in the style cannot pair with it, that lead to its acceptance as compatible. This feature makes the genetic involvements of the SSI model much complex than with the GSI model [46].

1. A SxSy plant cannot self when *Sx* and *Sy* are both equivalent.
2. However, when *Sx* is dominant over *Sy*, *SxSy* can self-pollinate.
3. The mate between ♀ *SySz* and *SxSy*♂will be inter-incompatible (IIC) with Sy=Sx and inter-compatible (ICO) with *Sx*>*Sy*. Consequently, dominance relationships favours cross-compatibility, but in the other direction ♀*SxSy* and *SySz*♂ with *Sy*=*Sz* the cross will be IIC whatever the dominance relationships between *Sx* and *Sy*.
4. As the intensity of dominance may vary between pair wise combination of S-alleles from 0 (no dominance) to 1 (full dominant), consequently the intensity of self-pollination varies from 0 (no self-pollination) to 1 (self-pollination equivalent to free pollination).
5. The self-incompatibility behaves as a quantitative trait and consequently, one must define a threshold to rank varieties between self-incompatible (S-I) and self-fertile (S-F) [47].
6. Others involvements will be considered in the fertilization section.

Several recent reviews on GSI and SSI models can be consulted by readers [49, 45].

The GSI model infers that the pollen must germinate and start pollen tube growth before undergone PCD if it is S-I or IIC. Pollen grain germination observations sustain this view, except in the *Papaver* model where the incompatible pollen grain does not germinate. The SSI model infers the expression of the SRK-kinase is constitutive in the style and thus the pollen tube may not grow longer and disappears fast. The recognition of the pollen tube and its growth should enable to state whether the reaction is incompatibility versus compatibility. In practice, it is not so obvious due to all pollen grains do not behave as the same way [50, 51, 52, 53].

6.3. Methods to estimate pollen stigma interaction

The microscope method to follow pollen tube growth till the ovule requires specialization of peoples to make observation, it is time consuming and thus expansive [51, 8, 10]. For seed plants such as sunflower self-sterility is measured by surrounding one head with a pollen-proof paper bag and leaving another head free pollinated. The ratio of both fruit sets enables to estimate S-I versus S-F [54]. For the olive the paper bag method runs efficiently, but the number of flowers under the bag should be consistent with significant statistics on fruit set and the paper pollen proof should be of good quality [55, 56]. In fact, the main concern for researchers is the source of the pollen, which can be in the orchard or much far and thus the pollen requests to be moved. Dry pollen appears to loose germination ability fast, whereas after freezing it may be kept over months [57].

Many studies have been undergone around the stage of the style waiting for the pollen and on the pollen waiting to land on the stigma. One original study by [58] dealt with calcium behaviour in the stigma, the style and around the embryonic bag. The calcium accumulation is probably correlated with the pistil maturity and receptivity.

6.4. Advantages and drawbacks

The advantages of the paper bag method are that several controls can be performed in one orchard, and thus, the number of parameters under variation is much estimated. However, to move pollen requests several controls to verify whether the pollen does not suffer from transportation. It is not so easy to verify and thus this point enhances the number of request controls. Villemur recommended freezing pollen, others prefer to move pollen in paper bag, or to move branches. All these methods are not neutral versus results and may explain the wide range of answers on crosses between olive varieties.

In any way the self-incompatibility is quantitative and thus requests to define thresholds to rank varieties. Probably, literature records on self-fertility and self-incompatibility on the one hand and inter-compatibility and inter-incompatibility between given pairs of varieties diverged due to position of thresholds that are not given in records. To be accurate it is required to add the threshold.

6.5. Unravelling S-alleles composition in the olive

Breton and Bervillé [46] have attempted to fill the gap for some varieties between published tables on successes and failures on self-fertility and self-incompatibility and for mates of pairwise combination of varieties containing + and – or 0, by translating the signs by S-alleles. Obviously, the attribution is relative as long as the alleles have not been sequenced. Moreover, before reaching a satisfying model we followed plenty of wrong tracks. As an example, we firstly supposed that self-fertile varieties (*Bouteillan, Cornicabra,* and *Verdale de l'Hérault*) contains one null-allele (S0) as in *Prunus* to lead to self-fertility.

Moutier's and co-workers have made the work to find better pollinisers [55, 56] – the word pollinators is devoted to animals (insects, mammals, and birds) that pollinate plants – to *Lucques* and *Olivière*. These varieties do not produce any pollen grain and should be pollinated by other varieties to ensure fruit set [25]. Consequently, the mates were not equilibrated between varieties, but due to the number of varieties involved (13), we considered this set of varieties pertinent to start the translation of + and – in S-alleles. Moreover, the series of varieties that did not mate *Lucques* was informative to state that the varieties should carry one allele at least in common to *Lucques*.

Male → / Female ↓	MS	*Amygdalolia*	*Cayon*	*Amellau*	*Tanche*	Cornicabra	*Salonenque*	*Bouteillan*	*VerdaleH*	*Arbequina*	*Aglandau*	CMS	*Picholine*	*Grossane*	Manzaniilla	Cailletier
Lucques		-	+	-		-	-	-	-	-			-	-	-	+
Amygdalolia		SI		+		-										⊕
Cayon		+	SI			+		+		-			-	-	+	
Amellau			+	SI					+				-			
Tanche			+		SI			⊙			-		⊙	-	-	
Cornicabra		-				SF										
Salonenque							SI			⊕	+		-	-	+	
Bouteillan			+		⊕			SF	⊕	+	-		+	+	-	
VerdaleH								⊙	SF				+	+		
Arbequina			-				⊙		+	SI			-	-	⊕	
Aglandau			+					-		+	SI		+	⊙	-	⊙
Olivière		-	+				-	-	-	+		nd	+	-	-	
Picholine			-		⊕		-	+	+	-	+		SI	-	⊕	
Grossane			-				-	+	+	-			-	SI	-	
Manzanilla			+				+	-		⊙	-		⊙	-	SI	
Cailletier		⊙									⊕					SF

For displaying symmetry in the table, *Lucques* and *Olivière* being male sterile their column as male is left empty.

Table 1. Rewritten from N. Moutier et al. [55,83] and Musho[8] with symmetric lines (female) and columns (male). All +- were converted into +. Cross in one direction only+, -; ⊕, ⊙: Crosses in two directions but dissymmetric fruit sets ; Symmetric fruit sets - -; or + + MS= male sterile; CMS cytoplasmic male sterile. Cases with dissymmetric fruit sets were pink circled. N.d. means not determined. SI and SF means self-incompatible and self-fertile, respectively.

Thus, the code we choose resulted from Gerstel' experiments [50] on Guayule (*Parthenium argentatum,* Asteraceae). Mates between two varieties involved 4 S-alleles coded *R1* to *R4*. However, there are 6 possible combinations of these 4 alleles leading to draw an hexagon with the 6 pairs at the 6 poles. The first step was to define the dominance relationships between the four alleles satisfying the + and 0 cases from the data table [8, 9, 55, 56]. To do this step we based on pair wise combination of mates which leads to opposite fruit sets in both directions of crosses. The other important information comes from the variety *Lucques,* which has little efficient pollinisers and therefore was hypothesized carrying 2 S-alleles almost equivalent (*R2*=*R3*), but dominant over *R1* and *R4*.

We constructed table 1 from Moutier's data to separate all mates that gave similar fruit sets (either +,+ or -,- in both directions) and those that gave opposite fruit sets (either +, - or -, + in both directions). Surprisingly, the same varieties were involved in differential fruit sets, this infers their genetic composition for the S-alleles had specific features that remain to be solved.

Based on the Gerstel's model the differences of fruit sets in reciprocal mates were due to the recessive alleles (here *R1* or *R4*) which are present in the pair [50]. *Cayon* was the only variety that mates *Lucques*: we attributed the pair *R1R4* with a priori *R1*=*R4*. Furthermore, we considered the pair *Manzanilla - Picholine,* we attributed the *R1*-allele to both because they gave reciprocal differences with *Cayon* that carries *R1* (it explains reciprocal differences) and *R2* to *Manzanilla* infers *R3* in *Picholine* because they cannot mate *Lucques* (*R2R3*). Once done this first translation, all the others result from this one for the four alleles *R1* to *R4*. Obviously, we were surprised to find *Tanche* with the same alleles as *Lucques* and that the four S-alleles explained the behaviour of eleven of the thirteen varieties.

Varieties studied[12]	Other studies	S-S/S-F	S-locus	Best Polliniser
Manzanilla, Amygdalolia	*Manzanillo*[51-35], *Manzanilla de Sevilla*[10], Picual[35], *Pendolino* [35], *Amygdalolia* [53]	S-S	*R1R2*	*R4R4-R5R5*
Picholine, Arbequina	*(Barnea, Corregiola)*[10]	S-S	*R1R3*	*R4R4, R5R5*
Cayon	*Corniale*[8], *Koroneiki*[9,10]	S-F	*R1R4*	*R5R5*
Grossane		S-S	*R1R5*	*R4R4*
Lucques, Tanche		S-S	*R2R3*	*R1R1,R4R4, R5R5*
Bouteillan, VerdaleH, Olivière, Cornicabra	*Kalamata*[9, 35], *(Mission, Kalamata, a-Pendolino, a-Kalamata, Verdale, UC13A6, King-Kalamata, Katsourella)*[78]	S-F	*R2R4*	*R1R1-R5R5*
Aglandau		S-S	R2R5	*R1R1*
Belgentier[13]	*Sevillano*[24]	S-F	*R2R6*	*R1R1, R4R4*
Amellau	*Ascolano-Tenera,Frantoio*[35]	S-F	*R3R4*	R1R1
Salonenque		S-F	*R3R5*	*R1R1-R4R4*
Cailletier	*Cailletier*[8]	S-F	*R4 R5*	*R1R1*

Table 2. Summary of *S*-alleles attributed to 39 olive varieties

To go further with the other varieties *Aglandau, Grossane,* and *Salonenque,* we attributed a new S-allele *R5* and we computed all dominance relationships with the other alleles to fit experimental data. Rapidly, we realized that the model was predictive. To attribute S-alleles to *Aglandau,* because the mate *Lucques* x *Aglandau* is lacking we considered the cross *Tanche* x *Aglandau* that failed and revealed that *Aglandau* carries *R2,* consequently *Aglandau* is *R2R5.* *Picholine* and *Grossane* gave similar patterns with all varieties except with *Aglandau* that sustains they carry *R1R3* and *R1R5,* respectively. *Salonenque* and *Grossane* have also similar

http://dx.doi.org/10.5772/55312

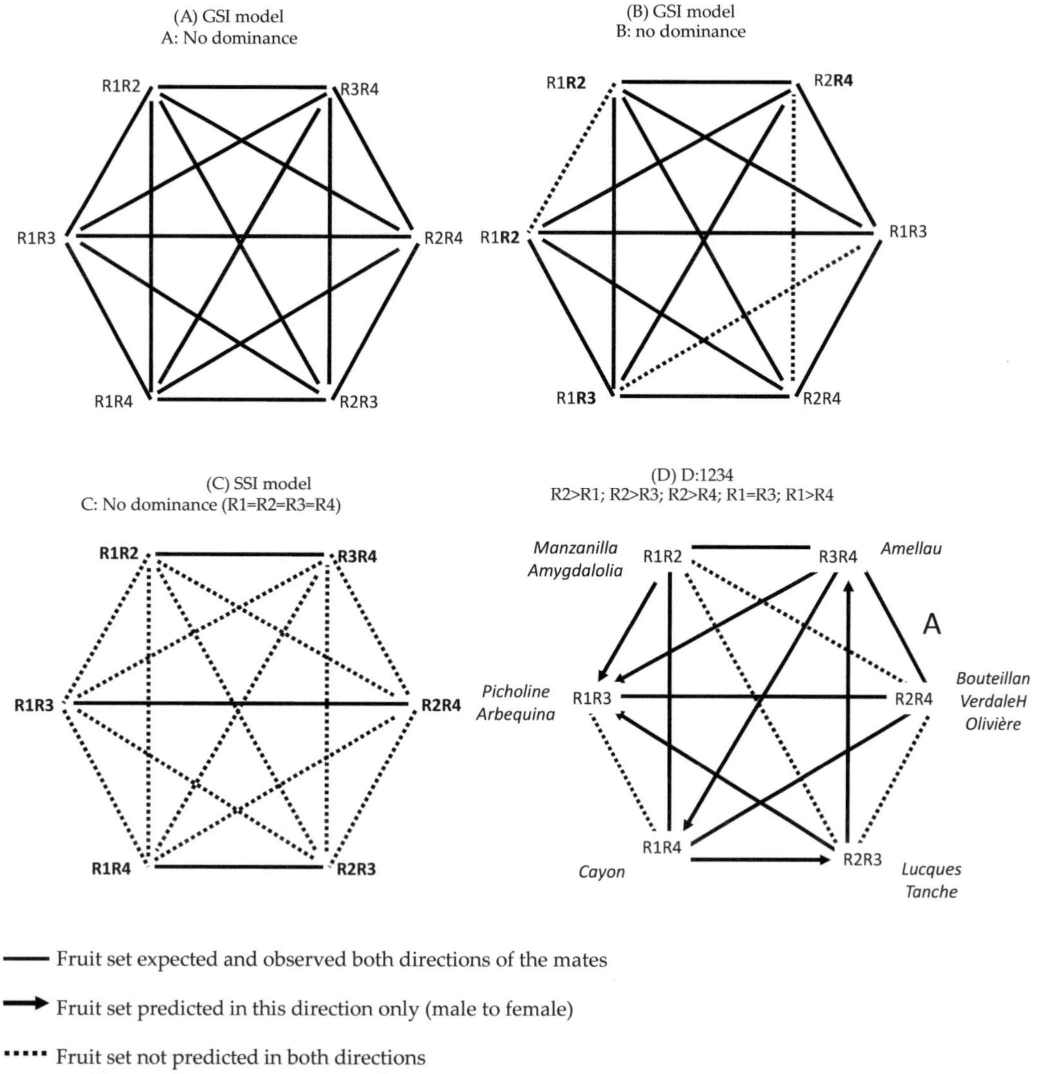

Figure 1. A: Prediction of fruit set for the six pair wise combinations of the R1, R2, R3 and R4 S-alleles in the frame of the GSI model. B: examples of mates that succeed or failed in the frame of the GSI model. C: Prediction of fruit set for the six pair wise combinations of the R1, R2, R3 and R4 S-alleles in the frame of the SSI model without dominance relationships between the S-alleles. D: application to data from Moutier et al.12

patterns except with *Cayon* and *Manzanilla*, which have *R1* in common, thus *Salonenque* carries *R3R5* and *Grossane R1R5*. Even if *Cailletier* has not enough mates, because it mates *Lucques* it cannot carry *R2* or *R3*. However, *Cailletier* gives reciprocal differences in fruit set with *Aglandau* (*R2R5*), consequently *Cailletier* should carry a recessive allele in front of R5. *Salonenque R3R5* cannot self-pollinate as intensively as *Cailletier*, thus we decided to attribute *R4R5* to *Cailletier*. It mates *Amygdalolia* (*R1R2*) [8] that is in agreement with *R4* and *R5*. *Cailletier* is *Taggiasca* in Italy. We should apply the parsimony principle to not proliferate S-alleles. Musho [8] has also shown that *Belgentier* can mate *Manzanilla* as *Sevillano* [51] although *Belgentier* cannot mate *Lucques*. All together these data show that *Belgentier* carries *R6* with *R6>R2*, and this allele is also present in *Sevillano*.

The model was constructed on crude data provided by different authors without taking account that in controlled crosses fruit sets result from both self-pollination and cross-pollination. Depending on the pair of S-alleles the self-pollination rate varies. Now, it is possible to experiment more accurately taking into account the self-pollination rates determined under paper bags and to deduce this rate from fruit sets (Table 2, Figure 1).

7. Interests of the results

Breton & Bervillé [46] have listed some other cultivars (see Table 5 in [46]).Wu [59] provided a complete diallel design for 5 varieties, but they carry only 3 pair wise combinations of S-alleles, however, they enabled to us to attribute *R1R2* to *Pendolino* and *Picual R3R4*. Many other publications provide mate results between olive varieties. However, authors did not provide enough information on mates (threshold, fruit sets,) making difficult the translation of the data in S-alleles.

Probably the list of varieties with deciphered S-alleles will grow rapidly for the benefit of researchers and olive growers, firstly, to enable diallel design between compatible varieties to check the importance of other factors (coincidence in blossoming, stigma receptivity, ovule receptivity longevity, other compatibility troubles,..) and to check at the style level whether all compatible pollen grains may germinate or if only some may germinate. These questions raise from eventual competition between pollen grains. Furthermore, the model enables to choose pollinisers a priori for a set of varieties.

Among about 20,000 pollen grains that land on a style many questions are still without answers. How many came from an inter-compatible tree [33, 60, 61, 62]. Now it is possible to design experiments to answer to such questions and to check whether competition between pollen grains does exist, not only between inter-incompatible and inter-compatible, but merely between inter-compatible pollen grains. Many observations sustained that only one pollen tube is detected in the style nearby the embryonic bag and only one reaches the ovule. This mechanism is not due to inter-incompatibility, but to some competition mechanism, which may screen among pollen grains based on still unknown parameters.

8. Fertilization

8.1. Strange features

The first feature at this stage is that even the distance between the style and the ovule is short, probably by less 2 mm, it is surprising that the pollen tube may reach the embryonic bag between 2 to 6 days after pollination [8, 9, 51, 36, 41]. In other species the pollen tube growth is fast as in *Petunia* fertilization occurs in the next two days and the style is very long (2-3 cm) in comparison to the olive. Probably, there are some regulation steps to unravel at this level.

The second main feature is the poor efficiency in fruit set in comparison to the number of hermaphrodite flowers [63, 64, 65]. Which is the mechanism which screens among the new fruits to cause the fall of some? Only nutritional resources allowance by the position in the twig, the branch, the side of the tree, and so on seem difficult to admit. We may suggest that the delay in fertilization may cause later fruit fall, and that the early fertilized ovules are more solid that those fertilized later.

8.2. Fruit fall

Fruit fall is erratic from early fertilization and for 6-8 weeks, thus it is important to notice fruit sets under paper bags after this period unless the risk is to overestimate fruit sets. Fruit falls seems programmed in the olive and thus we suggest a working hypothesis: is there linkage between the S-locus and lethal genes leading to seed abortion?

In most species with high heterozygosity levels revealed by many studies with molecular markers to study the neutral genetic diversity, self-pollination decreases drastically not only the vigour of the plants but also the fitness that is the number of descents per female (Maize, Sunflower, other out-crossing species). The self-incompatibility model in the olive admits self-fertility in most cultivars that may enhance homozygosity in progenies. Consequently, since we have not revealed yet homozygosity at the S-allele locus, we suggest this mechanism to eliminate these individuals.

9. Fruit development

We just address here which consequence may have the fertilization process on fruit development, but not all the aspects in quality of the compounds that did not result from the compatibility mechanism.

9.1. Different phases

The increase in size of the drupe is due to two different mechanisms. The young drupe accumulates water and cells are swollen, but also cells proliferate rapidly. The first sign in fruit differentiation is the hardening of the stone because sclerenchymous cells appeared. This step is very important about 50 days after fertilization because most nutritional resources are mobilized by this stage and all other parameters slowdown at this stage. Pulp development occurs then and oil starts to accumulate [66]. Oil content as oil composition is determined by the variety with little influence of the environment whereas other compounds (Sterols, phenols,) are much dependent of the environment [67].

9.2. Parthenocarpic fruits (Shotberry)

Self-pollination and cross-pollination modify the level of seedless fruits bore by most varieties [68]. The yield in shotberries is due to self-incompatibility and or inter-incompatibility and thus it is a sign that pollination is not optimal in the orchard. Some varieties display commonly different fruit size that result from fertilized and shotberries.

9.3. Stone and embryo number

The stone contains one embryo (seed), but some may contain no seed and some contains two seeds. Studies have shown that the average weight of fruits with two seeds is higher that those with one seed and those are higher that those with no seed. The seed number is variable between varieties and may indicate more or less that not enough inter-compatible pollen has landed on the stigma [69, 70]. By choosing now a priori inter-compatible varieties, it becomes possible to determine whether this suggestion is true or not.

The embryo number is an important parameter to forecast whether the drupes will fall easily before harvest or not. The more attached drupes contain two embryos [70]. If so, to enhance efficient pollen grain number in a monovarietal orchard could be obtained by enhancement of the adequate polliniser number. In the frame of the SSI model, the deficit in efficient pollen is crucial and if adequate pollen is lacking the orchard will never produce optimal yield. [20] have suggested based on fruit sets that *Picual* pollination is deficient in solid orchards under study, but they experimented that enhancement of cross pollination did not increase fruit set.

We suggest to them to reconsider the situation by reckoning the trees producing efficient pollen grains in the frame of the SSI model. Because *Picual* is *R1R2*, based on [59] data it is probably a bit self-fertile (S-II 0.21), the rate could be calibrated in different environments (varied temperature, watering/not watering, solid: mixed orchards) in Andalusia. However, *Picual* cannot be pollinated by any pollen grain harbouring the *R2* product (we suggest to call it P2) on its surface – that means it was produced on a variety carrying *R2* (as *Manzanilla*) and *R1* (as *Picholine R1R3* and *Cayon R1R4*), but this infers R1>R4, unless *R1* is masked by a dominant allele as in *Salonenque R3R5--*, but *Picual* is mated by *Arbequina* (R1R3) and *Cornicabra* (*R2R4*). Even with dominance as *R2 >R1*, there is debate for *Manzanilla* that is given as self-fertile in Spain, but behaves as S-S in most other countries [71, 72] see discussion in [46, 73, 74]. Supposing it is the same *Manzanilla* carrying the same allele pair, [51]Bradley & Griggs (1963) have shown the effect of high temperature on the expression of S-I in this variety. This may explain the variation observed for a tree for the different quarters exposed to north and south as examples [37].

10. Storage compound accumulation

The different genetic origins of the flesh and the embryo make deep differences in compounds accumulated in these organs. The drupe tissues are of maternal origin and thus they all have the same genetic composition. Oil accumulated in the flesh has the same composition for all drupes of a variety (no genetic variance), but it may exist environmental effects. For phenol compounds the variation is important and is exploited through the appellations (AOP, DOP,...).

In contrast for the embryo, the genetic variability exists between fruits from two origins: firstly from the allele segregation of the mother plant, each ovule has a specific genetic composition, and second the diversity comes from the pollen grain genome, each pollen grain has also a specific genetic composition.

If the commercial olive oil should have between 55% to 85% of oleic acid, the oil in the embryo has usually less content in oleic acid (about 30%) and the major fat is the linoleic acid as in the sunflower oil. However, the fatty acid are not free but each esterifies a alcohol radical from glycerol and the different isomers are used to recognize appellations [68, 75].

At maturity the fruit composition are none influenced by the incompatibility system and are not addressed here.

11. Paternity tests used to unravel S-I mechanisms

Researchers have tried to determine the father (the tree giving the pollen) of the fruits of one variety to look for better pollinisers. The problem is present and the molecular markers, now common to identify olive varieties, could probably enable such approach. The method in the olive has been initiated by [76,77, 78, 79], and then other publications have tried to identify the father in the progenies of different varieties, in Spain [80, 81], Italy (Unpublished)and France [82] (AFIDOL, Unpublished).

The meaning of the method is to identify the father, which has given some specific markers not present in the female tree, by its molecular pattern discriminate among those from all other olive cultivars.

11.1. Molecular markers

The tools to discriminate the father among other olive varieties are microsatellite markers widely developed for olive variety differentiation. To do that, the method consists to isolate seeds from fruits harvested on the female variety, to enable embryos to germinate in a seedling. Then for each seedling the DNA is prepared and each DNA profile using a series of microsatellite markers is recorded.

11.2. Significance, physiology and genetics

The method infers that all the possible fathers around the female tree have been analyzed with the same series of molecular markers. Consequently, the method is restricted to favourable environment, with a limited number of putative fathers.

The method is based on the effective pollination and further fertilization of the female by the male, which introduce a deep bias in the results, because the position of the male tree in the orchard can make the pollination easiest (under wind direction as example) than for other trees.

11.3. Drawbacks

Consequently, all found fathers in the progenies are surely inter-compatible with the female tree, but in anyway there is no response for all trees that had not given progenies with the female tree. The analysis of paternity data is therefore a bit frustrating for researchers, due to all negative results have no meaning.

12. Main research tracks

Architecture key genes that direct the shape of the trees are of interest to understand the best way to prune the trees. Their identification requires a specific genetic design, namely, to set up a progeny of about 200 trees between two varieties that display different shape without pruning. For the olive tree this step requires several tens of years.

Unravelling pollen-stigma interaction remains the main objective to handle accurately the R-alleles. Several approaches should be performed simultaneously, to enhance the chances to succeed in discovering this genes. Probably, international programs will sustain such tracks for the benefits of all.

13. Conclusions

13.1. Self-sterility and allergens

Crossing data from olive allergens (ten families) and from R-allele types (ten RxRy combinations), we did not find correlation that infers probably the genes that directed both mechanisms are not linked in the olive genome.

13.2. Male sterility and self-sterility

Many olive varieties do not produce pollen (*Farga,* Spain; *Chemlal de Kabylie,* Algeria, Lucques, Olivière,...), and some are deficient in pollen production (*Tanche,* France). These varieties cannot be used as male and their uses in genetic analyses to determine which S-alleles they carry may leave some doubt on the results, unless the varieties are used in a large number of crosses as *Lucques* and *Olivière* by [55,56, 83].

13.3. Looking for new pollinisers

Mates are systematically done with the same variety as female that may decrease the efficiency in identifying the S-alleles it carries. Consequently, we recommend to design crosses in both directions as much as it can be done.

13.4. Self-sterility an inter-incompatibility

Although the genetic of the system is simple (one locus with a few alleles), the mode of expression of the SI trait is very complex leading to enable mates even when parents carry the same alleles. Future research in this direction is probably to isolate the genes and proteins involved in the mechanism [84]. However, experiments to breed the olive are required [85].

We highlight in this chapter that most consequences still unrelated between most olive fruit development troubles come from the complex reproduction system of the olive tree. More details for each point addressed could be found in [86].

Acknowledgments Thanks are due to Pierre Villemur and Nathalie Moutier for helpful suggestions and discussion. This work was supported by the project "Patermed" coordinated by Stephane Anglès UMR 7533 LADYSS in the frame of the call ANR Systerra 2011-2014.

14. Summary

The olive tree displays specific features from the production of different flowers either hermaphrodite or staminate to maturation of fruits. Both types produce pollen, but more or less efficient as for female recipient. All steps are examined from a physiological and a genetic point of view to enable readers to well comprehend the key steps in flower and fruit production. The genetic system underlying the compatibility between pollen and stigma is explained in the frame of a new model, as well as the consequences for looking to polliniser trees to most varieties. Consequences in fruit set, seed development as seed fall are examined. The olive fruit is made of two parts one of maternal origin, the flesh, whereas the embryo has different genetic origins that explain some features in oil composition variation in different environments. Several hypotheses are suggested to explain the different specific features of the olive tree.

Author details

Catherine Breton[1,3*] and André Bervillé[2*]

*Address all correspondence to: catherine.marie.breton@gmail.com catherine.breton@supagro.inra.fr

*Address all correspondence to: berville@supagro.inra.fr

1 INRA, TGU-AGAP, Équipe DAVEM, 2 place Viala, Bât 21, Montpellier, France

2 INRA, UMR 1097 DIAPC. 2 place Viala, Bât 33, Montpellier, France

3 Institut des Sciences de l'Evolution de Montpellier (ISE-M), UMR CNRS 5554 Place E. Bataillon, cc63, Bât 24, Montpellier, France

References

[1] Hartmann HT. 1950 Olive flower bud formation California, Agriculture, November, p4.

[2] Loussert R, G. Brousse. L'olivier. Chap 3, Maisonneuve et Larose, Paris, 1978, pp459.

[3] Guayneychya O. Contribution à l'étude des phénomènes de croissance et de développement de l'olivier. Thèse USTL, Univ-Montp 2, p154. 1983

[4] Moutier N Villemur P Calleja M. Un olivier, comment ça pousse ? In L'Olivier l'arbre des temps Eds C Breton & A. Bervillé, Quae, Versailles. 2012.

[5] Cuevas J, Polito, VS. The role of staminate flowers in the breeding system of Olea europaea (Oleaceae): an andromonoecious, wind-pollinated taxon. Ann. Bot. 2004; 93: 547-553.

[6] Besnard G., Baradat, P., Breton, C., Khadari, B., Bervillé, A.. Olive domestication from structure of oleasters and cultivars using nuclear RAPDs and mitochondrial RFLPs. Genet. Sel. Evol. 2001; 0: 1-19.

[7] Seifi, E., Guerin J., Kaiser, B., and M. Sedgley, Inflorescence architecture of olive. Scientia Horticulturae 2008 ;116: 273-279.

[8] Musho U.-B. Contribution à l'étude de la biologie florale de l'olivier Olea europaea L.: mise en évidence de cas de stérilité mâle et recherche de pollinisateurs. USTL-Montpellier. 1977.

[9] Agrolive Pendolino Flower www.hotfrog.fr/Entreprises/AGROLIVE.

[10] Ouksili A. Contribution à l'étude dela biologie florale de l'olivier Olea europaea L. de la formation des fleurs à la pollinisation effective. USTL, Univ-Montpellier 2, 143p. 1983.

[11] Lavee, S. and Z. Datt.. The necessity of cross-pollination for fruit set of 'Manzanillo'298 olives. J. Hort. Sci. 1978; 53:261-266.

[12] Hartmann HT. Studies on self- and cross- pollination of olives. California, Agriculture, March, p4; 1961.

[13] Hartmann HT, Opitz KW. Olive production in Callifornia. Division of Agricultural Sciences, University of California, California, U.S.A., 64 p. 1966.

[14] Ferrara G, Camposeoa S; Palascianoa M; Godini A. Production of total and stainable pollen grains in Olea europaea L Grana 2009; 46: 2, 85 — 90.

[15] Damialis A, Fotiou C, Halley JM., Vokou D. Effects of environmental factors on pollen production in anemophilous woody speciesTrees (2011) 25:253–264; 2011.

[16] Zafra et al. BMC Plant Biology 2010, 10:36 http://www.biomedcentral.com/1471-2229/10/36

[17] Orlandi F, Ferranti F, Romano B, Fornaciarai M. Olive pollination: flowers and pollen of two cultivars of Olea europaea. NZ J Crop Hort Sci. 2010 ; 31:159-168.

[18] Hannachi H., S. Marzouk 2012 Flowering in the wild olive (Olea europaea L.) tree (oleaster): phenology, flower abnormalities and fruit set traits for breeding the olive. African J. Biotechnology 2012; 11: 8142-8148.

[19] Mukonyi Kw, Kyalo NS, Lusweti AM, Situma c, Kibet s Olea europaea subsp. cuspidata in Kenya http://adanimaldesign.com/rfga/fruit/glos179.html.

[20] Saumitou-Laprade P, VerneT P, Vassiliadis C, Hoareau Y, de Magny G, Dommée B, Lepart J. A Self-Incompatibility System Explains High Male Frequencies in an Androdioecious Plant. Science 2010. 327: 1648 - 1650.

[21] Pacini E. Juniper BE.The ultrastructure of pollen grain development in the olive(Olea europaea) 1. proteins in the pore. New Phytol 1979. 85, 157-163.

[22] Cuevas J, Pinillos V, Polito VS. Effective pollination period for 'Manzanillo' and 'Picual' olive trees. J. Hort. Sci.& Biotech. 2009. 84(3):370-374.

[23] Pinillos V, Cuevas J. Open-pollination Provides Sufficient Levels of Cross-pollen in Spanish Monovarietal Olive Orchards. HortScience 2009. 44:499-502.

[24] Bellini E, Giordani E, Rosati A. Genetic improvement of olive from clonal selection to cross-breeding programs, Adv. Hort. Sci., 2008. 22(2): 73-86.

[25] Besnard G, Khadari B, Villemur P, Bervillé A. A Cytoplasmic Male Sterility in olive cultivars Olea europaea L.: phenotypic, genetic and molecular approaches. Theoretical and Applied Genetics 100 2000. 1018-1024.

[26] Hannachi H, Breton C, Msallem M, Ben El Hadj S, El Gazzah M, Genetic Relationships between Cultivated and Wild Olive Trees (Olea europaea L. var. europaea and var. sylvestris) Based on Nuclear and Chloroplast SSR Markers Natural Resources, 2010, 1, 95-103.

[27] Carrión Y, M Ntinou, Badal E. Olea europaea L. in the North Mediterranean basin during th Pleni-Glacial and the Early-Middle Holocene. Quaternary Science reviews 2010. 20:952-968.

[28] [28 Hamman-Khalifa A, Castro AJ, Jiménez-López JC, Rodríguez-García MI, Alché JD Olive cultivar origin is a major cause of polymorphism for Ole e 1 pollen allergen. BMC Plant Biol. 2010; 2008; 8: 10.

[29] Jiménez-LópezJ C, Rodríguez-García MI, Alché JD. Systematic and Phylogenetic Analysis of the Ole e 1 Pollen Protein Family Members in Plants Systems and Computational Biology– Bioinformatics and Computational Modeling Chapter 12 245-260; 2009

[30] Jaradat ZW, J, Al Bzour A, Ababneh Q, Shdiefat S, Jaradat S, Al Domi H. Identification of allergenic pollen grains in 36 olive (Olea europaea) cultivars grown in Jordan. Food and Agricultural Immunology 2011:

[31] Ribeiro H, Cunha M, Abreu I. A bioclimatic model for forecasting olive yield, J Agri Sci 147 2009; 647-656.

[32] Avolio E, Pasqualoni L, Federico S, Fornaciari M, Bonofiglio T, Orlandi F, Bellecci C, Romano B. Correlation between large-scale atmospheric fields and the olive pollen season in Central Italy Int J Biometeorol 2008; 52:787–796]

[33] Tombesi A., - Biologia fiorale edifrutticazione, pp. 35-55. -In: Fiorino P. (ed.) Olea. Trattato di olivicoltura. Edagricole, Bologna, pp. 461, 2003.

[34] Olmedilla, A., Symposium and Training Course IV: Challenge for Plant Breeding and the Biotech Response, AGRISAFE 203288, EU-FP7-REGPOT 2007-1. Granada, Spain

[35] Ganino T, Rapoport H. Fabbri F. A Anatomy of the olive inflorescence axis at flowering and fruiting(1). Flowers and pollen of two cultivars of Olea europaea, New Zealand Journal of Crop and Horticultural Science, 31:2, 159-168.

[36] Serrano I, Olmedilla A. Histochemical location of key enzyme activities involved in receptivity and self-incompatibility in the olive tree (Olea europaea L.) Plant Science 2012 IN PRESS

[37] Griggs WH, Hartmann HT, Bradley MV, Iwakiri BT, Whisler JE. Olive pollination in California. Division of Agricultural Sciences, University of California, California, U.S.A., Bulletin 1975. 869, 50 p.

[38] Wu H-M, Cheung AY. Programmed cell death in plant reproduction. Plant Mol. Biol. 2000. 44, 267–281.

[39] Franklin-Tong V E, Self-Incompatibility in Papaver Rhoeas: Progress in Understanding Mechanisms Involved in Regulating Self-Incompatibility in Papaver. In Self-incompatibility in flowering plants: evolution, diversity and mechanisms. Springer, Berlin, pp 237-258. 2008.

[40] Kubo K, Entani T, Takara A, Wang N, Fields AM, et al. Collaborative Non-Self Recognition System in S-RNAse–Based Self-Incompatibility DOI: 10.1126/science. 1195243, Science 2010. 330, 6005: 796-799.

[41] Serrano I, S. Pellicione, Olmedilla A. Programmed-cell-death hallmarks in incompatible pollen and papillar stigma cells of Olea europaea L. under free pollination, Plant Cell Reports 29 2006; 561-572.

[42] Bosch M, Franklin-Tong V E, Sef-incompatibility in Papaver: signaling to trigger PCD in incompatible pollen. J. Exp. Bot. 2007; 59 481-490.

[43] Kakeda K, Toshiro I, Suzuki J,Tadano H, Kurita Y, Hanai Y, Kowyama Y. Molecular and genetic characterization of the S locus in Hordeum bulbosum L., a wild self-in-

compatible species related to cultivated barley. Mol Genet Genomics 2008; 280:509–519.

[44] Sassa H, Kakui H, Miyamoto M, Suzuki Y, Hanada T. Ushijima K, Kusaba M, Hirano H, Koba T: S Locus F-Box Brothers: Multiple and Pollen-Specific F-Box Genes With S Haplotype-Specific Polymorphisms in Apple and Japanese Pear. Genetics 2007; 175:1869-1881

[45] Takayama S, Isogai A. Self-Incompatibility in Plants, Annu. Rev. Plant Biol. 56 (2005) 467–89.

[46] Breton CM, Bervillé AJ. New hypothesis elucidates self-incompatibility in the olive tree regarding S-alleles dominance relationships as in the sporophytic model. CR Biologies 2012 in press.

[47] Cuevas J, Rallo L. Response to cross-pollination in olive trees with different levels of flowering. Acta Horticulturae 1990286: 179-182.

[48] Cuevas J, Rallo L, Rapoport HF. Initial fruit set at high temperature in olive, Olea europaea L. Journal of Horticultural Science 1994; 69: 655-672.

[49] Franklin-Tong N. Self-fertilization: article in: Brenner's Encyclopedia of Genetics. V.E. Franklin-Tong (ed.) Self-Incompatibility in Flowering Plants – Evolution, Diversity, and Mechanisms. Publ. Springer-Verlag Berlin Heidelberg 2008; 2011.

[50] Gerstel DU. Self-incompatibility studies in Guayule II. Inheritance, Genetics 1950; 35 482-506.

[51] Bradley WH Griggs L. Morphological evidence of incompatibility in Olea europaea, Phytomorphology 1963; 13 141-156.

[52] Collani S, Galla G, Baldoni L, Barcaccia G. Proceedings of the 53rd Italian Society of Agricultural Genetics Annual Congress Torino, Italy – 16/19 September, 2009 ISBN 978-88-900622-9-2 Poster Abstract – 7.73 Self-Incompatibility In Olive (Olea europaea L.); 2009.

[53] Koubouris, G C. Genetic and environmental factors affecting fruit set in olive (Olea europaea L.) and study of incompatibility in molecular level. PhD thesis Chania (Greece); 2009.

[54] Nooryazdan H, Serieys H, David J, Bacilieri R, Bervillé A. Construction of a crop—wild hybrid population for broadening genetic diversity in cultivated sunflower and first evaluation of its combining ability: the concept of neodomestication. Euphytica, pp. 1-17, doi:10.1007/s10681-010-0281-1

[55] Moutier N. Self-fertility and inter-compatibilities of sixteen olive varieties. In: Vitagliano C, Martelli GP (eds) Fourth International Symposium on Olive Growing. International Society of Horticultural Science, Bari, Italy, pp 209-211; 2000.

[56] Moutier N, Garcia G, Féral S, Salles JC. La maîtrise de la pollinisation en vergers d'oliviers, Olivæ 2001 ; 86 35-37.

[57] Alba V. ; V. Bisignano; E. Alba; A. Stradis; G. Polignano Effects of cryopreservation on germinability of olive (Olea europaea L.) pollen Genetic Resources and Crop Evolution 2011;, 58 (7) 77-982.

[58] Zienkiewicz Z, Rejón JD, Suárez C, Castro AJ, Alché J D, Rodríguez-García MI.Whole-Organ analysis of calcium behaviour in the developing pistil of olive (Olea europaea L.) as a tool for the determination of key events in sexual plant reproduction(2012BMC Plant Biol. 2011; 11: 150.

[59] Wu SB, Collins G, Sedgley M. Sexual compatibility within and between olive cultivars. Journal of Horticultural Science & Biotechnology 2002; 77 665-673.

[60] El-Hady S, Eman LF, Haggag MM, Abdel-Migeed M, Desouky IM. Studies on Sex Compatiblity of Some Olive Cultivars, Research Journal of Agriculture and Biological Sciences 2007; 3 504-509.

[61] Mehri H, Mehri-Kamoun R, Msallem M, Faïdi A, Polts V. Reproductive behaviour of six olive cultivars as pollenizer of the self-incompatible olive cultivar Meski. Advances in horticultural science 2003; 1 42-46.

[62] Selak G, Perica S, Goreta Ban S, Radunic M, Poljak M. Reproductive Success following Self-pollination and Cross-pollination of Olive Cultivars, Hort Science (2011) in the press.

[63] Arzani K, Javady T. Study of flower biology and pollen tube growth of mature olive tree cv. 'Zard'. Acta Horticulturae 2002 ; 586: 545–548.

[64] Kitsaki CK, Andreadis E, Bouranis DL. Developmental events in differentiating floral buds of four olive (Olea europaea L.) cultivars during late winter to early spring. Flora 2010: 205: 599-607.in Nutritional and Proteomic Profiles in Developing Olive Inflorescence Eds Christina K. Kitsaki, Nikos Maragos and Dimitris L. Bouranis

[65] Bassam F, Alowaiesh I, Harhash M. Flowering, pollination and fruiting of some varieties. Scientia horticulturae 2011; 129!;213-219

[66] Breton, C M, Souyris I, Villemur P, Bervillé A., Oil accumulation kinetic along ripening in four olive cultivars varying for fruit size. OCL 2009; 16 1-7.

[67] Pinatel C, Petit C, Ollivier D, Artaud J, Outil pour l'amélioration organoleptique des huiles d'olive vierges, in OCL 2004 ; 11 N°3.

[68] Koubouris CG, Metzidakis I, Vasilakakis M... Influnce of cross poliination on the development of parthenocarpic olive(Olea europea) fruits (shotberries). Expl Agric.: 2009; 46 67-76.

[69] Farinelli D, Boco M, Tombesi A. Results of four years of observations on self – sterility behaviour of several olive cultivars and significance of cross – pollination. Pro-

ceedings of Second International Seminar on "Biotechnology and quality of olive tree products around the Mediterranean Basin – Olivebioteq 2006", 5 – 10 November 2006, Marsala – Mazara del Vallo Italy, Vol. I: 275-282 Alcamo.. 2006.

[70] Farinelli D, Pierantozzi P, Palese AM. Maternal cultivar and pollinator effects influence on fruit drupe weight and seed number in the olive (Olea europaea L.). Submitted.

[71] Cuevas & Polito 1997Cuevas J? ? Polito VS. Compatibility relationships in "Manzanillo" 284 HortScience 1997. 32(6) 1056-1058.

[72] Villemur P, Musho US, Delmas JM, Maamar M, Ouksili A. Contribution à l'étude de la biologie florale de l'olivier (Olea europaea L.): stérilité mâle, flux pollinique et période effective de pollinisation, Fruits 1984. 39 467–473.

[73] Vossen P., 2007 http: //cesonoma.ucdavis.edu/files/27184.pdf.

[74] Vossen P., Primary world olive oil cultivars including several California table varieties for comparison, http://cesonoma.ucdavis.edu/files/27184.pdf. (2007).

[75] Ollivier D, Artaud J, Pinatel C,.Durbec JP, Guérère M. Differentiation of French virgin olive oil RDOs by sensory characteristics, fatty acid and triacyl glycerol compositions and chemometrics. Food Chemistry 2006; 97 382-393.

[76] Mookerjee S, Guerin J, Collins G, Ford C, Sedgley M. Paternity analysis using microsatellite markers to identify pollen donors in an olive grove Theor Appl Genet 2005; 111 1174–1182.

[77] Guerin J, Sedgley M. Rural Industries Research and development Corporation, rirdc.infoservices.com.au CrosS-pollination in Olive Cultivars Publication 07/169. 2007.

[78] Seifi E. PhD Adelaide University, Australia. 2009.

[79] Seifi et al. 2011], Seifi, E., J. Guerin, B. Kaiser and M. Sedgley, 2011 Sexual compatibility and floral biology of some olive cultivars. New Zealand Journal of Crop and Horticultural Science 39: 141-151.

[80] Diaz et al. 2007; Díaz A, A Martín, P Rallo, D Barranco, R De la Rosa 2006 Self-incompatibility of 'Arbequina' and 'Picual' Olive Assessed by SSR Markers. J. Amer. Soc. Hort. Sci. 131(2):250–255. 2006.

[81] Cuevas J., A.J. Diaz-Hermoso, D. Galian, J.J. Hueso, V. Pinillos, M. Prieto, D. Sola, V.S. Polito, Response to cross pollination and choice of pollinators for the olive cultivars (Olea europaea L.) 'Manzanilla de Sevilla', 'Hojiblanca', and 'Picual', Olivae 85 (2001) 26-32.

[82] AFIDOL, Unpublished. internal report.

[83] Moutier et al. 2010. Moutier et al. (2006). Moutier, N., Terrien, E., Pécout, R., Hostalnou, E., and J. F. Margier 2006 Un groupe d'étude des compatibilités polliniques entre variétés d'olivier. Le Nouvel Olivier 51: 8-11.

[84] Breton CM, Bervillé AJ. Les recherches sur l'olivier. ? In L'Olivier l'arbre des temps Eds C Breton & A. Bervillé, Quae, Versailles. 2012.

[85] Fabbri et al. 2008.Fabbri, A., Lambardi, M., and Y. Ozden-Tokatli, 2008 Olive Breeding in Breeding Plantation Tree Crops: Tropical Species Ed. S. Mohan Jain and P. M. Priyadarshan, DOI 10.1007/978-0-387-71201-7_12.

[86] Therios I; Olives. in crop production Science in horticulture series. Eds Jeff Atherton and Andalun Rees; 2009.